BMP TECHNOLOGY IN URBAN WATERSHEDS

Current and Future Directions

EDITED BY
Richard Field
Scott D. Struck
Anthony N. Tafuri
Michael A. Ports
Michael Clar
Shirley Clark
Betty Rushton

Published by the American Society of Civil Engineers

Library of Congress Cataloging-in-Publication Data on file

Published by American Society of Civil Engineers
1801 Alexander Bell Drive
Reston, Virginia 20191
www.pubs.asce.org

ISBN 0-7844-0872-6
Manufactured in the United States of America.

Contents

Preface

The purpose of this book is to provide information on various best management practice (BMP) options. BMP designs as well as guidelines for monitoring, selecting, implementing, and costs in urban watersheds are presented.

The urban area in the United States increased approximately five-fold since the mid-1940s and decreased our watersheds' capacities to: mitigate floods, and protect water quality and aquatic health through removal of excess contaminants before runoff enters receiving waters. Urban development also changes the hydrologic regime which can result in habitat loss and other ecological impairments due to relatively frequent, intense, and longer bankfull events, lower baseflows, and increased stream-channel erosion.

Cost-effective structural and nonstructural BMPs are being emphasized to reduce impacts from urbanization on wet-weather flow (WWF) quality and quantity. Urban wet-weather stressors can emanate from combined-sewer overflow (CSO), sanitary-sewer overflow (SSO), and stormwater.

Removal of WWF pollutants using control technologies can be very costly; whereas using BMPs can offer a less costly way to abate stormwater effects. During the relatively short period that BMPs have been applied for water-quality improvement, researchers have collected a substantial amount of data in an attempt to determine performance and develop and demonstrate process options.

This book does not intend to specify actual BMP selection. It is a compilation of the best presentations from the 2004 EWRI World Water and Environmental Congress' BMP Technology Symposium: Current and Future Directions, Salt Lake City, Utah. They were then expanded and improved in an attempt to provide information for selection of BMPs for a stormwater management program. It serves as an information source to individuals and agencies on the current state and future direction of BMPs in urban stormwater management.

Richard Field, D.WRE, P.E., Chair
BMP Technology Symposium: Current and Future Directions Book Committee 2004 EWRI World Water and Environmental Congress

Scott D. Struck, Ph.D., Co-Chair
BMP Technology Symposium: Current and Future Directions Book Committee 2004 EWRI World Water and Environmental Congress

Acknowledgements

The International Symposium titled **BMP Technology: Current and Future Directions** was held at Salt Lake City, Utah from June 28 to July 1, 2004. The symposium was sponsored and organized by the U. S. Environmental Protection Agency and the Urban Water Resources Research Council. It was held in conjunction with the World Water and Environmental Congress. The Congress was sponsored and organized by the Environmental and Water Resources Institute of the American Society of Civil Engineers. The objective of the symposium was to bring together the best researchers and the cutting-edge practitioners to present and discuss the current and future directions for the control and treatment of urban stormwater runoff.

The subjects covered by the published symposium proceedings are international in scope and as far reaching as the subject of urban stormwater management. The full range of technical issues, including BMP design, placement within a watershed, performance, inspection, modeling, monitoring, operation, maintenance and rehabilitation are covered. Also covered were legal issues, economics, multi-purpose design concepts, public participation and education, as well as low impact development. A total of ninety three papers were selected, peer reviewed, published in the proceedings, and presented at the symposium. The best papers and presentations from the symposium proceedings have been selected by the chairs of the individual sessions for inclusion in this publication. Thus, the best papers and presentations are expanded and combined into coherent chapters. Each chapter has been subjected to a rigorous peer-review process.

Many individuals were responsible for the success of the Symposium, including the symposium steering committee:

Richard Field, Committee Chair, US EPA, Edison, New Jersey
Scott Struck, Committee Co-chair, US EPA, Edison, New Jersey
Michael A. Ports, Black & Veatch, Kansas City, Missouri
Anthony Tafuri, US EPA, Edison, New Jersey
Michael Clar, Eco-Site, Inc., Ellicott City, Maryland
Betty Rushton, Southwest Florida Water Management District, Brooksville, Florida

And the symposium book committee:

Richard Field, Committee Chair, US EPA, Edison, New Jersey
Scott D. Struck, Committee Co-chair, US EPA, Edison, New Jersey
Michael A. Ports, Black & Veatch, Kansas City, Missouri
Anthony N. Tafuri, US EPA, Edison, New Jersey
Michael Clar, Eco-Site, Inc., Ellicott City, Maryland

Betty Rushton, Southwest Florida Water Management District, Brooksville, Florida
Shirley Clark, Penn State University-Harrisburg, Middletown, Pennsylvania

In addition special thanks are due to the sponsoring organizations:

Urban Watershed Management Branch of the Water Supply and Water Resources Division, National Risk Management Research Laboratory, Office of Research and Development, United States Environmental Protection Agency

and

Urban Water Resources Research Council
Environmental and Water Resources Institute
American Society of Civil Engineers

Chapter 1

U.S. EPA's Urban Watershed Research Program in BMPs and Restoration for Water Quality Improvement

Sally Guitierrez, Director, Water Supply and Water Resources Division, National Risk Management Research Laboratory, U.S. Environmental Protection Agency

Introduction

Urban land area in the U.S. more than quadrupled between 1945 and 1997, increasing from 15 to 64 million acres, primarily from conversion of land that was previously forested, pastured, and ranged (USDA, 2000). During this same period, paved road mileage nearly tripled (USEPA, 2000). This urbanization, with the accompanying expansion of impervious surfaces, significantly increased the nation's total amount of developed and paved surface area; reduced the amount of wetlands and other undeveloped land; and, decreased our watersheds' ability to mitigate floods, facilitate sediment replenishment, and protect water quality and aquatic health by removing excess nutrients and other chemical contaminants before runoff enters receiving waters. Urban development also causes changes in the hydrologic regime, which can result in habitat loss and other ecological impairments due to more frequent and longer bankful events, lower baseflows, and increased stream channel erosion (Roesner et al., 2001). Significant decreases in biotic measures of aquatic ecologic health also become apparent at low relative impervious areas (Center for Watershed Protection, 1994).

Urban areas are often proximal to highly-valued recreational and environmentally-significant waters, critical estuarine and coastal areas, as well as lakes and reservoirs that supply many drinking water systems. The water quality impacts associated with this increasing urbanization are evident in the impairments facing many of these waters. The EPA's biennial report, The National Water Quality Inventory (USEPA, 2002a), indicates that the States collectively believe that about 40% of assessed U.S. streams, lakes, and estuaries fail to meet one or more designated uses. As in previous inventories, pollution from urban land, transported by precipitation and runoff, is identified as a leading source of impairment, and many of the common stormwater-associated stressors are identified as the leading causes of impairment, including: siltation, nutrients, bacteria, metals, and oxygen-depleting substances.

During the past thirty years, the U.S. EPA established several regulatory programs to address the various point and nonpoint sources of urban runoff. After passage of the Clean Water Act (CWA) in 1972, water pollution control efforts focused primarily on wastewater discharges from point sources consisting of municipal treatment plants, industrial plants, and combined sewer overflows (CSO). To bring attention to the latter, the EPA issued a National CSO Control Strategy on August 10, 1989 reaffirming CSOs as point-source discharges subject to CWA National Pollution Discharge Elimination System (NPDES) permit requirements (Federal Register, 1989). Although the Strategy successfully focused attention on CSOs, it fell short in resolving many fundamental impediments to its implementation. In mid-1991, the EPA began a process to accelerate CSO strategy implementation. The initiative resulted in a CSO Control Policy which, in addition to providing guidance to NPDES permitting and enforcement authorities, also provided municipalities, especially those that are financially disadvantaged, the flexibility to consider the site-specific nature of CSOs and determine the most cost-effective means of reducing pollutants to meet the requirements of the CWA (Federal Register, 1994).

Far less emphasis had been placed on diffuse source pollution, i.e., pollution from contaminants picked up and carried into surface waters by stormwater runoff. To address the numerous environmental stressors associated with this pollution source, the CWA was amended in 1987 to establish NPDES stormwater discharge requirements. To implement these requirements, the EPA published the "Phase I" Stormwater permit program to address certain stormwater discharge categories associated with industrial activities and large storm sewer systems (located in 842 municipalities with populations of 100,000 or more) on November 16, 1990 (Federal Register, 1990). On December 8, 1999, the EPA promulgated "Phase II" stormwater regulations (Federal Register, 1999) expanding the list of regulated communities to include small systems in Bureau of Census-defined urbanized areas. Phase II requires NPDES permits and stormwater discharge controls in the regulated communities.

The EPA's approach, as outlined in the Phase II regulations, is to control the stormwater stressor load using best management practices (BMPs) augmented by a combination of natural and assisted restoration within the receiving water. The regulations rely heavily on a combination of structural and nonstructural techniques (i.e., BMPs) to mitigate the stormwater effects on receiving waters. Structural BMPs are designed to trap and detain runoff to settle or filter out the constituents before they enter receiving waters, while nonstructural BMPs control pollutants at the source to prevent or reduce contamination of stormwater runoff.

There are many types of structural BMPs available for stormwater runoff control and storage-treatment, and new treatment devices are frequently being introduced; however, there is a limited understanding of their effectiveness, their lifetime cost, the appropriate techniques to design or position them in the watershed, or their impact on receiving water after installation. The understanding of nonstructural BMPs is even more limited.

Objectives

The EPA's Urban Watershed Management Research Program (UWMRP) believes that efforts accumulating additional data measuring BMP-specific performance as relative stressor reductions will add little to the knowledge of BMPs, the internal physical and chemical mechanisms that control stressor and multi-stressor removals, or the ability to transfer knowledge to other BMP sites. Alternatively, establishing techniques that enable communities to predict the effluent quality *can* be translated to any urban installation. Providing tools and techniques that allow community-based environmental planners to select cost-effective techniques and practices that can restore and protect the receiving water body from stormwater-associated stressors on a watershed scale and within a predictable time is the UWMRP's overall objective for water quality research.

Approach

The UWMRP is emphasizing cost-effective technologies, and structural and nonstructural BMPs that reduce the impact of wet-weather flow related pollutants from *urban* areas. In the urban environment, wet-weather pollution can emanate from CSOs, sanitary sewer overflows (SSOs), and stormwater runoff. The UWMRP has a long history of involvement and expertise in CSO and SSO control as well as in stormwater runoff control. Early work in control technologies focused on developing and demonstrating: advanced collection system design alternatives; high-rate treatment technologies; techniques to optimize flow capture and treatment through in-sewer storage, flow routing, and wastewater treatment plant operation; and, infiltration/inflow identification and elimination technologies.

BMPs provide a mostly passive, and less costly, means to mitigate the effects of stormwater on receiving waters and have become a ubiquitous part of the landscape. Historically, engineers installed these controls for runoff volume (flood) control, and have only recently extended the application to include water quality considerations.

During the comparatively short period that BMPs have been applied to water quality, researchers have collected a substantial amount of data in an attempt to evaluate their performance. In addition, government agencies and professional organizations have published numerous guidance documents to assist in their implementation. Despite this substantial effort, fundamental questions on BMPs remain unanswered. The UWMRP is in the process of addressing these questions as part of its research program.

How do BMPs work?

In recent efforts using BMPs to control and treat stormwater runoff, the UWMRP concluded that, at least for the short term, a traditional chemical engineering approach will provide the greatest flexibility in implementing controls across the various national ecoregions. This approach hypothesizes that largely uncontrollable *but predictable* ambient conditions control the physiochemical processes within the BMP. These processes control any intra-event and inter-event transformations affecting the storm-introduced load and, in turn, define the effluent characteristics.

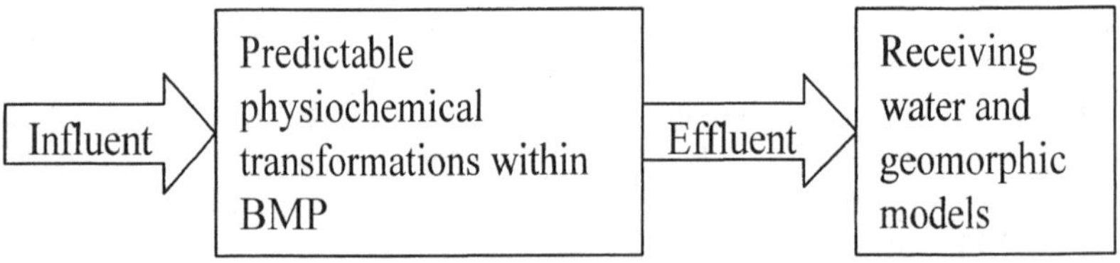

Once identified and understood, models can be developed that reasonably estimate the effluent load, regardless of the characteristics of the drained watershed. Receiving water models with appropriate spatial and temporal resolution can then address the effluent as an "intermittent point source" that discharges only for periods following a rain event with predicted stressor concentrations. Such water quality models exist--Total Maximum Daily Load (TMDL) developers routinely incorporate point sources in establishing load allocations.

The UWMRP identified nutrients, sediments, toxics, flow, and pathogens as the stressors of primary. For BMPs, the influent quantity and quality are the parameters of interest. Years of hydrologic and hydraulic modeling have yielded many alternate methods of differing sophistication levels that reliably predict the time-dependant influent flow rates. The influent flow rate is rainfall-driven and strongly affected by landscape characteristics (e.g., relative impervious area, topology, geologic properties) with most runoff events associated with more common small storms. The UWMRP demonstrated that the influent quality, expressed as a mean event concentration of a flow-weighted composite sample, for stable urban landscapes varies with both season and land use for macro nutrient species (Nietch *et al.*, 2001) and microbiological indicators (Selvakumar *et al.*, 2004). The annual means are typical of existing data sets and show year-over-year consistency. For other stressors of interest (e.g., heavy metals, suspended solids, organic carbon), the UWMRP has compiled less complete information.

Stressor reduction in BMPs commonly relies on particle control (i.e., filtration and settling processes) making the relative portion of the total stressor load that is particle associated a key consideration. The literature often links phosphorous removal, for example, as paralleling the suspended solids reductions. The UWMRP data support the hypothesis that the relative fraction of particle-associated phosphorous (i.e., total phosphorous - total dissolved phosphorous) varies seasonally. Given this variation, it is reasonable to expect seasonal performance differences in total phosphorous reduction linked to the ability of the BMP to trap the available particle-bound phosphorous fraction. These variations also help potentially explain the large range of total phosphorous reductions by BMPs reported in the literature.

Effluent hydrology is a function of outlet design and water depth that is, in-turn, defined by the time-dependant influent flows and control structure geometry (surface area changes with depth above the discharge control). In contrast, effluent concentrations are currently unpredictable, but, given the functional models outlined above, can be made predictable with an understanding of the transformations, if any, that occur within the structure. For some stressors, the UWMRP has a sufficient understanding of the controlling functions to begin developing, at least on a rudimentary level, the controlling functions that operate on the influent to control the effluent. For example, gravity-driven settling, a well-understood process, controls sediment reduction in wet ponds. Given the appropriate information on the effective spherical equivalent particle-size distribution, particle density as a function of effective spherical diameter, internal turbulence levels, influent total suspended solids concentration, and BMP geometry, the sediment effluent concentration can be predicted. The likely particle size distribution is reasonably predictable (Pitt, *et al.*, 1995) and serves as the basis for some existing models. Further, given the above information and information about water level relative to the outflow control, it is possible to predict the particle size distribution of the sediment in the effluent that will ultimately deposit in the streambed, a characteristic believed to influence the observed biotic integrity in streams (Paul and McDonald, 2003). Similarly, bacteria decay is generally accepted to follow a first-order process. Given a knowledge of the influent concentrations of organism groups of interest (measurements used to legally establish acceptable water quality), internal mixing characteristics, and ambient conditions controlling the decay rates, it should be feasible to predict the effluent quality.

The physiochemical transformation processes are complicated for other stressors. The nitrogen cycle, for example, outlines a complex, shifting equilibrium. Other researchers have completed work on the reaction pathways, the associated equilibrium, and kinetic constants. As with the information on bacterial decay constants, however, the information is often not specifically associated with stormwater runoff, often inconsistent, and sometimes contradictory. For pragmatic application to BMP design, it is unlikely that the full collection of equilibria be established. An a priori knowledge of the dominant influent and effluent constituents, coupled with the existing work, particularly the research supporting existing water quality models, will enable the researcher to focus on the necessary measurements related to the controlling reaction pathways.

Internal BMP hydraulics control the effective time that stormwater is within the BMP. The traditional engineering equivalent is the mean hydraulic residence time. Generally, this value represents the average time a given volume of water remains in the control structure. Developing an understanding of methods to predict the mean hydraulic residence time based on influent hydrology, inlet and outlet structure design, and control volume geometry will be critical in establishing the likely effluent quality and compliment the particulate upflow velocity approach. Academic researchers have started investigating this issue in a non-steady state, variable-volume system and some BMP field work has also been conducted (e.g., Kadlec and Knight, 1996). Others, Driscoll *et al.* (1989), for example, have statistically analyzed the long National Weather Service data record to provide a sound basis for inter-event rain duration and other climatic considerations. Their findings will simplify investigations. The water quality within control structures that retain water between storm events follows diurnal patterns and certain parameters (e.g., temperature, dissolved oxygen, pH, and oxidation reduction potential) can shift widely during these daily swings. For some stressors, e.g., metals, under some environments (e.g., iron and silica rich conditions) equilibria, controlled by Eh-pH conditions, constantly shift from soluble to partially soluble and condition-insoluble species. The USGS identified these diurnal patterns in natural waterways for zinc, cadmium, manganese, and arsenic (Nimick *et al.*, 2003; USGS, 2003). Investigations of natural in-stream cycling shows 300% concentration variation for dissolved zinc (Jones *et al.*, 2004). It is thus reasonable to expect similar fluctuations in the dissolved metal concentrations within the BMP. This shifting can partially explain the range of reported concentration reductions in BMP effluent relative to the influent concentration. Clearly understanding the effect of the predictable ambient condition fluctuations on the stressors is necessary to understand the likely effluent concentrations.

Research on chemical kinetics is best performed in the well-controlled laboratory environment that provides the conditions necessary to support higher accuracy and precision. Conversely, evaluation of the applicability of developed models must be undertaken at the field scale. Field scale, however, does not offer the independent variable control needed to develop a model and intermediate, pilot-scale research offering a balance of control and scale is indicated. Collectively, this information suggests that the UWMRP must undertake research tasks at multiple scales from bench through field scale.

How much do BMPs cost?

Cost will always be a consideration in the efforts to protect and restore water quality. For a BMP to pass the needed feasibility tests and be commonly installed, it must have technical worth *and* be cost effective. The watershed-wide BMP installation is a constrained-optimization problem. Watershed planners have constrained financial resources that limit the number, location, and construction schedule. These constraints conceptually allow the planners to define the feasible efficiency frontiers representing the greatest available load reduction for a fixed investment. The predicted lifetime cost of the BMP divided by the predicted lifetime stressor removal (both calculated with the same maintenance practices) is one measure of the BMP cost effectiveness. Without a generally accepted technique to estimate the total BMP cost, the local manager cannot define the efficiency frontier. The UWMRP completed preliminary internal reviews of the costs associated with urban BMPs (Raghaven *et al.*, 2001) and sponsored some extramural research keying on the obvious information gap of land (opportunity) cost (Heaney, 2002). The UWMRP has also teamed with extra agency experts to help guide the development of a whole-life cost model through cooperative efforts with the Water Environment Research Federation (WERF), the American Water Works Association Research Federation (AWWARF), and the United Kingdom Water Industry Research Ltd. (UKWIR) (WERF, 2004). The model will likely include considerations of both the capital and the operational costs based on current design and maintenance practices.

Where should BMPs be placed?

Assuming methods have been successfully identified to predict the effluent quality and cost associated with individual BMP operations, there still exists a need to understand how the individual reductions accumulate to reduce the watershed-wide stressor concentrations and how the reduction affects the ecology of the receiving water. On a pragmatic level, watershed planners need to know where to install the BMPs within the financial constraints to most directly affect the degraded water body and meet TMDL goals. The USGS regression-based research (e.g., SPARROW), clearly shows that stressors introduced at different locations within a watershed effect the receiving water differently (Alexander *et al.*, 2004). This leads to a constrained-cost optimization problem where the planner attempts to generate a maximum water body improvement for a fixed investment.

Typically available modeling systems include watershed loading, receiving water response, and partial representation of management practices. Noticeably lacking are comprehensive modeling systems for evaluating the location, type, and cost of wet-weather flow BMPs needed to meet water quality goals. To meet the need for place-based evaluation techniques that can be used to cost-effectively select and place BMPs to achieve a desired water quality goal, the UWMRP is developing a framework consisting of useful decision-making tools to facilitate placement of BMPs at strategic locations in mixed land use urban watersheds. This framework will use urban hydrologic/hydraulic/water-quality routing principles and will include appropriate spatial and temporal scales; watershed and receiving water hydrologic/hydraulic and water quality modeling based on integrated data collection; and, the evaluation of alternative solutions based on cost effectiveness/optimization.

The modeling approaches will consider degrees of urbanization, nonstructural and structural BMPs, retrofitting, and new development design.

Do BMPs substantively affect water quality?

Of the main monitoring approaches currently used to assess BMP effectiveness, those that involve sampling of the receiving water, e.g., upstream/downstream of a single BMP or within a watershed containing multiple BMPs, are among the least practiced. The numerous complexities associated with determining the effects of BMPs on receiving waters range from selecting an appropriate, implementable, and affordable/measurable metric to selecting appropriate receiving water monitoring sites, in particular, for watershed-level assessments. At present, the professional community lacks a clear consensus on a single technique or approach to declare that a given action improves the water quality condition within a given regional ecology and as a consequence, there is no clear measure of success.

With respect to the selection of an appropriate metric, a growing segment of the professional community appears to support a biotic measure as the integrator of all stresses influencing the community structure. However, neither selecting that biotic metric nor establishing the causal links necessary to define the controls needed to meet a given biotic goal are apparent on the horizon. For example, establishing measurements of the relative taxa richness for a water body could be a tool to measure water quality changes. Yet this metric does not clearly define the required landscape actions that will allow natural or human-assisted processes to reestablish the desired biotic condition measured by taxa richness.

The UWMRP research on watershed-wide BMP effectiveness monitoring is working on several scales from meso-scale to the full watershed. This will develop field monitoring guidelines to determine receiving water BMP effectiveness. Integrating site-specific BMP processes with respect to pollutant attenuation into a watershed-level management system for larger-scale water quality protection/preservation is a leading question among the watershed management community. The UWMRP is addressing this experimentally by outfitting a large, mixed-land use watershed in Clermont, OH with monitoring-based field sites from stream catchment headwaters to large multi-use water bodies. Controlled meso-scale experiments will assist in defining the linkage of stressor induced individual behavioral changes to changes at the community level of ecological organization. It is hypothesized here that these behavioral changes can be used as an integrative indicator of community responses at the larger scale. In a related study, a novel watershed management classification system is being developed to help watershed managers select appropriate stream and/or lake monitoring sites to determine and track the effectiveness of existing watershed management practices.

While the large-scale biomonitoring systems are undergoing development, deployment and testing, smaller-scale monitoring is being conducted. In particular, the role of small channel processes in ameliorating downstream nonpoint source pollution loads is being investigated in both controlled meso-scale and small-catchment field scale studies to develop specific, simple monitoring guidelines integrating biogeochemical parameters, community structure, and hydrogeomorphic features of small channels draining catchments with different levels of planned BMP/natural attenuation landscape features. More explicit representation of small-stream processes and pollutant/water quality parameter concentrations related to biotic responses will increase the understanding of the role these ecosystems play in sustaining small-scale subwatershed ecology and in reducing downstream nonpoint source pollutant loads in larger watersheds for better TMDL calculations, source water protection, and the subsequent implementation of watershed management strategies. Without an integrative and mechanistic framework for sequentially linking load reductions to water quality parameters to stream biota, attempts to measure the BMP effectiveness, including stream restoration, on ecological condition will fail. Until there is a better means of linking stressor loading characteristics to in-stream biological response, watershed planners will continue to design management and monitoring strategies throughout a watershed with, at best, a conceptual basis for how they affect a desired biotic endpoint. Products from this effort will provide the basis for turning this conceptual understanding into working models for better watershed management.

Accomplishments

Using control technologies to remove pollutants from WWF can have significant associated costs. BMPs offer a much less costly approach, however, the menu of structural and nonstructural BMP options is extensive and there is little consistency in the amount or quality of design, performance, and cost data for each of the many practices currently in use. For this reason, the EPA is committed to providing guidance to the regulated communities that will assist in both developing and implementing these practices as part of their stormwater management programs.

Two recent guidance documents produced by the UWMRP include a state-of-the-practice report describing the various BMP options; the design considerations involved; and, the general guidelines for selecting, implementing, and monitoring BMPs to reduce pollutants in urban stormwater (USEPA, 2004a); and, a three-volume manual that addresses BMP design in greater depth (USEPA, 2002b). Based on currently available information, the former report provides support to local decision makers on the selection, design and implementation of BMPs. A framework of material is presented on considerations for BMP design and monitoring, costs, use, and placement of multiple BMPs, and other methods that can contribute to pollutant reduction in stormwater and the optimization of receiving water quality benefits.

The three-volume design manual also addresses factors that should be considered in the BMP design, however, this document includes more detail in providing specific design guidance for the three most commonly used structural treatment BMPs, i.e., ponds (detention/retention), vegetated biofilters (swales and filter/buffer strips), and constructed wetlands. The state-of-the-practice for these BMPs is such that existing design data vary widely. The UWMRP manual provides a practical BMP design approach that incorporates pollutant removal and flood control. Included in the manual is a spreadsheet tool (Integrated Design and Assessment for Environmental Loadings (IDEAL)) that can assist planners and regulators in examining the hydrology, sedimentology, and water quality for BMPs. A summary of this document was published in 2002 (USEPA, 2002b); the full manual will be available in the fall of 2004.

Under EPA's Phase II Stormwater Regulations, stormwater runoff must be controlled from construction sites larger than one acre. Onsite or adjacent sediment barriers, such as straw bales or ditch checks, and sedimentation basins that trap eroded sediment can be effective if appropriately designed and maintained. At small development sites (i.e., 1 to 5 acres) spatial constraints are likely to limit the practicality of sedimentation basins. The UWMRP is leading an effort to develop a model supporting filter fence design to capture sediment in construction site runoff (USEPA, 2004b). A process-based model was calibrated under field conditions and used initially as a simulation tool to select alternative designs under a more extensive set of conditions than could be tested empirically. After screening with the model, specific conditions were field tested using synthetic rain generators. The model was used to develop standards covering design, installation, and maintenance. In addition, a user-friendly design aid is being developed that extends the process-based model into a spreadsheet tool that can predict site-specific design parameters.

A conclusion from the EPA/ASCE Best Management Practice study is that "There is a great need for consistency with regard to the constituents and methods utilized for assessing BMP effectiveness" (Strecker *et al.,* 2001). An ongoing UWMRP project is using recommendations from this study and others to establish monitoring approaches that can lead to a better and more consistent BMP performance characterization. A main issue is the apparent inconsistency in report percent removals. This approach can result in some BMPs appearing to perform better than others even though effluent quality may be statistically equivalent. A UWMRP project is using well-defined, low-cost influent and effluent monitoring (e.g., flow, suspended solids, temperature, and conductivity) to evaluate the performance of an existing BMP. Actual performance, measured by influent and effluent sampling is being compared with expected performance based on original design criteria to determine if the design and performance goals are met. The study also exemplifies that, when comparing BMP performance, original design criteria must be considered since BMPs with differing design objectives may not be directly comparable. The research on BMP monitoring approaches will result in greater consistency with regard to the constituents and methods used for assessing BMP effectiveness as load or concentration changes (USEPA, 2004c).

To improve our understanding of the relationships between BMP design parameters and effluent quality, the UWMRP is attempting to define and model the internal mechanistic processes of selected BMPs. Physical, chemical, and biological processes, including infiltration, residence time, settling, partitioning, and microorganism die-off (Selvakumar and Borst, 2003), are being studied in BMP mesocosms representing wet ponds and constructed wetlands. This research is examining: the effects of BMP type and season on constituent retention, inter-event water column chemistry, and pore water chemistry; the relationship between mesocosm constituent removals and processed-based algorithm predictions; and, surrogate parameters for BMP performance assessment. These models will be available to be used in TMDL development.

The influent data that will be used in the development of the internal mechanistic models discussed above were collected year-round from three different urban land uses. The results, which indicate which land-use area and runoff from which season resulted in the highest pollutant (microorganisms, nutrients, and metals) concentrations will additionally help watershed managers target BMPs to land uses and/or seasonal runoff associated with the greatest pollutant loads (Nietch *et al.*, 2001, Selvakumar and Borst, 2004). Another UWMRP study is determining the distribution of particle-bound phosphorus and nitrogen species throughout the suspended solids size range typically found in urban stormwater (Madge, 2004). An analogous effort examining the particle size range associated with heavy metals is complete. This data will determine the particle sizes associated with nutrient and metal effluent concentrations. BMP design improvements based on the results of this study will improve the capture of particulate-bound stressor, reduce stressor discharge into receiving water bodies, and thus reduce water quality impairments. Phosphorus and nitrogen dynamics in a wetland treatment system for septic tank overflow are also being assessed in a cooperative effort with the University of Cincinnati (Nietch and Brown, 2004a). The project will determine this technique's applicability for septic wastewater treatment in mixed land use watersheds.

Having the ability to determine BMP effluent quality is a critical element for developing watershed-wide models that link effluent concentrations to receiving water quality. These models are often the key components of TMDL determinations and other watershed management efforts. The UWMRP is developing an Integrated Watershed-Based Stormwater Management Decision Support Framework (ISMDSF) to support the management of WWF by strategic placement of BMPs in mixed land use urban watersheds (Lai and Fan, 2003). This decision tool will facilitate the selection and placement of the most cost-effective BMPs at strategic locations based on integrated data collection and hydrologic, hydraulic, and water quality modeling to achieve the desired water quality objectives.

The EPA's Stormwater Management Model (SWMM) is one of the most successful models produced for the water environment. SWMM is widely used throughout the world to analyze quantity and quality problems related to stormwater runoff, combined sewers, sanitary sewers and other drainage systems in urban and non-urban areas. It is considered by the engineering community in the U.S. to be the "de facto" model for comprehensive hydrologic/hydraulic/water quality analysis associated with urban sewer systems. From 1988-1999 approximately 5000 copies of the model were distributed by EPA and others. The SWMM users group has about 10,000 subscribers. Because of its strength and popularity, SWMM has been incorporated as is, or with improvements, by vendors into their software packages for commercial distribution. These packages include PCSWMM (Computational Hydraulics International, 2004), XP-SWMM (XP Software, 2004), and MIKE-SWMM (Danish Hydraulic Institute, 2004). SWMM is a dynamic rainfall-runoff simulation model, used for single-event or long-term (continuous) simulation of runoff quantity and quality from primarily urban areas. It allows for hydrologic computations including rainfall, snowmelt, soil infiltration, depression storage evaporation, and groundwater flow. It solves the complete dynamic equations of motion including simulation of backwater, surcharging, pressure flow, looped connections, and other hydraulic complexities. Currently, the UWMRP is leading a collaborative effort to produce a new version of SWMM that is compatible with current computational technology and a system that is more accessible for updating by the current generation of modelers (USEPA, 2004e). The new version, run under Microsoft Windows, uses object-oriented C code and includes a graphical user interface. It also includes a translation program to convert SWMM4 input files to SWMM5 input files (Rossman *et al.*, 2003, 2004; USEPA, 2004d).

SWMM5 will continue to be enhanced to include: bed-load transport in conduits, snowmelt improvement, groundwater/irrigation improvements, water quality routing/fate, a pollutant source generator, BMP unit processes, and diffusion wave/implicit dynamic routing methods. This work is expected to be completed in 2006. In addition, the UWMRP plans to apply SWMM5 to an urban watershed to evaluate its usefulness in meeting TMDL requirements for flow and several water quality stressors. This work will involve a cooperating watershed management agency that is actively engaged in WWF data collection and is willing to share the data for model calibration and application.

Research to better characterize the effects of BMPs on receiving water quality represents a critical step in our ability to assess BMP performance. The UWMRP is collaborating with Clermont County, OH on a research effort coupling experimental and monitoring data on hydrological, biogeochemical, ecological, and landscape processes to better characterize the effects of mixed land use BMPs on receiving water quality (Nietch, *et al.*, 2004b). The UWMRP will use a BMP landscape-level characterization within a GIS-based simulation modeling framework to suggest an optimal watershed size range for studying the effect of existing BMPs. Afterwards, a novel watershed management classification system will be developed to help resource managers to select appropriate stream and/or lake monitoring sites to determine and track the effectiveness of watershed management projects. A report has been produced that defines some of the preliminary GIS protocols and capabilities of a BMP geodatabase for watershed management (Wang *et al.*, 2004).

The development of BMP geodatabase technologies is paramount to tracking and assessing BMP implementation projects under the TMDL program. Methods for citing existing BMP ponds and riparian zones were developed and novel BMP-specific landscape characterizations were proposed. A related effort on small-channel carbon and nutrient dynamics for post BMP implementation assessments is also being conducted and will provide specific field monitoring guidelines for receiving water-specific BMP effectiveness determination. This work differs from other stream-related BMP effectiveness work in its focus on making in-kind measurements in small-channels draining catchments with different levels of planned BMP/natural attenuation landscape features with linkage to meso-scale, controlled experimental efforts.

Also in the area of monitoring receiving water quality, the UWMRP is in the process of developing, deploying, and testing meso- and large-scale biomonitoring systems to define the linkage of stressor-induced individual behavioral changes to changes at the community level of the ecological organization. The plan, outlining the proposed multi-scale monitoring and experimental assessments and the work in improving watershed water quality management is complete (Nietch, *et al.*, 2004b). The meso-scale studies will improve our understanding of the role these ecosystems play in sustaining small-scale subwatershed ecology and in reducing downstream nonpoint source pollutant loads in larger watersheds for better TMDL calculations, source water protection, and the subsequent implementation of watershed management strategies. On a broader scale, the project linking land-use practices and watershed quality will establish larger-scale water quality biological-based monitoring systems in coordination with small-scale stream channel monitoring in BMP-characterized catchments. Hence, a central database will be developed to allow for real-time water quality parameter linkage across a drainage network. In combination with the results from the meso-scale experiments, this data management system can be used in follow-up studies and within a hydrological modeling framework, integrate watershed responses to management actions more effectively than can be achieved with existing water quality databases.

Outreach

An integral part of the UWMRP's activities involves transfer of technology to the user community. The UWMRP maintains a website (www.epa.gov/ednnrmrl/) devoted to its WWF-related research. This site provides project summaries and links to ORD's project database; full-text versions of 35 project reports, 2 books, and 23 journal articles; the program's bibliography of more than 400 publications; a glossary of more than 150 WWF terms; links to other WWF sites; and WWF-related seminar and workshop notices. The UWMRP staff has also authored or coauthored *Water Environment Federation's* annual literature review on urban WWF research, and has additionally produced a compilation of their reviews, available from the website. In addition, staff has co-edited the multilingual glossary of more than 850 commonly used terms in urban drainage that is designed to facilitate international technical communication on this subject (Ellis *et al.*, 2004).

As an additional part of its outreach efforts, in recent years the UWMRP has published several books in the area of stormwater characterization and management. In 2002, UWMRP supported the development of a handbook entitled, "Stormwater Effects Handbook: A Toolbox for Watershed Managers, Scientists, and Engineers"

(Burton and Pitt, 2002). This book provides a logical approach for an experimental design to determine if stormwater runoff is causing adverse effects and beneficial-use impairments in local receiving waters. It addresses both adverse effects on aquatic life and human health, and considers exposures to multiple stressors such as pathogens, chemicals, and habitat alteration. In 2003, the UWMRP published a book entitled, "Wet Weather Flow in the Urban Watershed: Technology and Management", which contains chapters from many international experts on subjects that include the management of WWF and solids, the characteristics of urban stormwater runoff pollutants, source loading and management modeling, emerging stormwater treatment technologies, beneficial use of urban stormwater, low-impact development, and geomorphic considerations in stream protection and restoration (Field and Sullivan, 2003a).

Another UWMRP book, written entirely in-house, entitled "Management of Combined Sewer Overflows," summarizes the EPA/ORD's research involvement in developing countermeasures for CSO pollution (Field, Sullivan, and Tafuri, 2003b). The book is a reference for the user community faced with the challenges and mandates to combat urban wet-weather induced water pollution. It covers the gamut of engineering requirements, from pollution problem assessment and associated tools to management and control planning and design. Material is presented on source management and control technologies, treatment and solids management, disinfection, and effective control systems. The UWMRP staff also co-edited a book published by Technomic, entitled, "Innovative Urban Wet-Weather Flow Management Systems," that outlines the principals of sustainable urban water management and describes innovative methods to improve these systems, including innovative sewer design and monitoring systems, high-rate treatment, and stormwater reuse (Field, *et al.*, 2000c).

Future Direction

Ultimately, the UWMRP research targets the protection and restoration of aquatic life through mitigation of stress from urban WWF. This represents a fundamental sustainability question of human interaction with the environment. Reliably predicting BMP effluents is a necessary but not sufficient prerequisite to reaching this goal. Simply installing the controls that meet specified effluent requirements does not assure the protection of aquatic ecology.

The next stage of the BMP research, field demonstration of the model capability, will follow two tracks. The first is developing guidelines for new BMP installation. This process will follow traditional engineering processes of specifying the required effluent to meet water quality requirements and estimating the probable influent from landscape characteristics and local climatology records. The UWMRP will cooperate with others to incorporate the necessary design parameters into the BMP construction and facilitate the model demonstration. With the design constraints, it is possible to estimate the life-cycle costs and evaluate the project feasibility. The demonstration process will simultaneously apply the model to established controls to predict effluent characteristics. Direct comparison of the effluent characteristics with the predicted values will serve to validate the models. These demonstration and verification efforts will follow a traditional case-study approach. The UWMRP, in cooperation with other EPA laboratories, will couple these case studies with in-receiving water biotic measures of aquatic health. The aquatic health demonstration and monitoring programs will require several years with periodic biotic assessments to develop recovery rates. The ongoing research investigating the small-channel in-stream processes is laying the foundation for these programs that will shift the water quality from a process dominated by engineering intervention to a sustainable integration of humans in the environment.

Evaluating nonstructural BMPs requires a working integration of the BMP model and the watershed model. It is conceptually possible to make the evaluation without the model (e.g., paired watershed studies), but the cost and time requirements are prohibitive. Most communities undertaking significant nonstructural programs have BMPs in place or are installing the controls. The concurrent actions of structural and nonstructural BMPs confounds the data making the evaluation inherently problematic. When the model connections are available, the UWMRP will advance to nonstructural BMP evaluations.

References

Alexander, R. B., Smith, R. A. and G. E. Schwarz. 2004. Estimates of diffuse phosphorus sources in surface waters of the United States using a spatially referenced watershed model. *Water Science and Technology*. 45(3): 1-10.

Burton, G. A., and R. Pitt. 2002. Stormwater Effects Handbook: A Toolbox for Watershed Managers, Scientists, and Engineers. Lewis Publishers. Boca Raton, FL. 929 p.

Center for Watershed Protection. 1994. *The Importance of Imperviousness*. Watershed Protection Techniques, 1(3).

Computational Hydraulics International. 2004. PCSWMM.

Danish Hydraulic Institute. 2004. MIKE-SWMM.

Driscoll, E. D., Palhegyi, G. E., Strecker, E. W., and P. E. Shelly. 1989. Analysis of storm event characteristics for selected rainfall gauges throughout the United States. Report prepared for EPA, Washington DC. November 1989.

Ellis, B., Chocat, B, Fujita, S., Rauch, W., Marsalek, J. and R. Field. 2004. *Urban Drainage: A Multilingual Glossary*. IWA Publishing, London, UK.

Federal Register. 1989. National CSO Control Strategy. Volume 54: 37370. August 10th, 1989.

Federal Register. 1990. Stormwater Phase I Rule. Volume 55: 47990. November 16, 1990.

Federal Register. 1994. Combined Sewer Overflow Control Policy. Volume 59(75): 18688-18698. April 19th, 1994.

Federal Register. 1998. National Pollutant Discharge Elimination System--Proposed Regulations for Revision of the Water Pollution Control Program Addressing Storm Water Discharges. Volume 63(6):1536. January 9, 1998.

Federal Register. 1999. National Pollutant Discharge Elimination System--Regulations for Revision of the Water Pollution Control Program Addressing Storm Water Discharges. Volume 64(235): 68721-68770. December 8th, 1999.

Field, R., Heaney, J. P. and R. Pitt, eds. 2000c. *Innovative Urban Wet Weather Flow Management Systems*. Technomic Publishing, Lancaster, PA. 535p.

Field, R. and D. Sullivan, eds. 2003a. *Wet Weather Flow in the Urban Watershed*, Lewis Publishers, Boca Raton, FL. 382 p.

Field, R., Sullivan, D. and A. N. Tafuri, eds. 2003b. *Management of Combined Sewer Overflows*, Lewis Publishers, Boca Raton, FL. 238 p.

Heaney, J. P., Sample, D., and L. Wright. 2002. Costs of Urban Stormwater Control. US EPA Report No. 600/R-02/021. US EPA, Cincinnati, Ohio.

Jones, C. A., Nimick, D. A. and B. McCleskey. 2004. Relative effect of temperature and pH on diel cycling of dissolved trace elements in Prickly Pear Creek, Montana. *Water, Air and Soil Pollution*. 153: 95-113.

Kadlec, R.H. and R.L.Knight. 1996. *Treatment Wetlands*. Lewis Publishers, Boca Raton, FL.

Lai, F. H. and C. Fan. 2003. An Integrated Decision Support Framework for Placement of BMPs in Urban Watersheds. In: Proceedings of Water Environment Federation TMDL 2003 Specialty Conference. Chicago, IL. November 16-18th, 2003.

Madge, B. 2004. Analysis of Particulate Bound Nutrients in Urban Stormwater. Poster at: US EPA 2004 Science Forum: Healthy Communities and Ecosystems. June 1-3rd, 2004. Washington, DC.

Nietch, C. T., Borst, M., O'Shea, M. L., Selvakumar, A., and C. Casciolini. 2001. Estimating Diffuse Stormwater Nutrient Loads from Suburban Landscapes to the Navesink Estuary, New Jersey. In: Proceedings of the 16th Biennial Conference of the Estuarine Research Federation. November 4-8, 2001. St. Pete Beach, FL.

Nietch, C. T. and D. Brown. 2004a. Assessment of Nitrogen and Phosphorous Dynamics in the Grailville Wetland Treatment System for Septic Tank Overflow. US EPA NRMRL, Water Quality Management Branch Internal Planning Document.

Nietch, C. T., Allen, J., Haught, R. and J. Goodrich. 2004b. Integrated Research Plan for Linking Land Use Management Practices and Watershed Water Quality in Southern Ohio. US EPA NRMRL, Water Quality Management Branch Internal Planning Document.

Nimick, D.A., Gammons, C.H., Cleasby, T.E., Madison, J.P., Skaar, Don, and C. M. Brick. 2003. Diel cycles in dissolved metal concentrations in streams--Occurrence and possible causes. *Water Resources Research*. 39(9):1247.

Paul, J.F. and M. E. McDonald. 2003. Empirical, Geographically-Based Water Quality Criteria Development Using Conditional Probabilities: A Proposed Approach. Draft for internal review, National Health Effects and Exposure Laboratory, Research Triangle Park, NC. April 2, 2003.

Pitt, R., Field, R., Lalor M., and M. Brown. 1995. Urban stormwater toxic pollutants: assessment, source and treatability. *Water Environment Research*. 67(3).

Raghavan, R., Koustas, R. and S. L. Liao. 2001. Cost Estimating for Best Management Practices. In: Proceedings of the ASCE Urban Water Resources Research Council and Engineering Foundation Conference. August 19-21, 2001. Snowmass, CO.

Roesner, L, Bledsoe, B., and R. Brashear. 2001. Are Best-Management-Practice Criteria Really Environmentally Friendly? *ASCE Journal of Water Resources Planning and Management*. 27(3):150-154.

Rossman, L.A., Dickinson, R.E., Schade, T., Chan, C.C., Burgess, E., Lai, F-H. and D. Sullivan. 2003. Redevelopment of the US EPA Stormwater Management Model. In: Proceedings of WEFTEC 2003. October 11-15, 2003. Los Angeles, CA.

Rossman, L.A., Dickinson, R., Schade, T., Chan, C., Burgess, E. H., Sullivan, D. and F-H. Lai. 2004. SWMM 5 - the Next Generation of EPA's Storm Water Management Model.

In: *Innovative Modeling of Urban Water Systems, Monograph 12* (W. James, ed.), CHI, Guelph, Ontario.

Selvakumar, A. and M. Borst. 2004. Land Use and Seasonal Effects on Urban Stormwater Runoff Microorganism Concentrations. In: Proceedings of the World Water and Environmental Resources Congress 2004. June 27 - July 1, 2004. Salt Lake City, UT.

Selvakumar, A. and M. Borst. 2003. Microorganism Die-Off Rates in Urban Stormwater Runoff. Presented at: WEFTEC 2003 76th Annual Technical Exhibition and Conference. October 2003. Los Angeles, CA.

Strecker , E. W., Quigley, M. M., Urbonas, B. R., Jones, J. E., and J. K. Clary. 2001. Determining Urban Storm Water BMP Effectiveness. *Journal of Water Resources Planning and Management.* 127(3):144-149.

U.S. Department of Agriculture. 2000. *Agricultural Resources and Environmental Indicators*. Economic Research Service, Resource Economics Division.

U.S. Environmental Protection Agency. 2000. Our Built and Natural Environments, A Technical Review of the Interaction Between Land Use, Transportation and Environmental Quality. US EPA Report No. 231/R-00-005, Washington, DC.

U.S. Environmental Protection Agency. 2002a. *The Quality of Our Nation's Waters: 2000* US EPA Report No. 841/R-02-001. Office of Water, Washington, D.C.

U.S. Environmental Protection Agency. 2002b. *Considerations in the Design of Treatment Best Management Practices (BMPs) to Improve Water Quality.* US EPA Report No. 600/R-03/103. Office of Research and Development, Cincinnati, OH.

U.S. Environmental Protection Agency. 2004a. The Use of Best Management Practices (BMPs) in Urban Watersheds, in preparation.

U.S. Environmental Protection Agency. 2004b. Development of an Innovative Filter Fence for Sediment Control to Address Inefficiencies with Current Practices at Construction Sites, in preparation.

U.S. Environmental Protection Agency. 2004c. Performance Evaluation of Best Management Practices Using Monitoring and Sampling Data, in preparation.

U.S. Environmental Protection Agency. 2004d. Stormwater Management Model Redevelopment Project, in preparation. http://www.epa.gov/ednnrmrl/swmm/index.htm

U. S. Geological Survey. 2003. Diurnal Variation in Trace-Metal Concentrations in Streams. USGS Fact Sheet No. FS-086-03. December 2003.

Wang, X., Nietch, C. Huang, H., Goss, M. Hoffa, S. and J. Hamaker. 2004. Adopting the Problem Based Learning Approach in a GIS Project Management Class. A presentation at the Association of Collegiate Planning Schools Annual Conference. October 21-24, 2004. Portland, OR.

Water Environment Research Foundation. 2004. Post Project Monitoring of BMPs/SUDS to Determine Performance and Whole-Life Costs. WERF Stock Number 01CTS21T. Alexandria, VA.

XP Software. 2004. XP-SWMM.

Chapter 2

Milwaukee's Integrated Approach, A Watershed Moment

Kevin Shafer, Milwaukee Metropolitan Sewerage District

Introduction

Recently, the City of Milwaukee, Wisconsin embarked on a long- term planning effort based on the Environmental Protection Agency's watershed approach. The use of this integrated approach was prompted by the understanding that if every sanitary sewer overflow was stopped tomorrow, non-point pollution would still cause beaches to close and make fish unfit to eat. The Milwaukee Metropolitan Sewerage District (MMSD) launched this effort to develop an approach to reduce non-point water pollutants. To insure success, MMSD believes it must increase public awareness and public education of the sources of these pollutants in order to reduce non-point pollution. The use of good stormwater management principles by the general public is a major component of this effort.

For these principles to be effective, the public must be informed about how water becomes polluted and what manmade influences induce flooding. This public education must start with the basic understanding of what happens to a drop of water when it lands on an individual's property. Knowing how that drop of water becomes polluted, how it can cause flooding, and how it can cause sewer overflows is critical to any stormwater management program. Public outreach, public education, and public collaboration are the tools to making the leap to less flooding and cleaner waterways; stormwater management cannot be effective without it.

Background

The MMSD provides wastewater collection, treatment services and flood management services to approximately 1.1 million citizens. The MMSD service area includes 28 separate communities within 420 square miles. Of this 420 square mile total, 397 square miles are a separated sewer system and 23 square miles are a combined sewer system. Wastewater from this area is collected and treated at two wastewater treatment plants, Jones Island and South Shore. The plants process approximately 200 million gallons per day of wastewater discharging their effluent into Lake Michigan. Similarly, stormwater runoff in the Milwaukee area drains through 7 major watersheds, all of which discharge into Lake Michigan.

As with other major metropolitan areas in the United States, Milwaukee continues to wrestle with the effects of severe wet weather events inducing combined and/or separate sewer overflows. These overflows go into the area's river system which drains into Lake Michigan. Lake Michigan is also the drinking water supply for much of the Milwaukee region. In the early 1990's, a cryptosporidium outbreak in Milwaukee resulted in several deaths. This outbreak created a heightened sense of awareness to Lake Michigan's water quality. All of these issues led to intense public scrutiny by MMSD's stakeholders with the end result being a public demanding zero sanitary sewer overflows.

An integrated long-term planning effort completed in 1994 included the construction of a 405 million gallon deep tunnel system to temporarily store sanitary and stormwater flows. Since completion of this effort, the MMSD has seen the number of occurrences of separate sewer overflows fall from an average of 25 per year to 1.3 per year and the number of combined sewer overflows fall from an average of 50 per year to 2.3 per year. Yet, the Milwaukee region is still recording increased beach closings due to high bacteria counts, days when swimming in area waterways was not advisable and times when the fish caught in area waterways were not safe to eat. To help determine why this was happening, MMSD commissioned a study with the University of Wisconsin's Great Lakes Water Institute to perform DNA fingerprinting of the lake water at our beaches. This study showed that non-point pollution from beach runoff was the source of 80% of the bacteria that were closing Milwaukee's South Shore beach. This report, along with others performed nationally; show that non-point pollution was the number one cause of these water pollution concerns. Yet, the general public in Milwaukee still demands zero overflows.

Integrated Watershed Management

Due to these concerns, Milwaukee has embarked on the next generation of watershed management. Public outreach, education, and understanding are the basic building blocks this next generation will be built upon. This approach, called the Water Quality Initiative, uses the Environmental Protection Agency's watershed approach to deal with water quality issues at their source. The watershed approach analyzes pollutants found in the waterway and determines the sources of the pollutant. In this manner, stormwater runoff from agricultural and urban land forms is analyzed uniformly with sanitary sewer overflows and wastewater treatment plant effluent. Prioritizing the impacts of each of these sources and focusing resources on the high priority areas will begin to show real water quality benefits in a timely and cost-effective manner.

One element of the Water Quality Initiative is to initiate better management of stormwater runoff. The most important component of stormwater management is implementation guidance and public acceptance. To this end, the MMSD utilized a multi-faceted approach to dealing with stormwater management. The elements of this approach include:

Implementation of a regional flood management program to reduce existing and future flooding
Adoption of a Regional Stormwater Runoff Rule
Implementation of a Regional Stormwater Management Strategy

Regional Flood Management Program

The regional flood management program was implemented in response to two 100–year storm events in 1997 and 1998 that caused major flooding throughout the region. In addition to the goal of protecting our customer's investments from flooding, MMSD determined that all flood management alternatives developed under this program would also have to improve the environment. We focused on striving to remove concrete lining from streams and rivers, improving the natural habitat as well as advancing the quality of life in the adjoining neighborhoods. The product of this effort was an aggressive $300 million program striving to remove every inhabitable structure from the 100-year floodplain by the year 2010. When completed, this structural approach to flood management will provide a 100-year level of protection for the future land uses projected for the year 2020.

To address flood management beyond the year 2020, MMSD embarked on an innovative approach in 2000 to preserving undeveloped flood plain properties. Two projects were created, the *Conservation Easement* project and the *Greenseams* project. These two projects involve working with local community groups, municipalities, and others to acquire properties that are either riparian to the waterways, or are identified as critical to protect against future flooding and improve water quality. In all, this project has identified over 7,000 acres in the Milwaukee area that contain the necessary conditions to provide future flood reduction benefits. Additionally, these properties will remain as open space. Restoring wetlands at these sites will not only help to reduce future flooding, but will also remove water pollutants. By removing development pressures at these critical sites, the *Greenseams* project provides an open, natural vegetated buffer area along the waterways, allowing a greater reduction in non-point pollution.

Regional Stormwater Rule

Concurrent to implementing the regional flood management program, MMSD began the process of developing a regional stormwater runoff rule to reduce the risk of future flooding from redevelopment and new development. The first step to developing such a rule is to determine what ordinances existed to minimize future increases in stormwater runoff. A review of local ordinances shows that while some communities do a very good job of managing increased stormwater runoff, others have virtually no controls to manage this. In order to "level the playing field," between the communities to control runoff from new development and from redevelopment, the MMSD worked with all of its community leaders to develop a regional stormwater rule. This rule was implemented in 2002. The rule states that for any development or redevelopment that adds more than one half of an acre of impervious area to a site, the runoff from that site must be limited to 0.15 cfs/acre for the two year storm and 0.5 cfs/acre for the 100-year storm. Each community was required to change their local ordinances within one year of rule implementation to comply with the MMSD rule.

Stormwater Best Management Practices

Beginning in 2002, MMSD initiated a series of stormwater demonstration projects to highlight the various benefits of properly managing stormwater runoff from individual parcels. These projects establish partnerships with communities, developers, not-for-profit organizations, and universities to design, build, and monitor various stormwater reduction practices. Initially, these practices included porous pavements, rain gardens, green roofs, rain barrels, and bioremediation sites. The first of these projects were built in late 2002 and throughout 2003. The projects popularity with MMSD's partners prompted MMSD to take the next step forward; creating a Strategic Plan for Stormwater Reduction. This Strategic Plan will illustrate the effectiveness of these stormwater reduction practices on a system wide basis. The strategic plan targeted practices that were cost effective, easy to implement on a wide-scale basis, simple and straight forward to maintain, and able to provide both water quality and volume/peak discharge reduction benefits. The plan was adopted by the MMSD in March 2004. The three objectives of the plan were:

1. Reduce impacts of stormwater on the District's sewer system. This objective is aimed at reducing operation and maintenance costs of the sanitary system, increase available storage for the sanitary system, and ultimately help reduce sanitary sewer overflows.

2. Provide environmental benefits related to stormwater reduction. This objective is targeted at improving water quality in the waterways, reduce erosion and scouring problems, as well as improve green space and habitat.

3. Enhance public education regarding stormwater reduction. This objective will improve public involvement and environmental stewardship.

MMSD evaluated several potential stormwater reduction practices such as downspout disconnection, rain barrels, cisterns, rain gardens, green roofs, rooftop storage, green parking lots, stormwater trees, porous pavement , inlet restrictors/pavement storage, bioretention, onsite filtering practices, pocket wetlands, French drains/dry wells, infiltration sumps, and compost amendments.

Each practice was evaluated according to the following factors:

Flow impacts
Environmental impacts
Implementation impacts
Function
Operational maintenance
Potential to promote public involvement and awareness
Capital costs
Cost per impervious acre served
Cost per gallon of stormwater reduced

Based on these criteria, the residential practices recommended for detailed evaluation were downspout disconnections, rain barrels, rain gardens, compost amendments, and stormwater trees. In many cases, it was envisioned that these practices would be used in combination. An example of this would be downspout disconnection being directed into a rain barrel, the rain barrel would then overflow into a rain garden. This approach would create a stormwater train of practices. In commercial or industrial settings, green roofs, rooftop storage, green parking lots, porous pavement, and bioretention were recommended for further analysis.

The District also performed comprehensive hydraulic modeling to evaluate the anticipated performance of the stormwater runoff reduction practices. A hydrologic computer model was used to investigate impacts on peak storm flow and annual flow volume within the separate sewer service area and the combined sewer service area. The modeling analyses indicated the following:

Stormwater runoff reduction practices are effective in reducing stormwater runoff within the separate sewer service area. However, some practices, such as downspout disconnections and rain gardens, could potentially increase infiltration and inflow to sanitary sewers. Other practices, such as green roofs, rooftop storage, and bioretention, would reduce runoff without increasing the risk of infiltration and inflow to sanitary sewers.

Within the combined sewer service area, most stormwater runoff reduction practices would significantly reduce stormwater flow to the combined system, which could result in reduced combined sewer overflows and a reduction in the amount of stormwater that would be treated at the District wastewater treatment plants.

The modeling analyses concluded that stormwater runoff reduction practices are fairly effective at reducing peak flows and providing volume reduction benefits. Within residential areas, downspout disconnections combined with rain barrels and/or rain gardens were found to be most effective. Within commercial or industrial areas, the most promising practices were bioretention, green parking, porous pavement, and green roofs. To minimize the potential increase in infiltration and inflow to sanitary sewers, more demonstration projects in the separate sewer area were recommended.

However, in the combined sewer area, residential practices should be initiated immediately.

One of the elements of the strategic plan was to identify the criticality of a public education program to support the rest of the plan. The purpose of the public education program is to determine and demonstrate the information and education needed to convince landowners and communities to implement stormwater runoff reduction practices; to increase participation (and the resultant effectiveness) in the pilot projects, and to develop and promote general awareness and knowledge of stormwater management issues. The vehicle used to achieve this is the *Roadmap to Stormwater Management*. The *Roadmap* will serve as the vehicle for the information and education needed to support the pilot projects. The Roadmap will contain fact sheets, brochures, and a stormwater poster which will be designed to work in concert. The information will be tailored to address the needs and concerns of each project and will be distributed in a variety of ways.

Community partnerships are critical to the long-term success of MMSD's stormwater runoff reduction program. The active participation of established and trusted community organizations offers the project an opportunity to cast a wide net of participants and shared information. The most natural place to begin was with organizations who the District had existing relationships with, either through technical projects or like-minded missions to improve the environment.

These partners will be approached based on the practices or projects selected for implementation and their anticipated level of interest. Those with high interest and level of commitment will be contacted first. In later phases of the project and in years to come, additional groups may be targeted. For this phase to be able to model the effects of the stormwater runoff reduction, there is a need for significant numbers of participants, thus reinforcing the need to go where success is most likely.

Conclusion

For the past 20 years, MMSD has led the long-term planning efforts to reduce sanitary sewer overflows. While this planning has been very successful, area waterways are still polluted. The public's expectations of MMSD are also very high. These expectations include their desire for an immediate end to combined sewer overflows and water quality that meets fishable, swimable standards. In order to address these concerns, MMSD's role has been expanded to try to address non-point pollution and one of the impacts of this pollution, beach closings. A new integrated planning effort based on the watershed approach is currently being implemented that will allow for public outreach, education and gather input from them. Using water quality as the basis for this planning effort is unique and will result in a new phase of projects, different from those typically part of a sewer utility's long term plan. Strong science, public involvement, and this integrated plan will change the way Milwaukee deals with its waterways. Truly, a watershed moment!

Chapter 3

Integrated Approach to Urban Stormwater Management

Tony H F Wong

Associate Professor, Department of Civil Engineering, Monash University
PO Box 60, Monash University, Vic 3800, Australia

Introduction

The pursuit of "sustainability" has emerged in recent years as a progression from previous "environmental protection" endeavors. The ambition of "sustainability" and "sustainable development" is to have lifestyles, and their supporting infrastructure, that can endure indefinitely because they are neither depleting resources nor degrading environmental quality. Land development activities that do not consider sustainability for a growing community can have detrimental impacts on the physical environment, including the health and amenity of water environments.

Water environments, such as waterways and coastal waters, and water supply catchments are key areas where urban development can have significant impacts. In Australia, Ecologically Sustainable Development (ESD) initiatives can be described as going beyond the protection of the environment from the impacts of pollution, to also protecting, conserving and restoring natural resources. Environmental sustainability can be described as a condition where there is a zero net environmental cost associated with development activities. While such ambitions may seem beyond reach, they set a challenge that can reap wide ranging benefits – environmental, social and economic – with each step towards the ultimate goal of sustainability.

The key objective of sustainable water resource management is to protect drinking water from contamination by stormwater and wastewater. Managing the impacts of urban development on the water environment must include attention to all three streams of the urban water cycle and necessitates an integrated approach that is now widely acknowledged as achieving water sensitive urban development through Water Sensitive Urban Design. Thus, while the key theme of this paper relates specifically to integrated urban stormwater management, it would be inappropriate to discuss this without first placing into context the overall issue of integrated urban water management.

Integrated urban water management has many dimensions including:

- Integrated management of the three urban water streams
  - Potable water
  - Wastewater
  - Stormwater
- Integration of scale
  - Individual allotments and buildings
  - Precinct
  - Regional
- Integration into the built form
  - Building architecture
  - Landscape architecture
  - Public art
- Integration of structural and non-structural initiatives
  - Policies underpinning the sustainable operation of structural measures
  - Physical and social sciences
  - Setting design objectives that are consistent with the capability of existing technologies (best practice)
  - Simplified and transparent assessment of compliance to design objectives

In Australia, Water Sensitive Urban Design (WSUD) has evolved from its early association with stormwater management to provide a broader framework for sustainable urban water management. It is a framework that provides a common and unified method for integrating the interactions between the urban built form (including urban landscapes) and the urban water cycle. In Australia it is increasingly practiced in new urban greenfield development areas and urban renewal developments linked to a broader ESD agenda. Key guiding principles of WSUD include:

1. Reducing potable water demand through water efficient appliances and seeking alternative sources of water such as rainwater and (treated) wastewater reuse, guided by the principle of "fit-for-purpose" matching of water quality and end uses.
2. Minimising wastewater generation and treatment of wastewater to a standard suitable for effluent re-use opportunities and/or release to receiving waters.
3. Treating urban stormwater to meet water quality objectives for reuse and/or discharge to surface waters.

4. Using stormwater in the urban landscape to maximise the visual and recreational amenity of developments.

These principles are equally relevant to achieving sustainable developments in other places with differing water infrastructure systems to the Australian context. In some countries, the provision of water services (water supply and wastewater disposal) remains separate institutional entities from that of stormwater management. Within this context, WSUD initiatives at the local development level in such countries may predominately be guided by principles 3 and 4 directed at the management of urban stormwater both as a resource and as a source of pollution of the receiving water.

Managing urban stormwater as a resource for non-potable domestic use (eg. toilet flushing) remains an important consideration in spite of ample availability of (treated) recycled wastewater, as the latter requires increased consumption of energy to treat wastewater to a level that is largely not necessary for such uses.

The urban environment generates many pollutants that are ultimately "washed-off" by stormwater and discharged to the receiving water. These pollutants can lead to significant environmental degradation, loss of urban waterways, estuaries, and bays. Improving stormwater quality to protect and improve the health of urban waterways is an essential element of sustainable urban development. The integration of stormwater pollution control measures into the urban landscape and built form provides opportunities to maximise the visual and recreational amenity of developments.

Other countries, such as Singapore, have established centralized systems for wastewater treatment and reuse. The Government of Singapore has taken on the responsibility of integrating the water supply and wastewater management functions through its NEWater Scheme. While largely to meet water demand for industrial use, NEWater is treated to a potable standard and is the single most transformative step in water resources management towards "closing the loop" in Singapore's urban water cycle. In Australia, such schemes are emerging at a regional level, most notably the Rouse Hill Reclaimed Water Scheme, where treated wastewater is reticulated back to residential area through a separate pipeline for garden irrigation, toilet flushing, and laundry use.

A Changing Attitude to Urban Drainage

The management of stormwater runoff in conventional urban developments has been driven by an attitude that reflects the view that stormwater runoff has no value as a useful resource, is environmentally benign, and adds little to the amenity (aesthetic, recreation, education, etc) of an urban environment. Consequently, conventional urban stormwater management has focused on providing highly efficient drainage systems to rapidly collect and remove stormwater runoff using a combination of underground pipes and linear "engineered" overland flow paths (often located along the back fence line of properties to keep them out of sight). These systems kept stormwater runoff "out of sight" and consequently "out of mind."

The increased rates of stormwater runoff associated with conventional urban development coupled with a dramatic increase in stormwater runoff volume and associated contaminants such as litter, sediments, heavy metals, and nutrients has caused significant degradation of the natural environments. Increased flow rates have led to erosion of natural channels leading to urban waterway protection works which are often associated with channel lining, straightening and piping. These have led to conveyance of the stormwater pollutants to receiving waters further downstream.

To broadly address the environmental imbalance caused by conventional urban development practices, changes to the underpinning philosophy that drives conventional urban planning and design is required. It is becoming well accepted that a more holistic and multi-disciplined approach is needed to ensure planning and design decisions are made with a reliable understanding of the environmental, social, and economic consequences.

As part of this new paradigm in urban management, the treatment of stormwater runoff can no longer be considered in isolation to the broader planning and design of the contributing urban area. Rather, stormwater management needs to be considered at all stages of the urban planning and design process to ensure that site planning, architecture, landscape architecture and engineering infrastructure is provided in a manner that is sympathetic to the stormwater treatment system and receiving environment. Similarly, the stormwater treatment system needs to be sympathetic to the requirements of each of the other urban infrastructure elements in order for the "whole" package to function as an ecologically, socially, and economically sustainable urban system.

In contrast to the narrow flood protection focus of conventional stormwater management, the WSUD approach requires that urban stormwater management practices address multiple objectives in an integrated manner. These include managing stormwater as a valuable resource, protecting water quality and ecological health of receiving waterways and water bodies, and managing the risk to public safety from flooding.

Many cities now aspire to re-naturalising urban waterways, through initiatives such as transforming concrete-line waterways to natural systems at significant cost. One such notable project is the Cheongeceong Project in Seoul, Korea where an entire road infrastructure has been removed to accommodate the re-establishment of an old watercourse (see Figure 1). Such projects pose design challenges associated with the need to preserve flood discharge capacity in the waterway for a largely urbanized catchment, while attempting to rehabilitate the waterway to attain a higher ecosystem value.

Figure 1. Re-establishing a natural watercourse through the City of Seoul [Cheongeceong Project]

There have been some innovative solutions that have combined geomorphic re-establishment of a natural channel and water quality improvement. These involved a combination of innovative hydraulic designs that separate a natural low flow channel from a highly efficient underground high flow conveyance system, in combination with catchment stormwater treatment measures retrofitted into the built environment.

The success of WSUD as an urban planning and design paradigm will rest largely on the ability of the urban design industry to provide engaging and informative landscape design solutions within the public realm. These solutions should tell a 'story' about the interactions that occur between the human built form and the natural environment. The use of innovative landscape elements that show the connectivity between human activity and the urban water streams are increasingly recognised as having a powerful influence on the consciousness of individuals and recognition of their role and responsibility in the protection and enhancement of our natural water resources. By contrast, there is a distinct lack of visual connectivity between human activity, urban water streams, and receiving natural waterways with the conventional "piped" stormwater system. Thus within the conventional urban setting it is difficult for individuals to see, and indeed understand, the impact of their actions on the health and sustainability of our natural water resources.

Cooperative collaboration between the urban design professions can achieve "smarter" and more sustainable urban areas where urban landscapes engage, inform, and influence human behavior for the benefit of the natural environment and the improvement of the social fabric.

Integration of Scale

Best practice in sustainable water management encompasses elements at the site, precinct and regional levels, which can be combined to provide a strategic framework for integrated catchment management. Many of the concepts and types of treatment measures adopted in formulating integrated urban water management plans are applicable at all three scales as illustrated in Table 1.

Table 1 Inter-relationship between site-precinct-regional WSUD initiatives

Local	Precinct	Regional
Planning		
▪ Allotment density and layout	▪ Local street layout	▪ Major road layout ▪ Public open space and multi-use corridor
Water Conservation		
▪ Rainwater Tank	▪ Stormwater Storages/Pond ▪ Aquifer Storage & Recovery	▪ Stormwater Storages/Pond ▪ Aquifer Storage & Recovery
Stormwater Quality		
▪ On-site retention (infiltration) ▪ Porous pavement (driveway) ▪ Sand filters ▪ Bioretention planters ▪ Raingardens ▪ Vegetated buffers	▪ Precinct-wide infiltration basins ▪ Porous pavement (public car parks) ▪ Sand filters ▪ Bioretention swales ▪ Bioretention basins ▪ Vegetated swales ▪ Urban forests ▪ Constructed wetlands	▪ Riparian buffers ▪ Natural channels ▪ Urban Forests ▪ Constructed wetlands
Stormwater Detention		
▪ On-site Detention	▪ Retarding basins ▪ Ponds	▪ Retarding basins ▪ Lakes

Many of the WSUD elements presented in Table 1 share a common design and operation approach and philosophy with their differences merely in their spatial scale.

In relation to urban stormwater management, good site planning practices apply across the three spatial scales in terms of allotment layout, street layout, and regional planning of roads and public open space and multiple use corridors. Similarly, rain gardens and bioretention systems are used as local allotment WSUD elements for treatment of roof and pavement stormwater runoff operate in a similar way as street-scale bioretention system and urban forest parks treating catchment runoff from a larger, precinct-scale catchment.

Scientific research has been able to show that the ratio between the required size of these systems and the stormwater runoff contributing area to ensure effective treatment to meet the best practice objectives are essentially the same. Through close collaboration with landscape architects and urban designers, it has been possible to incorporate many of these technologies into the urban form at a range of spatial scale. The range of scales of constructed wetlands and bioretention systems that have been typically implemented in recent projects are discussed further in the next section.

Integrating Stormwater Treatment Measures into the Urban Landscape
Two of the most common stormwater treatment technologies that can be readily integrated into urban design are constructed wetlands and bioretention systems.

- The use of *constructed wetlands* for urban stormwater quality improvement is widely adopted in many Australian cities, many of which have been successfully incorporated into the urban landscape. Design considerations include the interaction between the wetland hydrology and hydrodynamic behaviour with the various physical, chemical, and biological treatment processes. The operating conditions of these systems are stochastic in nature, with intermittent and highly variable hydraulic and pollutant loading. Research and on-going refinement to practice in Australia have provided a rigorous basis for sizing constructed wetlands for stormwater management and for its integration into landscape design (Wong and Breen, 2002).

- Constructed swales systems are commonly used for stormwater conveyance, especially in non-urban areas. Recent adaptations of swale systems for stormwater quality treatment are directed at promoting a higher degree of stormwater treatment by facilitating infiltration of stormwater through a prescribed soil media. These systems are referred to as *bioretention systems,* where a trench filled with a "prescribed" soil of known hydraulic conductivity is used as a filter media. Recent research and monitoring of field applications have demonstrated that they present the best "soft-technology" for removal of urban stormwater pollutants. They can be readily incorporated in streetscape and linear parks where space may pose a constraint to other stormwater treatment measures. When designed with appropriate soil media and planting, these systems can have a long-term capacity to assimilate heavy metals washed off urban catchments.

Local and Precinct Scale Constructed Wetlands and Bioretention Systems

Recent developments in stormwater treatment technology have been able to confirm the scalability of stormwater treatment technologies for application in small confined areas (Wong, 2001). Through close collaboration with landscape architects and urban designers, it has been possible to incorporate many of these technologies into the urban form at a range of spatial scales.

Local Scale- Public building forecourts and local streetscapes represent the connecting pathways between the buildings where we live and work and the areas of sub-regional and regional public open space where we interact and recreate at a local and regional community scale. In terms of stormwater runoff generation, public building forecourts and local streetscapes represent public realm areas located closest to the source of most urban stormwater runoff (i.e., from impervious surfaces associated with buildings and road pavements). Integration of stormwater management functionality within landscape elements associated with public building forecourts and local streetscapes allows for a number of key WSUD best management practices to be satisified, namely: collection and treatment of stormwater runoff at its source; first use of stormwater runoff for watering the landscape; and visual connectivity between the built form and the urban stormwater stream and the amenity of the local landscape.

Precinct Scale- Precinct scale public open space areas provide an opportunity to integrate stormwater collection, treatment and storage/re-use facilities within the overall landscape design of these areas. With competing uses for these spaces the scale and landscape form of the stormwater management systems needs to carefully consider the other uses of the park and their potential interaction with the stormwater management systems. Issues of public safety and aesthetic amenity are important design considerations requiring site analysis to determine site usage patterns, journeys, site lines, and existing landscape character in order to ensure an appropriate landscape form.

A range of constructed wetlands and bioretention systems at various scales that have been implemented in recent projects over the last three years are shown in Figures 2 and 3.

Hampton Park Wetland (Melbourne Water)

Huntingdale Rd Retarding Basin Wetland

Redan Creek Wetland

NAB Building Forecourt Wetland

Lynbrook Estate Wetland

Figure 2. Various scales of stormwater treatment wetlands recently constructed in Australian projects

Figure 3. Various scales of stormwater bioretention systems recently constructed in Austra
projects

Integration of Stormwater Management Measures into the Built Form

New initiatives involving collaboration with building architects have extended the application of WSUD into a new dimension related to their integration into the building form. Figure 4 illustrates a proposed new apartment building in Sydney that attempts to "close-the-loop" with respect to the indoor water budget. The building is to be "plumbed" differently such that there is a separate collection of the water from the showers, baths, and hand basins to deliver this greywater to a package treatment plant in the basement of the building. This treated greywater is then pumped to a header tank in the roof of the building and gravity fed to the toilets.

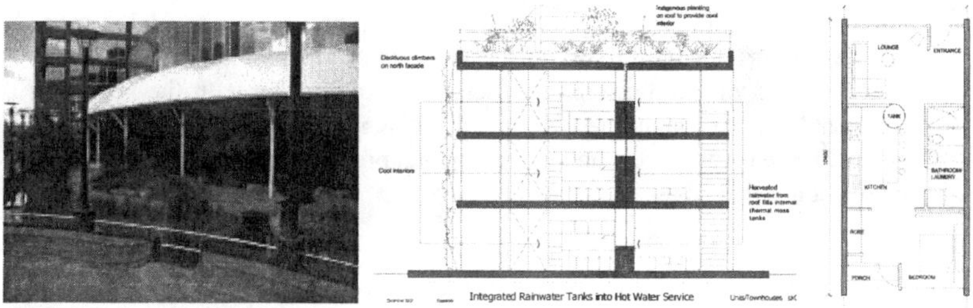

Figure 4. Integrating water management systems into buildings

A variation of this concept with particular emphasis on managing stormwater as a resource is the harvesting of roof stormwater runoff. This runoff could be treated with bioretention or constructed wetland systems laid out in a roof-garden and delivered to architecturally-designed rainwater tanks that are incorporated into individual apartments for toilet flushing. Contrary to the many misconceptions of roof top gardens, the entire roof space does not need to be fitted with vegetated areas. Often, the vegetated treatment areas need only to take-up 2% to 5% of the roof area to adequately treat stormwater runoff.

Figure 5. Rainwater tank configured as part of a wall

Integration of other stormwater treatment measures within building forecourts and atriums provide further opportunities for stormwater quality treatment and examples of these have been discussed in the previous section of this paper.

At a local domestic household, the use of rainwater tanks that are configured to form part of a boundary fence can result in significant optimisation of available space as housing allotment size decreases with increasing urban density (see Figure 5).

Integration of Structural and Non-structural Initiatives

Urban development is largely a private domain enterprise with government providing the framework within which land development is to take place. The most significant non-structural initiative is that of government policies on ESD, and in the case of urban water management, on WSUD. Policies at the state and local government levels often stop short of specifying performance objectives and this has led to confusion and uncertainty amongst stakeholders in the industry, rendering many of these policies being the subject of debate in their interpretations and hindering widespread implementation of WSUD practices.

Sustainable urban water management needs to be defined by a quantitative measure of performance that can be linked to objectives for sustainable environmental outcomes. These performance measures or design objectives of integrated urban water management strategies can be expected to evolve and to be adaptive as knowledge in sustainable urban water management itself is still an evolving field of multi-disciplinary sciences. In keeping with the principles of sustainable management, these design objectives should be risk-based that allows good balance between cost and environmental protection and allows flexibility for innovation and change.

Percentage reduction in per capita potable water usage and wastewater generated that is discharged to the water environment are fairly well accepted design objectives. Wastewater minimisation is linked to demand management and reuse of treated wastewater and is closely associated with the objectives of water conservation. Its key sustainability indicator and target are thus also closely aligned. In terms of stormwater quality, there is an on-going debate about whether the design objectives should be based on meeting probabilistic pollutant concentration levels or mean annual load reductions from that typically generated from urban environments (in the absences of any intervention measures). Irrespective of the form in which design objectives for integrated urban water cycle management are established, performance measures need to be separated from particular technologies that may be used to achieve performance (i.e., they need to be measures of the outcome not of a particular intervention).

Changes to policy and administrative framework are fundamental to underpinning the adoption of best practice in integrated urban water management. However, they can only be effective if the industry is equipped with the appropriate skill base and technology to implement WSUD strategies on the ground. Similar, regulatory authorities need to have the necessary tools to assess and approve WSUD proposals. Industry WSUD capacity building needs to encompass the full spectrum of industry and community engagement, from raising awareness of the pertinent issues of water sustainability amongst local communities and their political leaders to the business of engineering design and construction. Local experience with water sensitive urban design and availability of technical information is growing.

Conclusions

The concept of Water Sensitive Urban Design is based on formulating development plans that incorporate multiple stormwater management objectives and involves a pro-active process which recognises the opportunities for urban design, landscape architecture and stormwater management infrastructure to be intrinsically linked. WSUD espouses the need to integrate stormwater management into the planning and design of urban areas and applies across the entire spatial scale from a catchment-wide regional level to the precinct level to the local, building level. Opportunities for the innovative integration of stormwater management functions within contemporary urban landscape designs at a range of scales within the public realm and private buildings were presented in this paper. A shift towards "at source" stormwater management systems will further advance the development of innovative on-site, streetscape and precinct scale landscapes incorporating stormwater management functionality.

The challenge ahead is to continue the collaborative approach to urban design and to provide technically robust solutions that excite and engage communities and deliver improved environmental outcomes for today and future generations.

References

Wong, T.H.F. and Breen, P.F. (2002), *Recent Advances in Australian Practice on the use of Constructed Wetlands for Stormwater Treatment* proceedings of the 9th International Conference on Urban Drainage, Portland, Oregon, USA, 9-13 September 2002.

Wong, T.H.F. (2001), *A Changing Paradigm in Australian Urban Stormwater Management*, keynote address, 2nd South Pacific Stormwater Conference, Auckland, New Zealand, 27 – 29 June 2001.

Chapter 4

Best Management Practices (BMPs) and U.S. Environmental Protection Agency (EPA) Programs

Michael Borst, Ariamalar Selvakumar, William D. Shuster, Mary K. Stinson, Scott D. Struck, Michael A. Taylor, Hale W. Thurston, Joyce Perdek Walling

Introduction

This chapter contains three sections by U.S. EPA authors that reflect contribution of the Agency's research programs towards protection of watersheds through better understanding of capabilities and beneficial uses of BMP technologies. The topics presented here blend well with those presented by "outsiders," which proves that the EPA keeps up well with the entire scientific community of BMP experts. The three EPA sections range from a state-of-the-art review of BMPs' ability to manage pathogen contamination; a field study, which determined variability of pathogen concentrations in stormwater runoff depending on the land use and on the season of the year; and an ingenious, novel program of economic incentives for those who install BMPs on their properties for detention of excess runoff.

Managing Microbial Contamination in Urban Watersheds deals with pathogen contamination of urban watersheds, which is a serious problem with public health implications. In urban watersheds, wet-weather flow (WWF), such as combined sewer overflow (CSO), sanitary sewer overflow (SSO), and stormwater runoff, carry significant loads of fecal and pathogen contamination that should be prevented from entering an urban receiving water body. Most of the WWF are point sources, but some are diffuse sources. Importance of disinfection for point sources is stressed, because this practice stops pathogens from entering the watershed and its water bodies. Since issuance of the National CSO Control Policy of 1994, CSO has become the most frequently disinfected and studied type of urban WWF. Disinfection of SSO and collected, confined stormwater is also feasible and protects the watershed where implemented.

For management of diffuse pathogen sources in a watershed, structural and nonstructural BMPs are considered. Though effectiveness of structural BMPs for controlling pathogens has not yet been fully established, there are references with quantified pathogen removals. Nonstructural BMPs have been less studied, but they are effective in protecting the watershed by controlling pathogen sources. Most likely, diffuse sources of pathogens will continue to be managed by structural and nonstructural BMPs, and a rise in reporting on the effectiveness of these systems for pathogen reduction, positive and negative, is occurring.

Land Use and Seasonal Effects on Urban Stormwater Runoff Microorganism Concentrations provides actual data on concentrations of selected individual pathogens (*Pseudomonas aeruginosa* and *Staphylococcus aureus*) and several indicators (total coliforms, fecal coliforms, fecal streptococci, enterococci, and *E. coli*) at three different land-use designations (high-density residential, low-density residential, and landscaped commercial) in four seasons of the year. The pathogens and indicators concentrations were determined in stormwater runoff samples collected from separate storm sewers, each dominated by a single land use. As expected, organisms' concentrations from high-density residential areas were the highest. There was strong seasonal effect on organisms' survival at all land uses, with the lowest concentrations observed during winter. Collection of real data is important in selecting treatments, including BMPs, for a specific land use and, thus, achieving effective protection of the watershed and its receiving waters. Association of the collected data with specific land uses and climatic conditions makes it easily transferable to other geographical areas and applicable for use in models as well.

Trading Allowances for Stormwater Control: Hydrology and Opportunity Costs evaluates use of economic incentives for homeowners who install small water detention BMPs on their properties. This down-to-earth project involves quite complex research on the use of a tradable allowance system to promote placement of dispersed, small-scale BMPs. With the use of hydrologic modeling and cost information on each BMP, which include opportunity costs of land dedicated to this BMP, scenarios are investigated that include diverse price and spatial constraints on placement of this BMP. In many urban areas, excess stormwater runoff causes flooding, water pollution, groundwater recharge deficit and ecological damage to urban streams. An economic reward system for reduction of stormwater runoff by property owners is a worthy environmental program.

Managing Microbial Contamination in Urban Watersheds

Introduction

Discussed in this section are capabilities of disinfection for reducing microbial concentrations in point sources and best management practices (BMPs) for reducing microbial concentrations in diffuse sources (Stinson and Perdek, 2003). Further, out of many point and diffuse sources contributing to watershed contamination, only sources resulting from wet-weather flow (WWF) are discussed. This is because all components of WWF, such as combined sewer overflow (CSO), sanitary sewer overflow (SSO), and stormwater runoff, carry significant amounts of fecal and pathogen contamination that should be prevented from entering an urban receiving water body.

WWF point sources of pathogens can be effectively treated by disinfection, while diffuse sources are commonly managed with the use of structural and nonstructural BMPs. There is a difference between disinfection and a BMP. The effectiveness of a given disinfection technology to reduce microbial contamination is usually known, and treatment results can be predicted with reasonable certainty. On the other hand, the effectiveness of pathogen load reduction by a given BMP may be just an estimate (Field, 1996).

Use of structural and nonstructural BMPs is the common practice for managing stormwater runoff. The primary function of BMPs is reduction of flow volume, sediment, or nutrients, while reduction of pathogens is a secondary function and not always quantified.

It is feasible to disinfect stormwater and dry-weather urban runoff after its collection and confinement. Though infrequent, examples of either studies or implementation are presented (Pavia and Powell, 1968; SMURRF, 2000; Rasmus and Weldon, 2003).

Disinfection Technologies for Control of Pathogens

Since issuance of the National CSO Control Policy (U.S. EPA, 1994), which requires disinfection of CSO after primary clarification, CSO has become the most frequently disinfected component of WWF. Most WWF disinfection studies, with conventional and alternative technologies, have been conducted on CSO (U.S. EPA, 2002a). However, all components of WWF, such as SSO and stormwater runoff, carry significant loads of fecal and pathogen contamination that would be reduced by disinfection. Numerous factors need to be considered in discussing WWF disinfection, of which most important are the need for a high-rate disinfection process and the need for suspended solids removal prior to disinfection.

WWF disinfection can be achieved at shorter contact times, termed "high-rate disinfection" (U.S. EPA, 1979a; U.S. EPA, 1979b; Stinson *et al.*, 1998). Use of conventional contact time of about 30 min for disinfection of WWFs is extremely costly because of their relatively high flow rates and intermittently occurring volumes. High-rate disinfection is accomplished by: (1) increased mixing intensity, (2) use of higher concentrations of disinfectant, (3) use of chemicals or irradiation with higher oxidizing rates or microorganism-kill potential, or (4) combinations of these (Field, 1990). The use of increased mixing with any disinfection technology provides better dispersion of the disinfectant and forces disinfectant contact with a greater number of microorganisms per unit time.

WWF disinfection requires some form of filtration or sedimentation prior to introduction of disinfecting chemicals (U.S. EPA, 1973). High levels of particulate matter in WWF can provide a "shielding effect" in which particles present in the medium protect the microbes from the disinfecting agent (Sakamoto and Cairns, 1997).

WWF Disinfection Technologies - On a comparative basis, alternatives to chlorination (Cl_2) disinfection technologies, for example UV light irradiation and chlorine dioxide (ClO_2), generate significantly less toxic byproducts and residuals. However, chlorination/dechlorination is still predominantly used for WWF disinfection. Of the alternative technologies discussed here, UV disinfection has entered the market for WWF. Disinfection technologies are discussed below in diminishing order of their popularity for WWF treatment.

Chlorination/Dechlorination - Disinfection by Cl_2 has proven to be effective, and has been used for wastewater disinfection in the U.S. since 1855 (White, 1999). Chlorination/dechlorination is being used for WWF disinfection. Chlorine is readily available in several forms, is inexpensive, and effective against bacteria, though not fully effective against viruses and protozoa. There are health risks associated with Cl_2. Excess of free Cl_2 can cause chlorinated hydrocarbon formation, i.e., chloroform and trihalomethanes (THMs), which are suspected carcinogens. Chlorine residuals discharged to natural waters may be harmful to aquatic life. Use of Cl_2 solution in the form of calcium hypochlorite and sodium hypochlorite is preferred to the use of Cl_2 gas. There is high risk of gas leaks when transporting gaseous Cl_2. Aqueous Cl_2 is more expensive than gaseous Cl_2, but is a better disinfectant and, therefore, a changeover from gaseous to aqueous Cl_2 is encouraged. Addition of a dechlorination process to remove the excess of free Cl_2, with either gaseous sulfur dioxide or sodium bisulfite solution, reduces the health risks, but raises the cost of disinfection by about 30%.

Ultraviolet Light Irradiation - Since the early 1900s, UV light irradiation from mercury arcs has been recognized as an efficient disinfecting agent. UV disinfection has recently entered the commercial market for WWF disinfection, which is shown later in the implementation examples. At the germicidal wavelengths, within the range of 200 to 320 nanometers (nm), UV light disinfects water by altering the genetic material in microbial cells, preventing reproduction. UV light irradiation affects a wide range of microorganisms and does not generate known harmful secondary chemical byproducts (e.g., THMs). UV light irradiation is a physical procedure that does not alter the smell or chemical composition of water. UV disinfection for WWF requires relatively high level of physical pretreatment (with or without chemicals) to make UV light more effective for WWF (Field, 1996).

High variability in WWF flow rates influences UV disinfection effectiveness, because flow rate is a principal determinant of the dosage of UV light necessary for effective disinfection (Wojtenko *et al.*, 2001). High levels of suspended solids and minerals present in WWF also reduce UV light effectiveness. Also, fouling materials deposited on quartz sleeves decrease transmittance of UV light and thus its disinfection capability (Oliver and Cosgrove, 1975). Use of an in-place cleaning system can remove fouling materials from the quartz sleeves.

Chlorine Dioxide - Although it has not yet entered the market for WWF disinfection, the use of ClO_2 for WWF disinfection is a promising alternative to chlorination. The lack of any significant reactions of ClO_2 with water is the main reason for its biocidal effectiveness over a wide pH range. Chlorine dioxide is a powerful disinfectant and provides excellent disinfection at a fraction of the Cl_2 dosage, making it cost effective and relatively safe (Moffa, 1975; Aieta *et al.*, 1980). In addition, the low reactivity of ClO_2 with ammonia and reduced formation of halogenated organic compounds are other major advantages over Cl_2. However, presence of organic and inorganic impurities in water creates a large oxidation demand for ClO_2 and reduces its effectiveness (Katz *et al.*, 1994). Pretreatment of the wastewater by filtration or sedimentation increases effectiveness of ClO_2 disinfection (Stinson *et al.*, 1999). Chlorine dioxide disinfection is effective for Cl_2-resistant pathogens (e.g., viruses and protozoa) (Moffa, 1975).

The possibility of using a combination of ClO_2 and Cl_2 was investigated for municipal wastewater treatment by Katz *et al.* (1994). After adding both agents in equal amounts, improved disinfection efficiency was observed with all doses, and production of the byproducts, such as chlorite ion (ClO_2^-) and THMs, was greatly reduced. Chlorine dioxide used in combination with Cl_2 also resulted in a lower residual Cl_2 concentration (Moffa, 1975).

A major obstacle to the commercialization of ClO_2 disinfection for WWF is the necessity of its generation on site with the use of gaseous Cl_2, which cannot be transported through the urban areas. There is a new process of ClO_2 generation that uses $NaClO_2$ in the presence of UV light (Stinson *et al.*, 1998). In this process the transport and handling of gaseous Cl_2 is totally eliminated but this process is still under development and is not commercially available. Chlorine dioxide appears to have potential for becoming an effective Cl_2 alternative for WWF disinfection. Further investigations, however, are recommended to determine its effectiveness in a full-scale WWF application (U.S. EPA, 2002b).

Examples of Disinfection Studies and Implementation

Pilot Study at the 26th Ward WWTP Testing Facility in New York City - This pilot study demonstrated alternative disinfection technologies to hypochlorite disinfection for application at the Spring Creek CSO storage facility and at other CSO facilities as well. From December 1996 through March 1997, five technologies were tested: chlorination/dechlorination, UV, ClO_2, O_3, and electron beam irradiation (E-Beam). From August through November 1999, three technologies were tested chlorination/dechlorination, UV, and ClO_2, because these had shown potential for CSO applications.

All disinfection technologies were tested on the same primary effluent. Chlorination/dechlorination, ClO_2, and O_3 achieved targeted bacterial reductions of 3 to 4 logs. UV disinfection achieved 4-log bacteria reduction but leveled off when the suspended solids (SS) level in treated effluent reached 150 mg/L. While ClO_2 was superior in effectiveness and similar in cost to chlorination/dechlorination, the on-site generation technology for ClO_2 which requires gaseous Cl_2, is not acceptable for use in New York City. Ozonation was not cost effective for CSO applications. The upgraded Spring Creek facility, currently under construction, will continue to use sodium hypochlorite for disinfection with provisions to add dechlorination at a later date. Improvements will be made to increase disinfectant flash mixing and to automate hypochlorite feed and residual control (U.S. EPA, 2002c).

Continuous Deflection Separation, Fuzzy Filter and UV Treatment of SSO-Type Wastewaters - A pilot study was conducted at the Rockland County Sewer District Wastewater Treatment Plant in Orangeburg, NY from August 1998 to January 2001. The study evaluated high-rate solids removal technologies for treatment of SSO and CSO overflows followed by high-rate UV disinfection. Solids removal technologies were: Continuous Deflection Separation (CDS) and Fuzzy Filter compressed media filtration. The UV disinfection studies evaluated three different lamp systems: low-pressure lamps from Wedeco Ideal Horizons, medium-pressure lamps from Aquionics, and generic medium-pressure lamps.

The CDS process is capable of accomplishing approximately 10% total suspended solids (TSS) removals with a 1200-micron screen and 30% with a 600-micron screen. The CDS process provides protection of downstream filters or other pretreatment devices by removing debris and floatables. The Fuzzy Filter filtration process was effective in removing particles greater than 50 microns. In this application, the system is effective at 20% compression and at hydraulic loadings between 400 and 800 Lpm/m^2 (10 and 20 gpm/ft^2). At these conditions, TSS removals averaged approximately 40%.

The combined results generated with the three UV units indicate that a degree of disinfection with primary wastewaters can be accomplished by UV radiation. Reductions between 2.3 and 2.8 logs can be achieved at hydraulic loadings between 8 and 38 Lpm/kW of lamp input power (2 and 10 gpm/kW) based on the enumeration of blended samples. This is equivalent to approximately 3 to 3.5 logs when enumeration is conducted using standard analyses without blending samples. Doses greater than 40 mJ/cm^2 are required to achieve these reduction levels (U.S. EPA, 2002c).

Advanced Demonstration Facility (ADF) in Columbus, GA - Chlorination/dechlorination of CSO, along with several alternative technologies, were tested during 40 wet-weather events at a full-scale ADF in Columbus, GA. The CSO testing program at ADF was a part of a multi-year watershed study sponsored by the Columbus Water Works Agency with WWETCO, Inc. as the principal contractor and with the involvement of the Water Environment Research Foundation and the U.S. EPA. The ADF is comprised of multiple CSO technologies arranged as treatment trains: hydraulic controls, screening, vortex separation, compressed media filtration, and chemical disinfection using Cl_2 as sodium hypochlorite, ClO_2, peracetic acid, and UV disinfection. Disinfectants listed in order of their effectiveness were ClO_2, sodium hypochlorite, and peracetic acid, however all were capable of accomplishing a satisfactory disinfection. Chemical dosing under similar conditions required 12 mg/L of ClO_2, 15 mg/L of sodium hypochlorite, and 16 mg/L of peracetic acid.

As a result of the studies, an optimized model of the ADF facility was developed. The optimized facility includes two 32-ft diameter vortex separators with Cl_2 disinfection followed by dechlorination and 2,000 cubic feet of compressed media filtration followed by UV disinfection. The intercept capacity of the facility is 10 MGD and the recommended peak flow capacity is 90 MGD. Performance results of many studies and relative costs are given in a report from the studies (Columbus Water Works, 2001).

Washington, DC. Northeast Boundary Swirl Facility (NEBSF) (Disinfection Implementation) - The NEBSF, operated by the District of Columbia Water and Sewer Authority (WASA), provides treatment and disinfection for up to 400 MGD of CSO before discharging to the Anacostia River. The facility provides mechanical screening followed by three 57-ft diameter swirl concentrators. The effluent from swirl concentrators flows to a mixing chamber where sodium hypochlorite is added, usually at a dose of 5 mg/L. Sodium bisulfite is added at the end of the outfall for dechlorination, usually at a dose of 2 mg/L. Flows above 400 MGD are discharged untreated.

Samples taken during CSO events at the mixing chamber and at the river outfall are analyzed for enterococci and fecal coliform. Reported counts range from less than 10 MPN/100 mL to in excess of 250,000 MPN/100 mL. The high numbers are associated with events in excess of 400 MGD and represent blending of treated and untreated CSO. Annual operating costs for the NEBSF are estimated to about $230,000. The facility discharges on average about 100 times per year, with an average total volume of approximately 1,500 MG (Siddique, 2003).

Birmingham, AL. UV Disinfection at Peak Flow WWTP (Disinfection Implementation under Construction) - The Jefferson County Environmental Services Division for the City of Birmingham and about 20 neighboring communities are in the process of constructing a 350 MGD peak excess flow treatment facility. The new facility, named the Village Creek Peak Flow Wastewater Treatment Plant (PFWWTP), includes a pump station, with 360 MGD capacity, 20 surge basins with surface aeration for mixing (total capacity of 90 MG), granular, monomedia, deep bed filters with 350 MGD capacity, UV disinfection, and a 24 MW generating facility (primarily to power the pump station and UV). The UV system will have a total of 2,688 lamps at a peak power requirement of 7,526 kW. The total installation cost of the UV facility is estimated to be $13 million; the cost of UV equipment is about $10.7 million. Operating costs are not available (Chandler, 2003).

Oakland County, MI. Chlorine Disinfection at Acacia Park (Disinfection Implementation) - The Acacia Park CSO Retention Treatment Basin (RTB) is a four MGD basin that serves a combined area of about 818 acres. Disinfection is by sodium hypochlorite. The feed system provides a dose of 10 mg/L at a CSO flow rate of 426 MGD. There is no dechlorination. The disinfection target is a fecal coliform count of less than 400 cfu/100 mL at a total residual Cl_2 of 1.0 mg/L. Annual operating costs for the Acacia Park facility are estimated to be $120,000. The total volume of flow into the facility was estimated at 146 MG, with 88 MG retained and returned to the sewer system and 58 MG treated and discharged (Mitchell, 2003).

Bremerton, WA. UV Disinfection at CSO Treatment Facility

(Disinfection Implementation) - The City of Bremerton has recently constructed a CSO treatment facility that uses high-rate clarification, followed by UV disinfection, to treat flows up to 45 MGD. The facility uses a medium-pressure, high-intensity UV system that employs a total of 90 lamps. A 500 kW generator supplies power to the UV system as well as pumps, mixers, and other equipment. The clarification system uses a polyaluminum chloride coagulant. Bremerton installed a UV system at a a cost of about $600,000 to disinfect CSO discharges. The annual operation cost for the entire facility is estimated to be about $50,000 (Poppe, 2003).

New Orleans, LA. Stormwater Disinfection - New Orleans, LA

Evaluated a prototype disinfection facility for stormwater using sodium hypochlorite in the late 1960s and early 1970s; Pavia and Powell (1968), however did not adopt the practice permanently.

Santa Monica Urban Runoff Recycling Facility (SMURRF)_- Santa Monica's urban runoff recycling facility (SMURRF) project, completed in December 2000, in Santa Monica, CA, treats dry-weather runoff water from excessive irrigation, spills, construction sites, pool draining, car washing, the washing down of paved areas, and some wet-weather runoff. SMURRF treats an average of 0.5 MGD of the above urban runoff with solids and oil and grease removing technologies prior to UV disinfection. The treated runoff is reused for landscape irrigation and for flushing of toilets in dual-plumbed buildings. More information is contained on the Internet site: http://Epwm.Santa-Monica.Org/Epwm/Smurrf.html

Moonlight Beach Urban Runoff Treatment Facility - in the City of Encinitas, CA has been treating dry season urban runoff since September 2002. The facility accepts flows up to 150 gpm. The technologies used are filtration followed by UV disinfection. Coliform bacteria were reduced by over 99%. The facility does not operate during the wet season (Rasmus and Weldon, 2003).

Best Management Practices for Control of Pathogens in Urban Stormwater

Introduction

BMPs, have been used to control and manage the quality and quantity of urban runoff. BMPs can be structural and nonstructural. Structural BMPs can be farther divided into different groups and examples of such groups are given:

- Structures (wet ponds, dry ponds, constructed wetlands, filters, rooftop storage, and swales)
- Repair and upkeep of the stormwater and sewerage system (eliminating illicit connections, reduction of stormwater volume entering combined or sanitary sewer systems)
- Inflow reduction techniques (disconnection of roof leaders and redirection of area and foundation drains and basement sump pumps)

Nonstructural BMPs are generally good housekeeping practices or measures designed to institute good housekeeping for reducing or preventing pollutant deposition in a watershed, e.g., public education or regulation (Urbonas, 1999).

Structural BMPs - Effectiveness of structures such as wet ponds, dry ponds, constructed wetlands, filters, rooftop storage, and swales has been routinely measured as to their ability to reduce flow volume, sediment, or nutrients. Controlling pathogens or microorganisms also may occur but is a secondary goal for these BMPs and has been seldom measured. Some professionals believe that BMPs are ineffective for pathogen removal and should not be implemented for that purpose. This negative opinion is not supported by a wealth of data but rather by the scarcity of it.

Limited research has been conducted on the effectiveness of structural BMPs for controlling stormwater pathogen loads. Much of the existing information has been compiled by Winer (2000) and ASCE (2002) in U.S. EPA-sponsored projects. The data is compiled in a database format; therefore it is general in nature. Detailed information on a particular site is available from the original reference cited in the database. Reported fecal coliform removal efficiencies ranged from 99% at a wet pond in Ontario, Ca to –134% in a Fremont, CA wetland. These data show that while there are cases where microorganism reduction can be achieved to some extent by employing BMPs, BMPs also serve as environments where microorganisms are generated, presumably from increased wildlife populations and resuspension of bottom deposits (ASCE, 2002; Kurz, 1998; Winer, 2000). Variability of the data is caused by many factors including stormwater characteristics, BMP design, and environmental factors contributing either to microorganism die-off or surviva

Ponds and Wetlands - In some studies, better performance was observed in wet ponds over wetlands while other studies show that wetlands may provide advantages over ponds for indicator microorganism removals. One study found greater removal of thermotolerant coliforms, enterococci, and heterotrophic bacteria from stormwater in a wetland system (80-87%) than in a pond (–2-22%) (Davies and Bavor, 2000). The researchers attribute greater bacteria removal in the wetland to increased sedimentation aided by vegetation and increased removal of fine suspended particles (< 2 microns) with attached bacteria. Pond and wetland performance on microorganisms in sewage is an indicator of their_performance on stormwater. A wastewater treatment wetland removed 97-99.9% of fecal coliform and enterococci and 70% of coliphage (Stenstroem and Carlander, 2000). The investigators attribute the bacteria concentration reductions to the wetland's ability to remove suspended particles. Viruses have been shown to accumulate in wetland biofilms resulting in their removal from the effluent (Flood and Ashbolt, 2000).

Research by the University of Arizona on treatment of secondary sewage effluent in a duckweed-covered pond, a multi-species subsurface flow wetland, and a multi-species surface flow wetland reduced concentrations of *Giardia* cysts, *Cryptosporidium* oocysts, total coliform, fecal coliform, coliphage, and enteric viruses in secondary sewage effluent (Gerba *et al.*, 1999; Karpiscak *et al.*, 1996; Thurston *et al.*, 2001). Removal of the larger microorganisms, i.e., *Giardia* and *Cryptosporidium,* was the greatest in the duckweed pond, with sedimentation thought to be the primary removal mechanism. In contrast, the greatest removal of coliforms and coliphage occurred in the subsurface flow wetland, which may be related to the large surface area available for adsorption and filtration (Gerba *et al.*, 1999). When supplying potable water to a wetland at the facility, Thurston *et al.* (2001) showed that total and fecal coliform concentrations increased. In the summer, warm waters promote the growth of bacteria contained in the animal feces deposited in the wetland. Increased plant growth may increase root exudates, oxygen to the rhizosphere, and accumulation of organic matter, believed to increase microorganism growth (Thurston *et al.*, 2001).

Kern *et al.* (2000) conducted a seasonal effects study on dairy farm wastewater at a side-by-side wetland constructed at the Institute of Agricultural Engineering in Potsdam, Germany. The subsurface flow wetland with a horizontal water flow reduced fecal coliform densities by 99.3 and 95.8% in the summer and winter, respectively. The principal mechanism in eliminating fecal coliform seemed to be adsorption to soil particles followed by die-off and predation.

Karpiscak *et al.* (1999) studied the effectiveness of an integrated wastewater treatment facility consisting of solids separators, anaerobic lagoons, aerobic ponds and constructed wetlands, on dairy waste in Glendale, Arizona. In the aerobic pond, fecal coliform and *Listeria* concentrations decreased by 98.5 and 96.6%, respectively. Total coliform, however, increased by approximately 40%. Concentrations of all three organisms were decreased in the wetlands, total coliform by 79%, fecal coliform by 82.8%, and *Listeria* by 99.1%. Reductions are attributed to UV radiation, degradation of organic matter, solids settling, competition from other microorganisms, phytoremediation, and residence time.

Sand Filters - Sand filters operate by trapping suspended particles or adsorbing pollutants. Sand filters can be constructed in underground trenches or in above-ground, pre-cast concrete boxes. Advantages over ponds include lower areal requirements and the ability to install them out of public view (Kurz, 1998), which facilitate their use in ultra-urban environments where ponds are more difficult to site. Pathogen removal rates achieved by sand filters are comparable to those achieved by ponds and wetlands and the data are also variable (ASCE, 2002; Kurz, 1998; Winer, 2000).

Illicit Discharge Detection and Elimination - Improper connections to storm drainage systems convey contamination to receiving-water bodies. Sources of microbial contamination transported through this route include sanitary wastewater and septic tank effluent (Pitt *et al.*, 1993). The Clean Water Act requires separate municipal storm sewer system discharge permits to prohibit non-stormwater discharges into storm drains. U.S. EPA has an Internet site that presents information about illicit discharges, how specific municipalities are working to address them, and methods for identifying them:

http://cfpub2.epa.gov/npdes/stormwater/menuofbmps/illi_2.cfm (U.S. EPA, 2003).

Pitt *et al.* (1993) published an U.S. EPA User's Guide on investigating inappropriate pollutant entries into storm drainage systems available at:

http://www.epa.gov/ednnrmrl/repository/cross/cross.pdf. An update of this manual will be published in the near future and will include information on optical brightener monitoring, a quick and effective way for screening large watersheds for illicit wastewater connections.

Nonstructural BMPs - Nonstructural BMPs include institutional and educational practices whose goal is to change behaviors so that the amount of pollutants entering the stormwater drains and receiving waters are reduced (Urbonas, 1999). These common sense measures for addressing microbial contamination include limiting public and animal access to sensitive watershed or riparian areas, public education on the role of storm drains, erosion control, vegetative buffers, street sweeping, animal waste management, and pet waste or pooper-scooper ordinances. While quantitative data on nonstructural BMP effectiveness are limited, these practices have been shown to reduce receiving-water bacteria levels in rural and agricultural settings, primarily by controlling sources. Two examples of nonstructural BMPs for urban communities are dog parks and managing waste from resident Canada geese.

Dog Parks as BMPs - Waye (2003) cites the success of dog parks as BMPs. These parks should be located away from water bodies and provide fencing, public education on managing waste, and disposal bags and receptacles. Having a local community pet group take responsibility for a park and establishing the norm of picking up after one's own pets helps to ensure success of these parks.

Managing Waste from Resident Canada Geese - In recent years, resident Canada geese population in the U.S. has grown to 3.5 million (U.S. Fish and Wildlife Service, 2002). The problems encountered by local communities are the health and cosmetic problems associated with the fecal material generated, as well as the number of geese, and related traffic and safety concerns as these large birds cross traffic. Municipalities are instituting measures to protect public health from the impacts associated with this waste. The coastal town of Spring Lake, in New Jersey's Monmouth County, is experiencing high bacteria levels in a pond occupied by many Canada geese. During rain events, the pond overflows into the ocean, resulting in beach closures.

Colts Neck, another Monmouth County community, recently contracted with the U.S. Department of the Interior's Fish and Wildlife Service to asphyxiate Canada geese at local ponds. The local health officer defended the action based on the nuisance and potential health hazards posed by the geese droppings in and around the ponds (Jordan, 2003). Some citizens and animal rights advocates opposed the action and proposed alternatives. Waye (2003) names the possible alternatives identified by GeesePeace (www.geesepeace.org), including egg addling, vegetative barriers around water bodies, border collie patrols, goose repellants, and "no feed" zones.

Conclusions

All components of WWF, such as CSO, SSO, and stormwater, carry significant loads of fecal and pathogen contamination that should be prevented from entering urban receiving-water bodies.

WWF point sources of pathogens can be effectively treated by disinfection. For WWF two disinfection technologies are in commercial use: chlorination alone or followed by dechlorination, and UV irradiation. Other disinfection technologies for WWF are being developed. Though discussion of WWF disinfection is largely based on studies conducted on CSO, disinfection of SSO and of collected and confined stormwater is feasible, though seldom implemented.

The most important requirements for WWF disinfection are the need for high-rate processes and for suspended solids removal pretreatment.

WWF diffuse sources are commonly managed with the use of structural and nonstructural BMPs. The functions are reduction of flow volume, sediment, or nutrients and not reduction of pathogens. There are limited data on the effectiveness of structural and nonstructural BMPs for controlling pathogen loads.

Use of disinfection for reduction of pathogen contamination is effective and treatment results can be predicted with reasonable certainty. Use of structural and nonstructural BMPs is obscured by the highly variable and scarce performance data of these systems with respect to pathogen contamination. More research and quantitative data is needed on BMPs effectiveness in reducing microbiological loads in diffuse sources such as stormwater runoff.

With respect to protecting and improving receiving water quality, other options may need to be considered in addition to BMPs. These include incorporating a preliminary treatment step upstream of a structural BMP to create a treatment train or disinfecting the stormwater. Reducing runoff volume and source control are ways to decrease microorganism loads to receiving waters from urban runoff.

Land Use and Seasonal Effects on Urban Stormwater Runoff Microorganism Concentration

Introduction

Wet-weather flows, including stormwater runoff, combined sewer overflow, and sanitary sewer overflow, introduce chemical, physical, and biological stressors to receiving waters. The 2000 National Water Quality Inventory Report to Congress estimated that about 39% of assessed streams, lakes, and estuaries did not support designated uses such as fishing and swimming (U.S. EPA, 2002d). Leading pollutants include pathogens, nutrients, sediment, and toxicants. Runoff from both agricultural lands and urban areas is the primary source of these pollutants. Land use has been identified as an important parameter in predicting stormwater quality (Francy *et al.*, 2000; Mallin *et al.*, 2000).

This section hypothesizes that concentration of the identified microorganisms in the runoff differs between various urban land uses and seasons. Published data anecdotally support the hypothesis with some limited research directly supporting the supposition, although none examines a wide breadth of microorganisms. The literature also contains some contradictory examples. Some data from earlier studied outfalls were known to have cross connections. Other studies either did not report or did not test for cross connections. Most reported data, however, analyzed the receiving water, not the stormwater directly, forcing inferential analysis of the stormwater load. This emphasis on receiving water is consistent with the need to measure bacteria concentration in the regulated medium, and directly supports consideration of public health issues. Practical stormwater management, however, is unlikely to mitigate the load after it mixes with the receiving water, rather it works directly on the stormwater matrix.

Relationship with Land Use - Several monitoring programs have suggested a relationship between urban land use and the microbial concentrations in the stormwater runoff. While all of these results are suggestive, however, none of the programs were explicitly designed to evaluate the relationship of bacterial loads with the drained land use.

Olivieri *et al.*, (1977) reported concentrations of *Pseudomonas aeruginosa* (10^2 to 10^3 most probable number (MPN)/100 mL) and *Staphylococcus aureus* (10^1 and 10^2 MPN/100 mL) in stormwater runoff collected from residential areas in Baltimore, Maryland but with cross connections. A 1999 study to establish the source of unexpectedly high river and stream bacterial concentrations near Nashville, Tennessee showed fecal coliform, fecal streptococci, and enterococci concentrations directly related to the housing density, population, development, imperviousness, and apparent animal density. Surface runoff samples from more densely populated, sewered areas generally showed higher bacterial counts than runoff from less developed areas with septic tanks, suggesting that a relationship may exist between land use and potential bacterial loading (Young and Thackston, 1999). This study investigated the in-receiving water load and not the runoff concentrations. Similarly, Weiskel *et al.* (1996) reported a relationship for stormwater fecal coliform concentrations from impervious surfaces with the surrounding land use. The fecal coliforms yield from high-density residential areas, were significantly higher than those associated with nearby moderate-density residential areas, commercial areas, and low-density residential areas. Glenne (1984) and Chang (1999) independently concluded that population density, development age, and percent residential development may better predict bacteria levels in urban stormwater runoff than factors such as rainfall intensity and antecedent dry period.

Studies by Mallin *et al.* (1998; 2000) in North Carolina showed that fecal coliform abundance in estuarine creeks significantly correlated with watershed population, and even more strongly correlated with the percent of developed land within the watershed. However, watershed imperviousness was the most important anthropogenic factor associated with fecal coliform abundance in estuarine waters (Mallin *et al.*, 2000).

Examination of the relationship between bacterial concentrations in 136 stream-water samples collected in five important hydrologic systems in the United States and potential explanatory factors (mean annual basin temperatures, human population density, live stock density, stream flow, and land use) showed land use to be most significant (Francy *et al.*, 2000).

Relationship with Season - Evidence supports microorganism concentrations in stormwater varying seasonally. The earlier-mentioned Nashville, Tennessee study showed much higher summer fecal coliform counts than winter (Young and Thackston, 1999). These results are consistent with the data collected during the Nationwide Urban Runoff Program study, where fecal coliform densities in urban runoff during warmer months of the year were approximately 20 times greater than those found during cooler periods (U.S. EPA, 1983). Stormwater from city streets, a suburban business district storm drain, and a wooded hillside adjacent to a city park showed that peak total coliform, fecal coliform, and fecal streptococci densities occurred either in summer or autumn (Geldreich *et al.,* 1968). Stormwater samples collected in the residential and light-commercial portions of the Mt. Washington section of Cincinnati throughout the year showed total coliform, fecal coliform, and fecal streptococci densities were greatest in summer and lowest in winter (Evans *et al.,* 1968).

A study in two northwest Arkansas streams found significant seasonal influences on fecal coliform and fecal streptococci concentrations, with the highest concentrations occurring in summer (Edwards *et al.*, 1997). A study in Upvan Lake of Thane, India also found seasonal variation of bacterial population in the water body. Bacterial counts increased during summer months and decreased during winter months (Bagde and Rangari, 1999).

Objectives

As summarized above, suggestive evidence links concentration of microorganisms in urban stormwater runoff with season and land use (directly or indirectly). The objective of this study is to determine the concentrations of pathogens and indicator organisms from different land uses in different seasons and to investigate whether the concentration of pathogen and indicator organisms in the stormwater runoff varies between different land uses and seasons. Results will assist watershed managers to calculate organism load from stormwater runoff from different land uses and help to determine load allocations.

Experimental Design

The experimental design uses a nested hierarchical evaluation of three land uses with two outfalls in each land use. The design calls for at least two storm events in each season, analyzing multiple sub-samples at predetermined dilutions from flow-weighted composited samples. The minimum detectable difference is set at 1/3-log at the 95% level of confidence and 80% statistical power.

Site Description - This project was completed in the 246-km^2 Navesink River watershed (Figure 1). Discharging into Sandy Hook Bay, the Navesink estuary supports 2,300 acres of commercially-important shellfishing beds (NJDEP, 1999). The watershed drains a variety of land uses with variable population densities (Scro, 1993). The U.S. Geological Survey estimates the net mean annual flow from the river is about 2.3 m^3/s. The watershed is entirely within Monmouth County, New Jersey. The part of the county studied has separate stormwater and sanitary sewer systems. The county is divided into fourteen major land use/land cover categories generally paralleling U.S. Geological Survey classifications; e.g., residential with various levels of imperviousness, landscaped commercial/industrial, unlandscaped, agricultural, and commercial/industrial. The land uses selected for the study included high-density residential areas (65% imperviousness), low-density residential areas (17% imperviousness), and landscaped commercial areas. High-density residential areas consist of 1/8-acre lots, whereas, low-density residential areas consist of greater than 1-acre and up to and including 2-acre lots.

Organism and Indicator Selection - Two pathogens (*Pseudomonas aeruginosa* and *Staphylococcus aureus*) and five indicator organisms (total coliforms, fecal coliforms, fecal streptococci, enterococci and *Escherichia coli* (*E. coli*)) were selected for this study. *Pseudomonas aeruginosa* and *Staphylococcus aureus* were selected based on their abundance in stormwater reported in other studies (e.g., Milwaukee and Baltimore) (Oliveri *et al.,* 1977) and their association with diseases transmitted through water contact (e.g., skin, ear, and eye infections). Total coliforms, fecal coliforms, fecal streptococci, enterococci, and *E. coli* are commonly used or proposed bacterial indicators in water quality monitoring. Isolating and enumerating specific pathogens can be impractical in routine water quality monitoring leading to these surrogate analyses. Historically, total and fecal coliforms and fecal streptococci have served as the preferred indicators, but recent efforts are leading to recommendations to substitute enterococci and *E. coli* for water quality monitoring because of higher correlation with gastrointestinal illness (Gray, 2000).

Experimental Setup - Sewersheds selected drained small areas, were shown to be free of cross connections, and were within an urban watershed dominated by a single land use. It was necessary that the drained areas be near each other with outfalls situated above the expected high-water depth from storms and tides to avoid backflow and sample dilution, have safe, all-season public access, and provide reasonable equipment security. The outfalls had to provide conditions within the technical limitations of the sampling equipment (e.g., pipe diameter, suction lift of sampling pumps). Finding suitable locations that satisfy the criteria was difficult. Candidate outfalls were monitored for two weeks, generally following established procedures (Pitt *et al.*, 1993), before final selection to preclude the existence of dry weather flows that suggest inappropriate cross connections. Locations of the outfalls are shown in Figure 1 and the details are given in Table 1. The selected outfalls were near each other to reduce the variations introduced by geographic factors (e.g., antecedent dry period, rain duration and intensity, geology) recognizing that not all extraneous factors could be eliminated. The equinox and solstice dates divide the year into four seasons. Rain events are defined as rainfall preceded by at least 72 rain-free hours that produces enough runoff to allow collection of a sufficient sample for analytical needs.

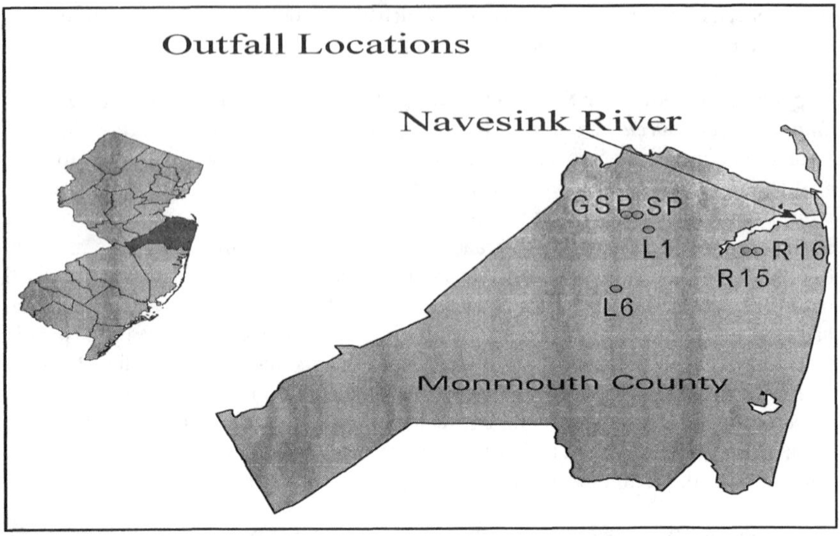

Figure 1. New Jersey County Map and Location of Stormwater Outfalls Selected for the Study within Monmouth County.

Sampling Equipment - Area-velocity flow meters (American Sigma, Loveland, CO) installed at each selected outfall recorded flow depth and velocity at six-minute intervals. Depth was measured using differential pressure (bubbler) or differential piezoresistive balanced bridge sensors. Twin 1-MHz piezoelectric crystals measure Doppler-based velocity. Internal electronics combine the measured values using the storm sewer pipe internal geometry to compute an associated flow rate. Instrument systems included ancillary measurements of temperature, specific conductivity, dissolved oxygen, and pH. Probes for these measurements were installed in 1-gallon plastic buckets positioned to collect the pipe discharge. Pre-rain event calibrations, following the manufacturer's outlined procedures, were completed before each rain event. A pre-calibrated automatic sampler (American Sigma, Loveland, CO) fitted with a single 10- or 20-L HDPE container was connected to each flow meter. Pre-calibrated tipping bucket rain gages (Onset Corp., Bourne, MA; American Sigma, Loveland, CO; and Environmental Sensors, Inc., Escondido, CA) installed near the outfall and away from obstructions recorded local rainfall in 0.2-mm increments. All field instrumentation was battery operated and generated delimited ASCII data files.

Table 1. Characteristics of Outfalls Selected for this Study

Outfall Name	Land Use	USGS Classification	Longitude	Latitude	Drainage Area (acres)	Outfall Diameter (in.)
R15	High-density Residential	65% imperviousness	74° 02' 37.4"	40° 21' 17.5"	6.04	15
R16			74° 02' 37.7"	40° 21' 17.2"	10.2	24
L1	Low-density Residential	17% imperviousness	74° 09' 48.9"	40° 22' 32.0"	2.69	18
L6			74° 11' 32.5"	40° 18' 45.4"	4.54	18
GSP	Landscaped	Approximately 15% of total area is vegetated	74° 10' 47.9"	40° 23' 17.9"	0.35	15
SP	Commercial		74° 10' 48.2"	40° 23' 15.2"	5.73	36

Note: Longitude and latitude of outfalls were measured using Trimble Geoexplorer II global positioning system. Land use is designated by Monmouth County, NJ. Drainage area was estimated from USGS topographic charts.

Generally following EPA guidance for collecting flow-weighted National Pollutant Discharge Elimination System (NPDES) stormwater samples (U.S. EPA, 1992), the meters triggered the automatic samplers when the measured flow water depth in the storm sewer reached 1 in. (2.54 cm). Upon triggering, the automatic sampler's internal peristaltic pump purged the 9.5-mm diameter pre-cleaned vinyl sample line before the initial sample collection. The pump transferred 1-L stormwater aliquots based on measured total flow until either the HDPE container was filled or the meter measured no flow for 3 hours. The samplers recorded the time of each aliquot addition to the composite sample. The samplers notified collection teams by cellular modem when collecting the first sample. The samples were recovered and immediately placed in a cooler with ice and driven to the EPA's laboratory for processing. The sampler collection bottles were decontaminated between storm events by thoroughly washing with non-phosphate detergent and rinsing with tap water. The container was then rinsed three times with (1/3 capacity of the bottle) reagent-grade water and air dried followed by a rinse with (1/3 capacity of the bottle) isopropyl alcohol and air dried.

Sample Analysis - The incoming samples were continuously stirred in the original 10- and 20-L container using an electrically-driven propeller (Stir Pak Mixer) for 3 to 5 minutes. After the premix period, but with the mixer still running, a siphon extracted 250-mL aliquots from mid-water height for analysis following Standard Methods (APHA *et al.*, 1998). In fewer than 10% of the events, the total elapsed time from the first aliquot collected through filtration marginally exceeded a 24-h total. These samples believed valid as independent experiments failed to detect changes in measured concentration with longer holding times (Selvakumar and Borst, 2004). Each sample was sequentially diluted with buffered water using three dilution factors based on previous analyses of samples from the same source. Sequential dilutions generally used at least 10 mL aliquots and always used at least 5 mL. Samples at each dilution were analyzed in quadruplicate. The dilution factors were selected to obtain the method-recommended colony count on at least one dilution set. All results were normalized to give concentrations in colony forming units (CFU) per 100 mL. Each analytical batch included laboratory blanks and positive controls. Blanks were run before and after each analytical set. Verification was performed on ten colonies for each organism. After incubation, the plates were manually enumerated.

Data Analysis - Concentrations from the analyses yielding plate counts in the method-specified range were used. If multiple dilutions produced counts in the desired range, all dilutions producing results within the range were used. If no dilution gave plates in the desired colony-count range, all countable plates were used. The data analysis excluded non-quantitative data (e.g., too-numerous-to-count or less than 1). The concentrations were $\log_{10}$- transformed before data analysis and were tested using standard analyses of variance (ANOVA) techniques at $\alpha = 0.05$ (i.e., 95% level of confidence). ANOVA was performed to test the statistically significant difference between concentrations in stormwater from different land use areas and seasons. The statistical analysis was completed using Statistica '98 (Statsoft, 1998).

Results and Discussion

Samples from 14 rainfall events were collected from June 22, 2000 to May 26, 2001. Precipitation in the sampled events ranged from 1.8 mm to 51.8 mm with rain intensities (cumulative rainfall/event duration) ranging from 0.9 to 15 mm/h. At least one sample per season was collected from all outfalls. Some samples were not collected because inlet tubes became clogged with vegetative material, ice or other debris or due to malfunctions in the flow recording or sampling equipment.

The runoff at the time of sampling was slightly acidic (pH averaged 6.77 $\pm$ 0.9) with a conductivity (0.19 $\pm$ 0.24 mS). Runoff temperature varied seasonally and averaged 15.1°C in spring, 20.3°C in summer and 13.1°C in fall. Winter temperatures were not measured due to freezing of probes. These characteristics fall within the range reported for stormwater (Makepiece *et al.,* 1995) and are similar to the collective national ranges reported in the NPDES database (Pitt *et al.*, 2003a).

The organism concentrations found in urban stormwater runoff from three different land uses are summarized in Table 2. The overall concentrations of total coliform (4.2×10^4 - 1.9×10^5 CFU/100 mL), fecal coliform (5.6×10^3 - 2.2×10^4 CFU/100 mL), fecal streptococci (3.5×10^2 - 3.2×10^3 CFU/100 mL), enterococci (1.0×10^3 - 6.6×10^3 CFU/100 mL), *E. coli* (1.5×10^3 - 8.5×10^3 CFU/100 mL), *Pseudomonas aeruginosa* (3.4×10^2 - 1.2×10^3 CFU/100 mL), and *Staphylococcus aureus* (4.6×10^3 - 1.8×10^4 CFU/100 mL) were found in the stormwater runoff. The concentrations in the stormwater runoff are similar to the collective national ranges reported in the NPDES database (Pitt *et al.*, 2003a) and by other investigators who demonstrated the existence of a potential health hazard in stormwater discharges (Olivieri *et al.*, 1977; Qureshi and Dutka, 1979). The results further show that urban stormwater runoff is a major diffuse source of pollution to receiving waters.

Relationship Between Concentration of Microorganisms and Land Uses - As shown in Table 2, the organism concentrations vary with land use. Statistically significant differences were found between land uses for all microorganisms studied except *E. coli*. Based on the stormwater data from a representative number of NPDES stormwater permit holders, Pitt *et al.* (2003b) found significant differences for land use categories. Except for enterococci and *E. coli*, the highest geometric mean organism concentrations were from high-density residential land use areas. These concentrations were higher than those associated with low-density residential and landscaped commercial land uses supporting the observations of other investigators (Weiskel *et al.*, 1996; Bannerman *et al.*, 1993; Mallin *et al.*, 2000). Since the outfalls were free of cross connections, the major sources of microorganisms in the stormwater runoff were most likely from the feces of domestic animals, insects, and wildlife.

The measured microorganism concentrations varied in samples collected from outfalls within a given land use (e.g., the concentrations from R15 and R16, each draining high-density residential areas, differ) as shown in Table 3. Area draining into outfall R15 includes a middle school, whereas area draining into outfall R16 is all houses.

Relationship Between Concentration of Microorganisms and Seasons - The microorganism concentrations varied significantly ($P < 0.05$) between seasons with the lowest generally observed during winter (Table 4). The mean concentrations vary within about an order of magnitude. The organism concentrations during the summer were not significantly different from those observed during fall and spring. These findings are consistent with others, who also reported that organism concentrations were higher during warmer months than in cooler months (Geldreich *et al.*, 1968; Evans *et al.*, 1968; Edwards *et al.*, 1997; Pianetti *et al.*, 1998; Badge and Rangari, 1999; Young and Thackston, 1999; U.S. EPA, 1983; Pitt *et al.,* 2003b).

The seasonal differences in organism concentrations have significant implication from the standpoint of stormwater treatment/storage. Since organism concentrations are higher during warmer months than during cooler months, any microorganism removal by structural BMPs must be based on warmer month concentrations. Otherwise, the results can underestimate the concentrations during warmer months when recreational uses such as swimming and boating increase.

Conclusions

Stormwater samples collected from storm sewers draining small municipal separate storm sewer systems shown to be free of sanitary-sewage cross connections within an urban watershed and dominated by a single land use were analyzed for pathogens (*Pseudomonas aeruginosa* and *Staphylococcus aureus*) and indicator organisms (total coliforms, fecal coliforms, fecal streptococci, enterococci, and *E. coli*). The concentrations found in the urban stormwater runoff are similar to those reported in the NPDES database. The runoff concentrations are uniformly higher than standards commonly established to protect public health, confirming the concerns associated with bacterial loading to recreational and drinking waters.

Other than *E. coli*, the microbial concentrations in stormwater runoff vary within and between urban land uses drained by the monitored sewer. Generally, the concentrations in runoff from high-density residential areas are higher than the concentrations in other tested land uses. The higher concentrations in the more impervious areas coupled with the increased runoff volume associated with these areas will lead to higher receiving water loadings. The results will help watershed managers target BMPs or treatments to land uses having higher runoff loads having greater impact on lowering the total load to a receiving water.

Microbial concentrations from outfalls within a given land use vary. The mean concentrations vary by less than an order of magnitude, typically less than ½-log, suggesting that the outfall selection (after cross connection screening) is not critical to estimate the local runoff loading from a given land use. Although concentrations vary among land uses, the relative differences are comparatively small with respect to the overall magnitude. These differences coupled with other modeling uncertainties do not warrant changing the existing aggregate approach to estimating the load from urban areas.

Seasonal concentrations in stormwater runoff vary within about an order of magnitude, with winter concentrations generally lowest.

Table 2. Microorganism Concentrations (95% level of confidence) by Land Use. Data includes analytical results from both outfalls for all seasons.

Organism	High-Density Residential		Low-Density Residential		Landscaped Commercial		Statistically Significant?
	Concentration (CFU/100 mL)	N	Concentration (CFU/100 mL)	N	Concentration (CFU/100 mL)	N	
Total Coliforms	1.1×10^5 - 1.9×10^5	147	6.4×10^4 - 1.3×10^5	74	4.2×10^4 - 7.9×10^4	112	Significant (P<0.001)
Fecal Coliforms	1.2×10^4 - 2.2×10^4	146	5.6×10^3 - 1.3×10^4	109	7.2×10^3 - 1.9×10^4	90	Significant ($P = 0.006$)
Fecal Streptococci	1.9×10^3 - 3.0×10^3	105	3.5×10^2 - 7.1×10^2	103	1.4×10^3 - 3.2×10^3	79	Significant (P<0.001)
Enterococci	3.2×10^3 - 5×10^3	72	1.0×10^3 - 2.2×10^3	80	4×10^3 - 6.6×10^3	73	Significant (P<0.001)
E. coli	2.0×10^3 - 3.9×10^3	130	1.5×10^3 - 5.9×10^3	64	3.1×10^3 - 8.5×10^3	98	NS ($P = 0.13$)
Pseudomonas aeruginosa	6.3×10^2 - 1.2×10^3	95	3.4×10^2 - 6.5×10^2	69	3.5×10^2 - 6.5×10^2	104	Significant (P< 0.001)
Staphylococcus aureus	1.1×10^4 - 1.8×10^4	98	6.5×10^3 - 9.8×10^3	92	3.2×10^3 - 4.9×10^3	111	Significant (P<0.001)

Note: N = number of data.

NS = not significant (P is set at ≤ 0.05)

Table 3. Microorganism Concentrations (95% level of confidence) by Outfall. Data includes all seasons.

Organism	Concentrations (CFU/100 mL)											
	High-Density Residential				Low-Density Residential				Landscaped Commercial			
	R15	N	R16	N	L1	N	L6	N	GSP	N	SP	N
Total Coliforms	7.9×10^4 - 1.6×10^5	83	1.2×10^5 - 3.2×10^5	64	1.0×10^5 - 2.4×10^5	33	3.5×10^4 - 1.1×10^5	41	2.8×10^4 - 5.4×10^4	45	4.7×10^4 - 1.2×10^5	66
Fecal Coliforms	7.9×10^3 - 1.4×10^4	101	2.5×10^4 - 9.1×10^4	45	5.6×10^3 - 1.9×10^4	61	3.5×10^3 - 1.2×10^4	48	6.8×10^3 - 2.8×10^4	50	4.7×10^3 - 1.9×10^4	40
Fecal Streptococci	1.3×10^3 - 2.3×10^3	69	3.7×10^3 - 6.2×10^3	36	3.5×10^2 - 1.0×10^3	47	2.8×10^2 - 6.9×10^2	56	2.9×10^3 - 6.3×10^3	35	6.8×10^2 - 2.2×10^3	44
Enterococci	3.9×10^3 - 6.8×10^3	44	1.9×10^3 - 4.0×10^3	28	7.9×10^2 - 2.8×10^3	38	8.5×10^2 - 2.6×10^3	42	3.6×10^3 - 6.3×10^3	36	3.5×10^3 - 8.5×10^3	37
E. coli	1.3×10^3 - 2.8×10^3	87	3.5×10^3 - 1.0×10^4	43	3.5×10^3 - 2.8×10^4	27	5.4×10^2 - 2.8×10^3	37	2.6×10^3 - 1.2×10^4	45	2.5×10^3 - 9.5×10^3	53
Pseudomonas aeruginosa	7.9×10^2 - 1.7×10^3	58	3.1×10^2 - 9.3×10^2	37	3.7×10^2 - 8.5×10^2	34	2.3×10^2 - 6.8×10^2	35	4.8×10^2 - 1.0×10^3	52	2.1×10^2 - 5.1×10^2	52
Staphylococcus aureus	6.2×10^3 - 1.1×10^4	56	1.9×10^4 - 4.0×10^4	42	6.6×10^3 - 1.3×10^4	40	5.6×10^3 - 9.3×10^3	52	2.1×10^3 - 3.7×10^3	56	4.4×10^3 - 7.6×10^3	55

Table 4. Microorganism Concentrations (95% level of confidence) by Season. Data include all outfalls.

Organism	Summer		Fall		Winter		Spring		Statistically Significant?
	Concentration (CFU/100 mL)	N	Concentration (CFU/100 mL)	N	Concentration (CFU/100 mL)	N	Concentration (CFU/100 mL)	N	
Total Coliforms	1×10^5 - 2.3×10^5	63	7.2×10^4 - 1.2×10^5	128	1.4×10^4 - 2.5×10^4	69	2.4×10^5 - 5.1×10^5	73	Significant ($P<0.001$)
Fecal Coliforms	2.5×10^4 - 8.2×10^4	95	6.7×10^3 - 1.1×10^4	111	2.3×10^3 - 4×10^3	87	1.5×10^4 - 2.8×10^4	56	Significant ($P<0.001$)
Fecal Streptococci	8.5×10^2 - 1.3×10^3	93	2.5×10^3 - 4.4×10^3	103	1.5×10^2 - 3.6×10^2	64	3.8×10^3 - 7.6×10^3	24	Significant ($P<0.001$)
Enterococci	1.4×10^3 - 3.4×10^3	48	4.7×10^3 - 7×10^3	74	5.8×10^2 - 1.2×10^3	60	7.8×10^3 - 1.1×10^4	44	Significant ($P<0.001$)
E. coli	8.4×10^3 - 3.3×10^4	69	1.5×10^3 - 2.4×10^3	91	3.5×10^2 - 1.1×10^3	59	5.6×10^3 - 1.1×10^4	68	Significant ($P<0.001$)
Pseudomonas aeruginosa	7.1×10^2 - 1.3×10^3	60	2.9×10^2 - 4.7×10^2	103	1.8×10^2 - 3.5×10^2	66	3.4×10^3 - 6.1×10^3	40	Significant ($P<0.001$)
Staphylococcus aureus	5.5×10^3 - 9×10^3	76	6.2×10^3 - 9.8×10^3	132	1.2×10^4 - 2.2×10^4	46	5.2×10^3 - 8.7×10^3	60	Significant ($P< 0.001$)

Note: N = number of data.

Trading Allowances for Stormwater Control: Hydrology and Opportunity Costs

Introduction

Proliferation of impervious surface coincident with urbanization and "sprawl" allows more stormwater from rain events to reach a stream faster causing higher peak flows that can lead to stream alteration and habitat degradation. Furthermore, where impervious surface prevents rainfall from infiltrating the soil, less water is available for groundwater recharge, which can also reduce stream base flow. Depending upon the land use in the watershed, nutrients and toxics can be scrubbed off roadways and parking lots. Excess stormwater runoff in a watershed can concentrate these non-point pollutants leading to offsite impacts, such as habitat alteration, increased toxic loadings, and other types of stream degradation.

An earlier study by Thurston *et al.* (2003), proposed a dispersed set of relatively simple stormwater management technologies to control the negative impacts of runoff for an urban area, and at a lower cost than large engineering infrastructural projects. The central premise of this work was that a market based tradable allowance mechanism for trading runoff reductions would be a practical and cost-effective method by which to assign dispersed runoff control throughout urbanized areas (while the term "tradable permits" is more common in the economics literature, we use the term "tradable allowances" to avoid any confusion with the permits used in US EPA's National Pollutant Discharge Elimination System - NPDES). Due to the primitive state of data and other information about the development and performance of tradable credit systems, the researchers added several caveats to their analysis, not the least of which was ignoring opportunity costs associated with the proposed best management practices (BMPs).

The pilot study presented here utilized new results regarding socioeconomic issues involved with trading stormwater volume, and introduced the utility of this arrangement with a specialized continuous hydrologic simulation model. Next, estimated were opportunity costs for the most prevalent land use and BMP type in the watershed selected for the pilot study. Specifically, hedonic modeling techniques were used to estimate residential homeowners' valuation of the land that makes up their parcel, and assume that the negative of this value represents a lower bound for the opportunity cost of a BMP.

Tradable Allowances: Theoretical Considerations

The cost effectiveness of the tradable allowance approach to pollutant reduction in air sheds is well established in the literature (Tietenberg, 2000; Baumol and Oates, 1975; Eheart, 1980), and in practice: the U.S. Acid Rain Program has used trading to achieve huge (SO_2) emissions reductions, the 1994 Regional Clean Air Incentives Market (RECLAIM) is used to control SO_2 and nitrogen oxides (NOx) in the Los Angeles, California area, and states in the northeast U.S. designed a regional NOx cap and trade program to control transport of ground-level ozone, where reductions began in 1999.

However, watersheds differ from air sheds in key aspects, like the confinement to a channel, non-uniform mixing and downstream accumulation of pollutants. These features present new challenges for the establishment of tradable allowance systems. Watershed trading is not a new concept; the US EPA's *Draft Framework for Watershed-Based Trading* (1996) provides an overview of some twenty tradable allowance programs across the United States. These mostly deal with the control of nutrients, and the application of trading to the control of stormwater is relatively recent.

The trading regime explored here requires detailed mapping information on individual properties including size, percent impervious surface, and soil type. In the past, these high information requirements made the type of program suggested in this study extremely difficult, if not impossible, to implement. Recent improvements in geographic information systems (GIS) technology, however, allow collection and management of large amounts of geographic data fairly inexpensively. Using spatially-explicit hydrologic simulation models, runoff from properties as a function of not only their physical attributes can be predicted. Also, different arrangements of parcels and their trading potential can be assessed. Since property rights at the parcel level are well delineated, runoff information at the parcel level makes allowance trading a technically feasible management alternative. By applying the standard assumption that property owners facing explicit costs for managing runoff from their properties will be cost-minimizers, the effects of changes in allowance prices and in land use (e.g., increased impervious surface) on runoff flow and stream loading in the watershed can be predicted.

Although lumped-parameter, event-based rainfall-runoff models have been used earlier to predict peak runoff discharge these models are limited in their ability to simulate the great number of factors that underlay a trading program for stormwater runoff. The inability to account for antecedent soil moisture conditions and its impact on runoff formation processes is a major source of error in hydrologic prediction. There are two major influences that regulate the potential improvement in stormwater management, namely, the existing landscape setting (e.g., soils, topography), and whether these attributes are both suitable for and adaptable to best management practices; and secondly, the type and number of trades made that result in the successful placement of stormwater best management practices. This type of trading system may not work for parcel-level, infiltration-type best management practices if soils are only slowly permeable by water, or if slopes are extreme. In this type of scenario, stakeholders would likely participate in paying for costs of downstream abatement or storage.

On the other hand, since trading credits are likely to be assigned on the basis of conditions for runoff abatement or generation; and after trading has started; best management practices will be implemented, changing again the landscape hydrology.

For example, given the placement of rain garden runoff interceptors, the parcel-level hydrology will be altered on the basis of runoff prevention. For both of these issues, some method of estimating the impacts of different trading scenarios must precede implementation. Ideally, this assessment would include separate models for a screening level and then continuous simulation of different trading arrangements. The screening-level model would likely be more an assessment of soils and topography, to determine the suitability for a site under consideration for implementation of a trading credit system for stormwater.

Although an expert-system for this type of assessment may not exist, one could easily be developed from widely available soil survey information. For this work, a commonly available model (Soil-Water Assessment Tool, USDA) has been adapted to assist in estimating the watershed-wide impacts of parcel-level shifts in hydrology. Since this model is parameterized at the parcel level, different trading scenarios can be implemented in a straightforward manner. This complex model will be the subject of a forthcoming publication that contrasts different trading ratios and other scenarios.

The trading program used in this study, as is the case with most water trading programs around the nation, is a "clearinghouse" approach. That is, instead of freely traded allowances on an open market, an authority for the management of surface water runoff in the area, such as a stormwater utility, sets the price of allowances, and this price is known to those participating. Parcel owners then choose to pay the fee or build a BMP to control their runoff, and within the watershed the responsibility can be traded as long as the aggregate level of reduction is met. It is assumed that the stormwater authority has determined the ecological limits of streams and sewers in the relevant watersheds, or as is the case in this study, imposes a reasonable *ad hoc* reference level (i.e., runoff in predevelopment conditions), and grandfathers each parcel owner an allowed flow of water from each parcel. Runoff volume in excess of the grandfathered amounts either must be controlled on site, or allowances for controlled water on other parcels have to be obtained in the allowance market.

If heterogeneous costs for abatement exist, there is an opportunity for some cost saving over a command-and-control program, for example. It is therefore crucial that all costs in the preliminary assessment are accounted for.

There are at least two components of cost facing each parcel owner: the construction and maintenance of BMPs and the opportunity cost of land taken out of other uses. Each of these depends on the volume of water the BMP is required to detain, Q, and the related square footage of land necessary to build the appropriately sized BMP, denoted $Q(sf)$. So, total cost, CT is given:

$$CT = const\,\&\,main\,\mathrm{t}(Q(sf)) + OpportunityCost(Q(sf))$$

and marginal abatement costs (MC) are given as:

$$\partial CT / \partial Q = MC ,$$

The marginal (or incremental) cost of abating a given unit of runoff *MC,* are assumed to increase monotonically with quantity abated such that:

$$\partial MC / \partial Q \geq 0 , \qquad (2)$$

This is primarily due to increasing opportunity cost of land. That is, although the specific abatement technologies may exhibit decreasing costs to scale over the relevant size of the BMP, it is assumed that an individual property owner faces increasing opportunity costs for land taken out of other uses and dedicated to BMPs. A negative sign on the quadratic term for land area in the hedonic price curve estimation confirms that this is the case with residential properties.

Opportunity Costs of BMPs

It has recently been recognized that when considering other-than-engineering solutions to stormwater management, the inclusion of other-than-engineering costs is imperative, and that these costs are usually not known. For example, Sample *et al.* (2003) note that in the vast literature on costs of facilities for stormwater detention and retention, including costs of length of pipes, manholes, etc., that data on BMPs are probably the least reliable. In an attempt to include opportunity costs in their life cycle cost analysis of BMPs, the authors focus on a case study found in the work on sewer design done by Tchobanoglous (1981). Sample *et al.* (2003) use the discounted stream of benefits, based on real estate appraisals, of land value forgone by dedicating land to BMPs. While this approach is unique, and considers the time dimension of the opportunity cost, a different approach based not on land value data from realtors, but extracted through hedonic analysis from market transactions was chosen for this study.

Hedonic models exploit the existence of heterogeneity in attributes that make up a given item that is traded in the market to estimate the portion each attribute adds to the value of the goods. A hedonic price function that includes various elements of properties sold in Hamilton County, OH in postal zip codes near the Shepherd Creek area needs to be estimated. Estimation of hedonic price function for housing is commonplace.

Econometrically, estimation is fairly straightforward. Co-variates usually include things such as number of bathrooms, fireplaces, the existence of central air conditioning, etc. Some neighborhood characteristics are usually also included such as school quality and proximity of parks. In this study, focus is on the value homeowners place on their yards, for it is this that is taken away if square footage is dedicated to an on-site stormwater detention facility such as a sand trench or rain garden.

The estimated Hedonic price function is:

$$p = h(z, \alpha) \qquad (3)$$

where p is the price of the house, z is a vector of attributes such as those mentioned above, and alpha is a vector of parameters that define the shape of the price function. Assumed is the standard approach that the household budget constraint is given by:

$$b = h(z) + x \tag{4}$$

where x is all other goods, and maximizing utility subject to the budget constraint implies optimal conditions for each attribute:

$$\frac{\partial u(x, z:\beta)}{\partial z_l} = \lambda \frac{\partial h(z)}{\partial z_l}, l = 1, \ldots L \tag{5}$$

If equilibrium conditions hold in the housing market the value owners place on the attribute "yard area" is expressed:

$$w\hat{tp} = \Delta z_l \frac{\partial h(z)}{\partial z_l} \tag{6}$$

where z_l is the co-variate "land" and the result is the amount of land dedicated to a BMP multiplied by, in the case of linear estimation of the hedonic price function, the estimated coefficient on the land variable. It follows that, if land is dedicated to a BMP, the negative of the above represents the opportunity cost of land in the BMP. In the case of the semi-log functional specification, the interpretation of the estimated coefficients is that they approximate the percent change in the property value due to a change in the given co-variate.

Data - Data on over 100,000 house sales from the period 1998 - 2001 were downloaded from the Hamilton County, OH, County Auditor's web site. These data are publicly available, and include over 100 columns of information, including sale price, parcel numbers, address, name of buyer, date of the sale, school district etc. Also included is information on attributes of the property: number of bathrooms, square footage, existence of central air conditioning etc. Only residential, non-vacant properties that were sold in zip codes adjacent to the pilot project site were used. Incomplete and obviously erroneous entries were culled, resulting in 24,218 observations. Variables used in ordinary least square (OLS) estimation are listed in Table 5, along with their means and definitions.

Table 5. Variables Used in Regression Analysis

Variable (n= 24218)	Mean	Definition
HALFBAT	0.44322	Number of half bathrooms
FULLBAT	1.5114	Number of full bathrooms
SFTFIN	1551.8	Area of the house finished in square feet
NETLOT	0.32064	Area of property alone (lot size minus footprintof house) in square feet
NETLOTS	0.75029	Net square footage of property squared
SCHRANK	15.159	School rank in Hamilton county based on statewide aptitude test pass percent*
FIRE	0.46061	Number of fire places
COND	3.4385	Condition rating assigned by realtor, 1-5
SALEPRIC	107,370	Sale price of the property in current dollars
Cool	0.72108	Dummy variable that takes the value 1 if central air conditioning, 0 otherwise

Results - OLS was used to estimate coefficients on the parameters described in the previous section. Two functional forms were employed, linear and semi-log, in which the dependent variable, sale price is logged, each with a quadratic term for square footage of lawn to capture decreasing marginal utility. Estimation results are presented in Table 6.

Tested was the choice of the linear versus semi-log specification using the Box and Cox (1964) test statistic under the null hypotheses that the two are equivalent,

$$l = \frac{T}{2}\left|\ln\left(\frac{SSE_L / \bar{y}_G^2}{SSE_{LL}}\right)\right| \sim \chi^2$$

where SSE_{LL} is the sum of squared errors from the log-linear model, and the term $SSE_L/\bar{y}_G^2$ is the sum of square errors from the application of least squares to the linear model with $(y_t / \bar{y}_G)$ as the dependent variable where

$$\bar{y}_G = \exp\left[\frac{1}{T}\sum_{t=1}^{T} \ln y_t\right]$$

in this case $SSE_L = 5.29 \times 10^{13}$, $\bar{y}_G = 91413.47$, and $SSE_{LL} = 4586.5$ which yields a value of the test statistic of 3922, and because

$SSE_L / \bar{y}_G^2 = 6341.0 > SSE_{LL}$

Table 6. Regression Results

Characteristic	Parameter Estimates (Standard Error)	
	Linear	Semi-log
HALFBAT	11168 (642)	0.11511 (0.00597)
FULLBAT	14065 (693)	0.11219 (0.00645)
SFTFIN	44.7 (0.745)	0.000256 (0.00000693)
NETLOT	0.278 (0.0137)	1.17×10^{-6} (1.28×10^{-7})
NETLOTS	-9.39×10^{-8} (7.88×10^{-9})	-4.44×10^{-13} (7.33×10^{-14})
SCHRANK	-1908 (56.1)	-0.0169 (0.000522)
FIRE	15800 (655)	0.155 (0.00609)
COND	6820 (416)	0.115 (0.00387)
COOL	8280 (723)	0.199 (0.00672)
CONSTANT	294 (2200)	10.4 (0.0204)
R^2	0.5464	0.4567
SSE	5.29×10^{14}	4586.5

the null hypothesis is rejected and the conclusion is that the models are not observably equivalent and that the semi-log model provides better estimates.

The semi-log specification estimates approximate percent changes in the housing price due to a change in the attribute level, in this study a net-square footage of land. While the current Shepherd Creek data set does not include house prices, the average of the effect on the population sample can be used to estimate the coefficients. This yields an approximate average of $0.13 with the semi-log model.

The Shepherd Creek Study Area

Cincinnati's Mill Creek and its tributaries occupy some 44,000 ha (170 square miles) in 23 municipalities in the counties of Hamilton and Butler, Ohio. Mill Creek is considered by the OEPA to be the most polluted waterway in the state (OEPA, 1994).

Stormwater runoff is a major direct and indirect contributor to the pollution of Mill Creek. Stormwater runoff loads the creek directly as rainfall dissolves and mobilizes detergents, phosphorous, fuels, and nitrogenous waste from parking lots, fertilized lawns, streets and agricultural areas. The contaminated runoff flows both overland and through storm drains into the creek.

One possible solution to the runoff problem currently under review for Mill Creek calls for a variety of engineering options which include the costly separation of sanitary and storm sewers in some areas, and in other areas the installation of large scale BMPs which would provide collection and high rate treatment on-site (Donovan, 1995). Another program calls for the construction of a 29 km (18-mile), 9.1 m (30-foot) diameter detention tunnel to be excavated 46 m (150 feet) below the surface at a cost of $800 to $850 million (the first estimate comes from Klepal (2001), "Giant Tunnel Under Mill Creek Proposed," and the larger estimate comes from more recent personal correspondence with an engineer from the Metropolitan Sewer District).

The dispersed-BMP solution investigated in this study would be facilitated by an incentive-creating tradable allowance plan. Two necessary conditions for tradable allowance regimes to be cost reducing are that: i) there be sufficient difference in abatement cost across parcel owners so that potential cost savings can be realized through market exchange of runoff control; and ii) the transactions costs of such programs be no greater than the gains achieved. For this test case in the Shepherd Creek, abatement cost is in fact heterogeneous, largely a function of substantial variation in parcel use, and soil type. Abatement costs, including the construction costs and the opportunity cost of land dedicated to the BMP, are the focus of this study. The actual transactions for a trading mechanism will be the focus of future research. With these conditions satisfied, a tradable runoff allowance (TRA) system can efficiently assign runoff control to dispersed locations, and may avoid the larger cost of centralized approaches.

Application - To employ the opportunity cost estimates in the formulation for this TRA program: individual parcel-owners excess runoff responsibility was estimated, a clearinghouse type market was established to provide incentive to trade, and opportunity costs of land in the BMP cost functions being considered by the homeowners were included.

The boundaries of the Shepherd Creek sub-watershed using established hydrology were delineated. The sub-watershed was divided into parcels, then characteristics such as soil type, land use and parcel boundaries were entered into an ArcView GIS project. Figure 2 shows the area in detail, illustrating the headwaters, parcels, different soils and impervious surfaces such as buildings, driveways, etc.

Mount Airy Forest was chosen as the section of Mill Creek for this study area because of the diverse land use, topography, and soil types. Mount Airy Forest is a protected area that occupies a large piece of the study area, and proposed and ongoing subdivision developments provide heterogeneous land use and the opportunity to research a dynamic system.

Figure 2. Impervious Surface in Shepherd Creek

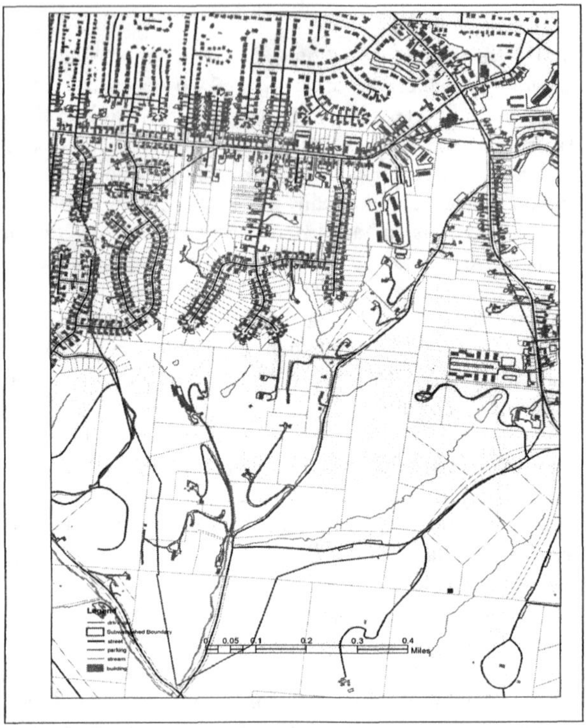

Mathematical calculation of the property owners' runoff responsibilities is implemented in Microsoft Excel, and then exported to Arc View for visualization. An ArcView representation of excess stormwater runoff occurring with the design storm in Shepherd Creek is shown in Figure 3.

Numerical Example - 420 parcels are used in the numerical example of the study. Most are single-family residential properties with some multi-family apartments and industrial land use. The soil is of hydrologicsoil group B, C or D and the terrain ranges from forested to grassy to impervious. Parcel owners in the case study area are assumed to choose the most suitable BMP for their land use and soil type from among a variety of BMPs that exist for different applications. Table 7 summarizes the commonly used BMPs. The reader should note that applicability and cost of BMPs varies with, among other things, type of soil in the proposed site.

Figure 3. Excess Stormwater Runoff

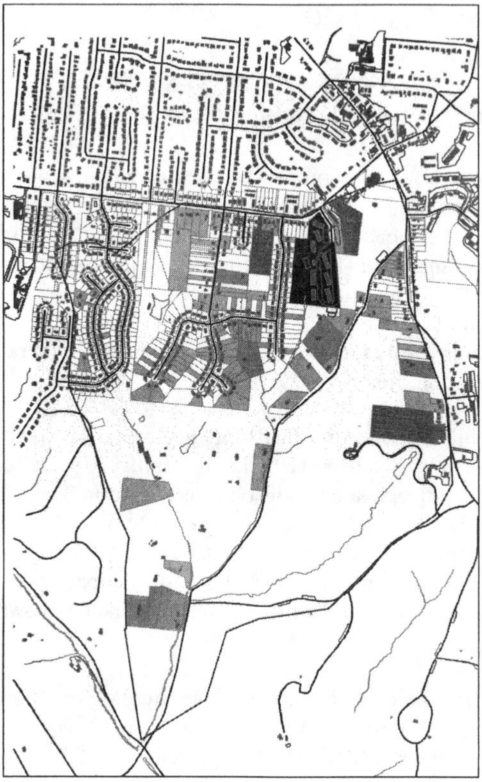

For the computational exercise, the type of BMP, concomitant detention cost (DC) functions and inverse cost functions, were assigned to parcels in the sub-basin based on land use and soil type as follows:

Parcels with residential land use and HSG B are assumed to employ sand filters. The cost function is: $DC_{ResB} = 26.6Q^{0.64}$, where Q is the quantity of stormwater runoff detained in cubic feet.

Residential parcels with HSG C or D are assumed to use grassy swales, the relevant cost function of which is $DC_{ResC} = 4.94Q + 0.278Q$. Where the second term is the estimated Opportunity Cost of land estimated with the linear specification. And $DC_{resc} = 4.94Q + 0.125Q$, again, the second term is the Opportunity cost of land, but in this representation using the average estimated impact from the Semi-log model.

Commercial (including multi-family land use) and industrial properties of HSG B are assumed to use infiltration basins or extended detention ponds; the pertinent cost function is: $DC_{ComB} = 15.3Q^{0.69}$

Commercial and industrial properties of HSG C or D are assumed to employ infiltration trenches with cost function: $DC_{ComC} = 157Q^{0.63}$

For this exercise, some properties, such as public schools and public rights of way, were not included. While these do contribute a significant amount of stormwater runoff and toxic loadings to streams, public agency responsibility was not included in the present analysis (The area of the roads in this watershed is 4% of the total land area. According to TR55, the contribution to the runoff volume from rain that falls directly on the roads is 15% of the total runoff volume. This is partially mitigated by extant requirements for BMP construction for public infrastructure. (CAGIS, personal communication, June 20, 2001)).

The suitability of specific BMPs for specific parcels is based primarily on Heaney *et al.* (2002) and Schueler (1987). Suitability is a function of site characteristics including service area and soil type, as shown in Table 3-3. It is important to note that assignation of BMP type is done partially due to geologic characteristics and space constraints, and partially for modeling convenience. Current ongoing work includes the use of an experimental auction approach to determine both willingness to pay (or accept) for BMP construction and actual homeowner's choice of BMP design itself.

To calculate the cost of a dispersed set of BMPs to store, on site, all of the excess runoff from a design storm event of three12 cm (1.23 in.), assigned was the appropriate BMP technology on a parcel-by-parcel basis in the small case area and solve each landowner's cost equation. This calculation presumes that the parcel owner is responsible for *all* the runoff over that which would result if the parcel were in its undeveloped state. The water detained with this constraint is 3477 m^3 (122,775 ft^3). Considering only construction costs of BMPs, this results in an average cost of \$5.40 per cubic foot of stormwater runoff detained via BMPs over all properties in the study area. If a policy causing all parcel owners to use BMPs to manage all excess SWR is implemented, the estimated opportunity cost of the land dedicated to the BMPs can be assumed, in this case \$0.27 per ft^2 using the linear specification and \$0.13 per ft^2 using the semi-log specification. The cost for watershed-wide detention rises to is \$9.22 per ft^2 and \$7.12 per ft^2, respectively. It can be assumed that costs would be higher for all price levels of allowances due simply to the increased cost of BMPs (recognition of opportunity costs), and this appears to be the case.

Table 7. Commonly Used BMPs

Name	Effective Area Served (Acres)	Indications (Restrictions and Strengths)	Cost Function (Q in ft^3)	Soil Permeability Ranges (minimum infiltration rate inches/hour)	Slope restriction (yes/no)
Wet Pond (Retention Pond)	>15	Large space requirement, high sediment input potentially problematic, has thermal impacts, depth limit	$C=6.1Q^{0.75}$ (<100,000 c.f. storage) $C=34Q^{0.64}$ (>100,000 c.f. storage)	Loamy Sand to Clay (>2.4)	No
Infiltration Basin	>5 <25	Cannot be near bedrock or foundations, uses more space than trench, can handle high sediment input	$C=15.3Q^{0.69}$	Sand to Loam (<.55)	Maybe
Extended Detention Pond	>15	Large space consumption, may not handle high sediment input	$C=10.7Q^{0.69}$	Sand to Sandy Clay (<.04)	No
Porous Pavement	<10	Has many restrictions (depth, slope, space, proximity to foundations/bedrock)	$C=4Q$	Sand to Loam (<.55)	Yes
Infiltration Trench	<10	Cannot be near bedrock or foundations, needs little space, can handle high sediment input	$C=157Q^{0.63}$	Sand to Loam (<.55)	Yes
Filter Strip/ Sand Filter	<10	Use in residential properties precludes other uses.	$C=26.6Q^{0.64}$	Various	Yes
Grassed Swales	<10	Small scale. Reserved here for Residential properties.	$C=4.94Q$	Various	No

In an effort to incorporate market incentives it can be assumed that the stormwater utility has the mandate to require either the building of a property level BMP or purchase an allowance, the proceeds of which may be used for stormwater detention capacity elsewhere. If this were the case the property owner would take into account the opportunity cost of his land. It has been modeled explicitly below.

Trading Allowances - It has been posited elsewhere (Thurston *et al.*, 2003) that it may be possible to organize a trading system under which parcel owners in a watershed trade stormwater detention capabilities based on their (differing) costs to do so, to arrive at a lower cost solution to the stormwater control problem at the watershed level. If one may choose between buying an allowance to let his stormwater flow, or build detention facility, it stands to reason he will weigh the costs of each alternative and choose the lesser of the two. A parcel owner would take into account the opportunity cost associated with the BMP as well as the construction cost when determining whether he would build detention capacity or buy a permit, through a clearinghouse or other market, from someone who does. It must be made clear here that only re*sidential* BMP opportunity costs have been estimated and included, and have not included opportunity costs for commercial properties, and yet those properties have been considered in the trading scenarios that follow. Therefore conclusions about trading behavior and subsequent costs and savings are tenuous at best. Essentially, it can be said, that opportunity cost for non-residential properties is zero. This may be true in the special cases of rooftop detention, under-parking lot detention, or commercial property where space dedicated to BMP would not otherwise be put into use.

If the allowance price is below the minimum (in this case about $2.50), the market is essentially a fund-raising mechanism only, because the cost of credits is low enough that even without the consideration of opportunity costs it is always cheaper to buy the allowance. This is the current system in most cities today; a fee for runoff is assessed and revenues are used by a stormwater utility to construct tunnels or other stormwater runoff infrastructure. But if the price of the allowances is raised such that it becomes binding, some parcel owners will buy the allowances and some will build BMPs on their property. Table 8 shows the per-unit cost of detention assuming some different prices for allowances when we assume 1) no opportunity cost of land for residential BMPs, 2) the opportunity cost estimated in the semi-log model, and 3) the opportunity costs estimated with the linear model.

Table 8. Per Unit Cost of Detention

Scenario	Opportunity Costs Considered ($/ft^2 land)	Cost per Cubic Foot Detained ($/ft^3)	Quantity Detained (ft^3)	Allowance Revenue ($)
	0.00	5.46	119,685	46,275
$15.00	0.13	7.34	114,625	122,250
Allowances	0.28	8.93	109,600	197,625
$8.00	0.00	5.28	118,111	37,312
Allowances	0.13	7.37	101,834	167,528
	0.28	11.88	66,531	449,952
$5.00	0.00	5.73	99,954	114,105
Allowances	0.13	12.97	44,414	391,805
	0.28	12.98	44,404	391,855
$2.50	0.00	N/A	0	306,937
Allowances	0.13	N/A	0	306,937
	0.28	N/A	0	306,937
Command	0.00	5.40	122,775	0
and Control	0.13	7.12	122,775	0
	0.28	9.22	122,775	0
18-mile Tunnel	N/A	8.93-12.64	122,775+	0

These calculations were made assuming that the parcel owner chooses to retain *all* the excess runoff, or to retain *none* of it, that is would buy allowances instead.

Conclusions

This study is a part of an ongoing effort to fully describe a realistic market for tradable stormwater allowances, the creation of which is intended to alleviate the problems caused by increased stormwater runoff in urban and urbanizing areas. As impervious surface grows relative to natural landscapes the problem will increase, and municipal authorities will look for ways to deal with it that are both practical from a political standpoint and ecologically effective. An allowance system that may be less costly than alternatives has been outlined, but a proof is needed that assumptions used were not unrealistic. To this end it has been shown here that inclusion of the opportunity cost of residential land dedicated to particular stormwater control practices increases the overall cost of stormwater control in a pilot study area. Using hedonic modeling techniques, the value parcel holders place on their land area has been estimated, this information has been used in the study's TRA.

The hedonic price function can provide a lower bound estimate of the opportunity costs associated with giving up an additional square foot of a landowner's yard. However, the private costs that a landowner associates with devoting a portion of his or her yard to a stormwater best management practice may be based on more than the loss of physical area. These more diverse values can be better captured through stated preference valuation methods.

The most popular stated preference method to estimate the total economic cost of a non-market good is the contingent valuation method (CVM). The CVM approach uses surveys to elicit participants' maximum willing to pay for a good or service in a hypothetical market. These methods are not without criticism. In particular, concerns exist with whether hypothetical markets can bring out actual statements of value. Recently, experimental auctions have been gaining support as vehicles for stated preference valuation because they show promise in overcoming the problems associated with hypothetical bias. Further research will include the collection of experimental auction data from homeowners to validate some of our assumptions about decision making we model, and the actual implementation of BMPs for the purpose of measuring effectiveness.

References

Aieta, E. M., Berg, J. D., Roberts, P. V., and Cooper, R. C. (1980). Comparison of Chlorine Dioxide and Chlorine in Wastewater Disinfection. Journal WPCF 52(4), 810-822.

American Society of Civil Engineers (ASCE). 2002. National Stormwater Best Management Practices (BMP) Database, http://www.bmpdatabase.org/

American Public Health Association, American Water Works Association, and Water Environment Federation. (1998). "Standard Methods for the Examination of Water and Wastewater," 20th Edition.

Bannerman, R. T., Owens, D. W., Dodds, R. B., and Hornewer, N. T. (1993). Sources of Pollutants in Wisconsin Stormwater. Wat. Sci. Tech., 28(3-5), 241-259.

Badge, U. S. and Rangari, A. K. (1999). Periodicity of Coliform Bacteria in an Aquatic Environment. Wat. Sci. Tech., 40(7), 151-157.

Baumol, W. J. and Oats, W. E. (1975). The Theory of Environmental Policy, Cambridge University Press, Cambridge

Chang, G. (1999). Personal Communication with Thomas R. Schueler. Austin TX Environmental and Conservation Services Dept., City of Austin, TX.

Chandler, H. (2003). Personal Communication. Environmental Services, Jefferson County. AL.

Columbus Water Works. (2001). Wet Weather Demonstration Projects: CSO Technology Testing, Source Water Assessment and Protection, Watershed Assessment and Management. Draft Report, http://www.cwwga.org/.

Davies, C. and Bavor, H. (2000). The Fate of Stormwater-Associated Bacteria in Constructed Wetland and Water Pollution Control Pond Systems. Journal of Applied Microbiology 89(2), 349-360

Donovan, S. (1995). “Sanitary Sewer Overflow Removal: The Approach Used by the Metropolitan Sewer District of Greater Cincinnati,” Proc., National Conference on Sanitary Sewer Overflows, Washington, DC.

Edwards, D. R., Coyne, M. S., Daniel, T. C., Vendrell, P. F., Murdoch, J. F., and Moore, P. A., Jr. (1997). Indicator Bacteria Concentrations of Two Northwest Arkansas Streams in Relation to Flow and Season. Transactions of the ASAE, 40(1), 103-109.

Eheart, J. W. (1980). “Cost-Efficiency of Transferable Discharge Permits for the Control of BOD Discharges,” *Water Resources Research*, 16(6), 980-986.

Evans, F. L., III, Geldreich, E. E., Weibel, S. R., and Robeck, G. G. (1968). Treatment of Urban Storm Water Runoff. Journal of WPCF, 40(5), R162-170.

Francy, D. S., Helsel, D. R., and Nally, R. A. (2000). Occurrence and Distribution of Microbiological Indicators in Groundwater and Stream Water. Water Environ. Res., 72(2), 152-161.

Field R. (1990). Combined Sewer Overflows: Control and Treatment, In: Control and Treatment of Combined-Sewer Overflows, Editor: P. E. Moffa. Van Nostrand Reinhold, New York, NY. 119-190.

Field R. (1996). Stormwater Pollution Abatement Technologies, In: The Control and Treatment of Industrial and Municipal Stormwater, Editor: P. E. Moffa. Van Nostrand Reinhold, New York, NY. 151-237.

Flood, J. A. and Ashbolt, N. J. (2000). Virus-Sized Particles Can be Entrapped and Concentrated One Hundred Fold Within Wetland Biofilms. Advances in Environmental Research 3(4), 403-411.

Heaney, J. P., Sample, D., and Wright, L. (2002). *Costs of Urban Stormwater Control* US Environmental Protection Agency, National Risk Management Research Laboratory, EPA-600/R-02/021.

Gerba, C. P., Thurston, J. A. Falabi, J. A., Watt, P. M., and Karpiscak, M. M. (1999). Optimization of Artificial Wetland Design for Removal of Indicator Microorganisms and Pathogenic Protozoa. Water Science and Technology 40(4), 363-368.

Geldreich, E. E., Best, L. C., Kenner, B. A., and Van Donsel, D. J. (1968). The Bacteriological Aspects of Stormwater Pollution. Journal of WPCF, 40(11), 1861-1872.

Glenne, B. (1984). Simulation of Water Pollution Generation and Abatement on Suburban Watersheds. Water Resource Bulletin, 20(2).

Gray, D. (2000). Qualitative Review of Epidemiology Studies, in Regional BEACH Program Conference – 1999 Proceedings: U.S. Environmental Protection Agency, Office of Water, EPA 823-R-00-003, February, 43-45 (http://www.epa.gov/waterscience/beaches/proceedings/Beach-99.pdf).

Jordan, B. (2003). Colts Neck Defends Decision to Gas 284 Canada Geese, Asbury Park Press, July 11, 2003. Asbury Park, NJ.

Karpiscak, M. M., Gerba, C. P., Watt, P. M., Foster, K. E., and Falabi, J. A. (1996). Multi-Species Plant Systems for Wastewater Quality Improvements and Habitat Enhancement. Water Science and Technology, 33(10-11), 231-236.

Karpiscak, M. M., Freitas, R. J., Gerba, C. P., Sanchez, L. R., and Shamir, E. (1999). Management of Dairy Waste in the Sonoran Desert Using Constructed Wetland Technology. Water Science and Technology, 40(3), 57-65.

Katz, A., Narkis, N., Orchansky, F., Fiedland, E., and Kott, Y. (1994). Disinfection of Effluent by Combination of Equal Doses of Chlorine Dioxide and Chlorine Added Simultaneously over Varying Contact Times. Wat. Res., 10, 2133-2138.

Kern, J., Idler, C., and Carlow, G. (2000). Removal of Fecal Coliforms and Organic Matter From Dairy Farm Wastewater in a Constructed Wetland Under Changing Climate Conditions. J. Envir. Sci. Health, 35(8), 1445-1461

Klepal, D. (2001). "Giant Tunnel Under Mill Creek Proposed," The Cincinnati Enquirer, 12 August.

Kurz, R.C. (1998). Removal of Microbial Indicators from Stormwater Using Sand Filtration, Wet Detention, and Alum Treatment Best Management Practices. Surface Water Improvement and Management (SWIM) Program Technical Report, Southwest Florida Water Management District, Tampa, FL.

Makepeace, D. K., Smith, D. W., and Stanley, S. J. (1995). Urban Stormwater Quality: Summary of Contaminant Data. Critical Reviews in Environmental Science and Technology, 25(2), 93-139.

Mallin, M.A. (1998). Land-Use Practices and Fecal Coliform Pollution of Coastal Water. Proceedings of the 14th Annual On-Site Wastewater Treatment Conference, North Carolina State University, Raleigh, NC.

Mallin, M. A., Williams, K. E., Esham, E. C., and Lowe, R. P. (2000). Effect of Human Development on Bacteriological Water Quality in Coastal Watersheds. Ecological Applications, 10(4), 1047-1056.

Mitchell, D. (2003). Personal Communication. Hubbell, Roth, & Clark, Inc., MI.

Moffa, P. E. (1975). Bench-scale High-rate Disinfection of Combined Sewer Overflows with Chlorine and Chlorine Dioxide EPA-670/2-75-021. U.S. EPA Office of Research and Development, Cincinnati, OH.

New Jersey Department of Environmental Protection. (1999). Initial Surface Water Quality Characterization and Assessment Report (Based on DEP Data) for Watershed Management Area 12. http://www.state.nj.us/dep/dsr/watershed/area12-report.htm

Ohio Environmental Protection Agency. (1994). "Biological and Water Quality Study of Mill Creek and Tributaries," Ohio Environmental Protection Agency Technical Report SWS/1993-12-9, Columbus, Ohio.

Oliver, B. G. and Cosgrove, E. G. (1975). The Disinfection of Sewage Treatment Plant Effluents Using Ultraviolet Light. Can. J. Chem. Eng., 53,170.

Olivieri, V. P., Kruse, C. W., Kawata, K., and Smith, J. E. (1977). Microorganisms in Urban Stormwater. USEPA Report No. EPA-600/2-77-087. July.

Pavia, E. H. and Powell, C. J. (1968). Stormwater Disinfection at New Orleans. In: Proceedings of the 41st Annual Conference of the Water Pollution Control Federation, Chicago, IL.

Pitt, R., Lalor, M., Adrian, D. D., and Field, R., and Barbe, D. (1993). Investigation of Inappropriate Pollutant Entries into Storm Drainage System: A Users Guide. Office of Research and Development, U.S. Environmental Protection Agency, Cincinnati, Ohio. EPA/600/R-92/238.

Pitt, R., Maestre, A., Morquecho, R., Brown, T., Swann, C., Cappiella, K., and Schueler, T. (2003a). Evaluation of NPDES Phase I Municipal Stormwater Monitoring Data. National Conference on Urban Storm Water: Enhancing Programs at the Local Level Proceedings, Chicago, IL, February 17-20.

Pitt, R., Maestre, A., and Morquecho, R. (2003b). The National Stormwater Quality Database (NSQD, version 1.0). 76th Annual Technical Exhibition and Conference, Water Environment Federation, Los Angeles, CA, October 11-15.

Poppe, J. (2003). Personal Communication. Wastewater Division Manager, City of Bremerton, WA.

Prince George's County, Maryland, Department of Environmental Resources. (2000). Low-Impact Development: An Integrated Design Approach, EPA 841-B-00-003, January.

Qureshi, A. A. and Dutka, B. J. (1979). Microbiological Studies on the Quality of Urban Stormwater Runoff in Southern Ontario, Canada. Wat. Res., 13, 977-985.

Rasmus, J. and Weldon, K. (2003). Moonlight Beach Urban Runoff Treatment Facility. Stormwater, the Journal for Surface Water Professionals, May/June, 12-22.

Sakamoto, G. and Cairns, W. L. (1997). UV Dose-response Determination: Collimated Beam vs. Pilot Studies. In: Proceedings of Hawaii Water Environment Association Conference, Honolulu, HI.

Sample, D. J., Heaney, J. P., Wright, L. T., Fan, C., Lai, F., and Field, R. (2003). "Costs of Best Management Practices and Associated Land for Urban Stormwater Control," *Journal of Water Resources Planning and Management*, 129(1), 59-68.

Santa Monica Urban Runoff Recycling Facility (SMURRF). (2000). http://Epwm.Santa-Monica.Org/Epwm/Smurrf.Html.

Schueler, T. R. (1987). Controlling Urban Runoff: A Practical Manual for Planning Designing Urban BMPs, Washington Metropolitan Water Resources Planning Board, Washington, DC.

Scro, R. (1993). Navesink River - An Assessment of Shellfish Water Quality and Nonpoint Source Pollution, Water Monitoring Project, Shellfish Growing Area 3, NJDEPE Division of Science and Research, CN 109, Trenton, NJ 08625-0409.

Selvakumar, A. and Borst, M. (2004). Effects of Sample Holding Time on Concentrations of Microorganisms in Stormwater. Water Environment Research, 76(1), 67-72.

Siddique, M. (2003). Personal Communication. CSO Control Program Manager, District of Columbia Water and Sewer Authority. Washington, DC.

Stenstroem, T. A. and Carlander, A. (2000). Occurrence and Die-Off of Indicator Organisms in the Sediment in Two Constructed Wetlands. Water Science and Technology, 44(11-12), 223-230.

Stinson, M. K., Field, R., Moffa, P. E., Goldstein, S. L., Smith, K. J., and Delva, E. (1998). High-Rate Disinfection Technologies for Wet Weather Flow. In: Advances in Urban Wet Weather Pollution Reduction. Water Environment Federation Specialty Conference, Cleveland, OH.

Stinson, M. K., Wojtenko, I., and Field, R. (1999). High-rate disinfection techniques for combined sewer overflow. In: Proceedings of the 26th Annual Water Resources Planning and Management Conference, Tempe, AZ.

Stinson, M. K. and Perdek, J. M. (2003). Chapter Three: Management and Control of Pathogens. In: Managing Urban Watershed Pathogen Contamination, EPA/600/R-03/111, EPA Office of Research and Development, Cincinnati, OH.

Tchobanoglous, G. (1981). Wastewater engineering: Collection and pumping of wastewater, McGraw-Hill, New York.

Thurston, J. A., Gerba, C. P., Foster, K. E., and Karpiscak, M. M. (2001). Fate of Indicator Microorganisms, Giardia and Cryptosporidium in Subsurface Flow Constructed Wetlands. Wat. Res., 35(6), 1547-1551.

Thurston H. W., Goddard, H. C., Szlag, D., and Lemberg, B. (2003). "Controlling Stormwater Runoff with Tradable Allowances for Impervious Surfaces," Journal of Water Resources Planning and Management, 129(5), 409-418.

Tietenberg, T. (2000). Environmental and Natural Resource Economics, Addison-Wesley, Reading, Massachusetts.

U.S. Department of Agriculture. (1986). *Technical Release 55: Urban Hydrology for Small Watersheds*, Engineering Division, Soil and Conservation Service, Washington, D.C.

U.S. Environmental Protection Agency (U.S. EPA). (1973). Microstraining and Disinfection of Combined Sewer Overflows - Phase II, EPA-R2-73-124 Office of Research and Development, Washington, DC.

U.S. EPA. (1979a). Combined Sewer Overflow Abatement Program Rochester, NY, Vol. II Pilot Plant Evaluations; EPA/600/2-79-031b. EPA Office of Research and Development, Cincinnati, OH.

U.S. EPA. (1979b). Disinfection/Treatment of Combined Sewer Overflows, Syracuse, New York. EPA/600/2-79-134. Office of Research and Development, Cincinnati, OH.

U.S. EPA. (1983). Results of the Nationwide Urban Runoff Program, Volume 1: Final Report. Water Planning Division, Washington, DC. December. NTIS Publication No. 84-185552.

U.S. EPA. (1992). NPDES Storm Water Sampling Guidance Document. EPA 833-92-001. Office of Water, Washington, D.C.

U.S. EPA. (1994). Combined Sewer Overflow Control Policy, 59 Federal Register 18688. April 19.

U.S. EPA . (1996) Draft Framework for Watershed-Based Trading, Office of Water, EPA 800-R-96-001, May.

U.S. EPA. (2002a). Management of Combined Sewer Overflow, EPA/600/R-02/xxx. DRAFT. Office of Research and Development, Cincinnati, OH.

U.S. EPA. (2002b). CSO Disinfection Pilot Study: Spring Creek CSO Storage Facility Upgrade. Research Summary, EPA/600/R-02/077. Office of Research and Development, Cincinnati, OH.

U.S. EPA. (2002c). Continuous Deflection Separation, Fuzzy Filter and UV Treatment of SSO-Type Wastewaters: Pilot Study Results, EPA/600/R-02/100. Office of Research and Development, Cincinnati, OH.

U.S. EPA. (2002d). National Water Quality Inventory 2000 Report. EPA-841-R-02-001. Office of Water, Washington, D.C.

U.S. EPA. (2003). Illicit Discharge Detection and Elimination, http://cfpub2.epa.gov/npdes/stormwater/menuofbmps/illi_2.cfm. Office of Water, Washington, DC.

U.S. Fish and Wildlife Service. (2002). Service Releases Draft EIS on Resident Canada Geese, http://southeast.fws.gov/news/2002/n02-000.html. Press Release, March 4, 2002.

Urbonas, B. (1999). Assessment of Stormwater Best Management Practice Effectiveness. In Innovative Urban Wet-Weather Flow Management Systems, by Heaney, J.P., R. Pitt, and R. Field, EPA/600/R-99/029, http://www.epa.gov/ednnrmrl/publish/book/epa-600-r-99-029/. U.S. EPA Office of Research and Development, National Risk Management Research Laboratory, Cincinnati, OH.

Waye, D. (2003). Removing Bacteria from Runoff, From Alum to UV - Draft. NPS News-Notes, Issue 73, Sept.-Oct.

Weiskel, P. K., Howes, B. L., and Heufelder, G. R. (1996). Coliform Contamination of A Coastal Embayment: Sources and Transport Pathways. Environmental Science and Technology, 30(6), 1872 - 1881.

Wolman, M. G. and Miller, J. P. (1960). "Magnitude and Frequency of Forces in Geomorphic Processes," *Journal of Geology*, 68(1), 54-74.

White, G.C. (1999). Handbook of Chlorination and Alternative Disinfectants. Wiley-Interscience Publication, John Wiley & Sons, Inc., New York, NY.

Winer, R. (2000). National Pollutant Removal Performance Database for Stormwater Treatment Practices, Second Edition, Center for Watershed Protection, Ellicott City, MD.

Wojtenko, I., Stinson, M. K., and Field, R. (2001). Challenges of Combined Sewer Overflow Disinfection with Ultraviolet Light Irradiation. Critical Reviews in Environmental Science and Technology, 31(3), 223-239.

Young, K. D. and Thackston, E. L. (1999). Housing Density and Bacterial Loading in Urban Streams. Journal of Environmental Engineering, 125(12), 1177-1180.

Chapter 5

BMPs in the Heartland: An Institutional Framework for Stormwater Management

Michael A. Ports, PE., D.WRE., F.ASCE and Tom Jacobs

Introduction

The particle size distributions of stormwater at different locations in an urban area greatly affect the ability of different source area and inlet controls in reducing the discharge of stormwater pollutants. A series of U.S. EPA funded research projects has examined the sources and treatability of urban stormwater pollutants (Pitt, *et al.* 1995). This research has included particle size analyses of 121 stormwater inlet samples from three states (southern New Jersey, Birmingham, Alabama; and at several cities in Wisconsin) that were not affected by stormwater controls.

Kansas City, Missouri traces its genesis back to the year 1821, when the Missouri Territory was admitted to the Union as a state. That same year, a Frenchman named Francois Chouteau traveled up the Missouri River from St. Louis and established a trading post on the south bank of the river. Other French families gradually joined him. From the beginning, the fledging settlement suffered from flooding. In 1826, M. Chouteau rebuilt his trading post on higher ground after a large, destructive flood. Now more than 1,800,000 people live in the Greater Kansas City Metropolitan Area. The area straddles the border between the states of Kansas and Missouri. Moreover, the area consumes part or all of eight counties and all or part of one hundred sixteen incorporated municipalities. By far the largest municipality is Kansas City, Missouri

(KCMO) with a population of approximately 450,000. Spread out over approximately 320 square miles, KCMO now includes parts of four counties and continues to grow in both population and area as new territory is annexed. Roughly half of KCMO lies north of the Missouri River and half south of it. The area south of the river generally is older and much of it is served by combined sewers. KCMO extends from the old downtown and industrial areas along the river southward some 18 miles characterized by newer and newer developments. The Northland generally is much younger in character and continues to undergo rapid urbanization.
More than a dozen of the smallest municipalities have a population of less than 500 people. Not surprisingly, the metropolitan region exhibits significant variability in terms of local government policies, regulations, and standards. The textured geographic, economic, and political terrain hampers the efforts of many communities in working together to address important regional issues such as stormwater management.

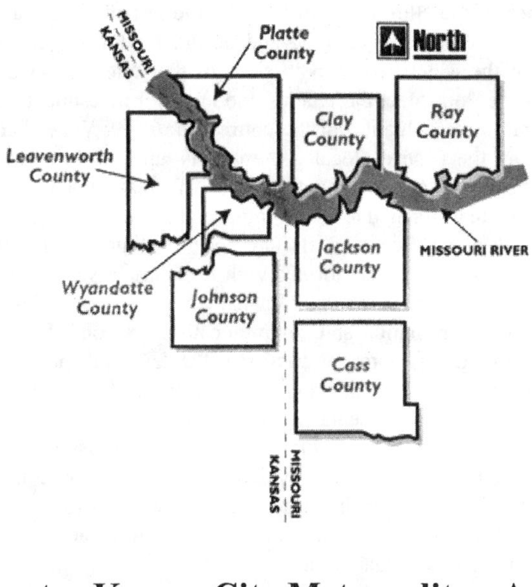

Greater Kansas City Metropolitan Area

Background

Despite the many hurdles, the community as a whole has made substantial progress. Numerous independent, yet related, initiatives have yielded positive results. Foremost among different initiatives is the development of regional design standards for stormwater management and for erosion and sediment control. Also, significant is the formation of the Jackson County Stormwater Committee, which provides a forum for all of the municipalities within that county to discuss issues of mutual interest. While successful within the county, the committee suffers from the same cross boundary problems.

From time to time, the Kansas City District, Corps of Engineers has provided a vehicle to facilitate cooperation and address cross boundary issues. A prime example of such a success is their Turkey Creek Flood Control Project. That one 13-square-mile watershed encompasses parts of two states, three counties, and no less than thirteen municipal governments and has had a history of significant flood damages over many decades. The Kansas City District has been successful in facilitating fundamental agreement among the various units and levels of government to develop a plan that when fully implemented will go a long way toward solving the major flooding problems in the lower portion of the watershed and enhancing the integrity and health of the main stream channel in the upper portion of the watershed. However, in spite of that success, the plan does not address in any comprehensive way stormwater management, receiving water quality, and habitat viability.

Another example may be found in the Blue River Watershed. Over the years, the Kansas City District has invested heavily in a series of flood control projects along the lower reaches of the Blue River in KCMO, the last of which are scheduled for construction soon. While the planning and design for the projects proceeded, the upper portions of the watershed on both sides of the state line have urbanized at a rapid pace. In order to protect the federal flood control investment, the Kansas City District has begun a re-evaluation of the entire Blue River Watershed. Preliminary discussions among the District, local governments, and the Mid-America Regional Council, have begun to explore the feasibility of creating a new, jointly funded position to coordinate watershed level programs and projects.

Within KCMO three significant efforts have been initiated. First, the Stormwater Coordinating Committee has been formed with formal representatives from all of the City agencies involved in stormwater management. It provides a needed and effective forum for coordination and communication. Second, KC-One project has been launched with much fanfare. Its mission is the development of a comprehensive city-wide stormwater management program. And third, the KCMO Water Services Department invited senior representatives from Johnson County Wastewater and the Wyandotte County Department of Public Works to discuss issues of mutual concern. The group agreed that the forum provided was valuable and, therefore, should continue to meet. Interestingly, it further was decided to expand the scope of the forum to include all aspects of drinking water, wastewater, and stormwater as well as invite more stakeholders throughout the region.

Developing Regional Design Standards

Working together, the community has developed shared resources to manage public works projects through the local chapter of American Public Works Association (APWA). The chapter was formed in 1963. In 1981, it produced a set of design criteria and specifications for storm drainage projects that were adopted by ordinance in most communities. They were updated in 1990 to incorporate lessons learned.

While the standards functioned well, evolving local priorities led the community to revisit stormwater management. As a result of severe flooding in October 1998, a strong regional consensus reframed the practice of stormwater management to address water and environmental quality, in addition to flood control and public safety. The rationale underlying the new approach includes substantial long-term savings in the development and maintenance of public infrastructure, protecting of natural resources, compliance with new federal water quality regulations, and improving the overall quality of life.

A "Water Resources Action Plan" developed in 2001 by area public works and planning directors throughout the metropolitan area called for more regional cooperation on watershed management through an institutionalized Water Resources Policy, Planning and Practices (WRP3) committee structure. A key short-term objective of the committee was to revise local stormwater planning guidelines and design standards. A broad-based Steering Committee was established in 2002 to guide the initiative. Financial support was provided by many local governments and by a grant from the Environmental Protection Agency. Staff from the Mid-America Regional Council, the regional council of governments and metropolitan planning organization, administered the project. The committee issued a request for proposals and selected a consulting team to do the work.

The Steering Committee tended to focus on the technical issues associated with analyzing, designing, and constructing stormwater management facilities. Representing a cross-section of the community, each individual participant had their own views on what needed to be done. Their views ranged across the entire spectrum of possibilities. However, everyone understood that a set of widely-accepted policies was needed to guide the development of criteria that would benefit the entire region. This led to a two-step process. Step one involved developing consensus on the policies. Then, step two involved developing the criteria to support the governing policy.

Developing Policy Consensus

The steering committee began by listing all of their ideas on what needed to be accomplished. Almost one hundred ideas were presented. Then a series of facilitated meetings were conducted with the steering committee to develop the governing policies. First the group reaffirmed its commitment to the following goal and objectives previously established through the Action Plan:

GOAL: provide tools and design standards for communities in metropolitan Kansas City and outlying areas to manage stormwater, protect local water resources, and conserve the quality of life through three intrinsically connected objectives:

- Reduce the risk of flooding
- Protect surface water quality and the environment
- Increase the amenity value associated with regional land and water resources through the design of multipurpose community projects like greenway trails.

Project participants further agreed upon a number of review criteria and overarching issues to guide project implementation. Review criteria sought to ensure that the project would be regionally consistent, achievable, effective and acceptable. Once the overall goal and objectives were defined, the next task defined several key issues that would be important for the proposed new standards to address, including

- Resource Protection
- Planning Integration
- Education
- Operations and Maintenance
- Inspection and Enforcement
- Inter-jurisdictional Cooperation.

The proposed goal, objectives, and criteria were presented to three outside stakeholder groups. The first was the WRP3 committee. The second group was the Board of Directors of the local APWA chapter. The third group of stakeholders consisted of approximately 75 individuals who attended the workshop advertised by MARC. The WRP3 committee offered two observations.

- An alternative to "conventional development" is needed.

- Public support for protecting natural resources, creating multipurpose projects, and water quality management is widespread.

Each workshop began with a summary presentation of the progress made in developing a goal, objectives, criteria, and primary issues. Facilitated discussion ensued to understand stakeholder views, identify areas of consensus and of divergent opinions. Small group discussions were held to brainstorm new ideas and issues. Then, summary presentations from each group were presented. Notes from each group were collected for reference.

The outside stakeholder process affirmed the general direction of the effort so far. The next task was to begin addressing the ideas and issues that were generated. The ideas and issues were grouped into eight major categories.

- Standards specific to new development or redevelopment
- System capacity requirements
- Hydrology
- Open channels
- Geomorphology and bioengineering
- Detention
- System components
- Water quality

Translating Policy into Standards

With widespread support, the team dove immediately and enthusiastically into addressing the pressing technical issues. However, it quickly became apparent that support for broad goals and objectives does not necessarily translate into widespread agreement on the specific and detailed technical issues and requirements needed to achieve the adopted goal and objectives. During initial meetings to address the specific technical issues, intense discussions ensued over virtually every aspect of stormwater management. Different, even competing, individual community values and approaches were defended by many, if not most, of the stakeholders.

In general, the results from step one directed the effort toward very aggressive resource protection measures such as strict stream setback requirements, specific requirements for maintaining or improving water quality, and requiring multi-purpose solutions. As the proposed measures were developed, resistance from stakeholders grew. Many were not comfortable with the notion of *pushing the envelope too far*. They were often heard saying that their governing body would not support some of the more aggressive measures to protect resources or improve water quality.

It became apparent that many such issues could not be resolved in a meeting of the full steering committee. While the detailed issues primarily were technical in nature, the real discussion focused on the policy underlying the technical issue. For instance, almost all agreed that stream setbacks are an excellent way to preserve natural resources. Almost all further agreed that, in general, the wider you can make them, the more beneficial they are. However, some stakeholders felt that their community would not support wide setbacks. Other stakeholders believed that they had the necessary political support. Such a divergence of positions could not be overcome solely with technical justification. It took intense negotiation between opposing positions to identify and develop specific criteria that both could support. Similar discussions were conducted for all work within the riparian zone, water quality requirements, maximum allowable runoff from development sites, and hydrologic methods.

Subcommittees were organized to conduct the intense discussions. Usually four to eight of the stakeholders volunteered for specific subcommittees because they had a keen interest in the subject area. The subcommittee process proved to be very productive. The project participants could work closely on the specific issues. As a direct result, a spirit of teamwork, cooperation, and conciliation became apparent. The detailed discussions gradually transformed from one characterized by two opposing viewpoints to another based upon common ground.

Subcommittees usually met in a series of meetings. Following each meeting, the project team conducted research to address questions, provide information on state of the art practices, and provide technical support. This allowed the committee to reach an informed consensus. The process culminated in a presentation by the project team and subcommittees to the steering committee. Questions arising out of the presentation, typically were addressed easily and quickly. New design criteria were written to supplement the 1990 criteria. Some of the major new criteria include:

- Maximum release rates limited to predevelopment rates for range of storms.
- Minimum setbacks for streams with 40 acres or more tributary area.
- A user-friendly guide to evaluating stream stability.
- Specific requirements for energy dissipation at outfalls.
- Requirements for locating stream crossings.

- Design criteria for grade control.
- Minimum recommended guidelines for maintaining runoff water quality from residential, commercial, and industrial sites.
- BMP design criteria tailored to the Kansas City metropolitan area.
- An innovative approach for choosing BMPs using TR-55 methodology.

Lessons Learned

The new design standards were adopted by the full local APWA membership in November 2003. Several local communities have already adopted the criteria and begun using them. Some are considering adoption or are in the process of adoption. Still other communities around the Midwest are reviewing the standards and how they might be tailored to meet their specific needs.

This project accomplished many of its original objectives; and, the design criteria are widely supported. However, the project required painstaking deliberation to achieve community support. The steering committee and project team learned that development of design criteria for multiple jurisdictions first required the determination of common ground. Once technical and political expectations were commonly understood, the group worked to assess how to effectively "raise the bar" for the community. The wide range of governing body concerns, technical abilities, and technical approaches confirmed the ultimate support necessary to develop and adopt the final criteria and methodologies.

KC-One

The effort to develop a comprehensive city-wide stormwater management program for KCMO has been dubbed KC-One. The mission of KC-One is to create one plan, one people, and one for stormwater management throughout Kansas City. The new program will clarify responsibilities within KCMO, establish priorities, provide timelines and schedules, and offer funding recommendations for comprehensive stormwater management.

From the outset in 1826, Kansas City has been beset by the problems associated with periodic flooding. More recently, major floods occurred in 1993 and again in 1998. Both events caused loss of life as well as significant property damage. And both events revealed that flooding problems plague most every neighborhood in KCMO. The NPDES municipal separate storm sewer system regulations continue to expand the responsibilities of KCMO in managing stormwater. Numerous public surveys indicate that the community supports a holistic approach to stormwater management that involves flood control, water quality management, natural resource protection, and minimizing the adverse impacts from new developments. Thus, KC-One offers a unique opportunity to shape the way that KCMO manages its watersheds and manages stormwater both now and in the years to come.

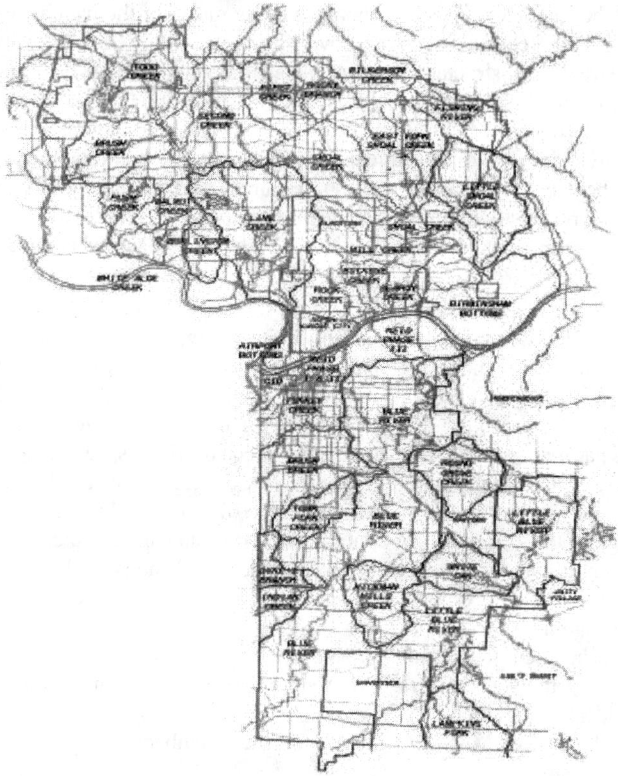

The KCMO Watersheds

The comprehensive plan will include recommendations in the following areas:

- Administration: define roles and responsibilities of City departments, staffing needs, and requirements for operation and maintenance.

- Policy: incorporate best management practices, stream protection, land use planning, and development plan approval.

- Public Involvement: listen and learn from the community to ensure that the plan is responsive to the needs of the stakeholders.

- Stormwater Master Plans: prioritize improvements in 35 watersheds and ensure technical consistency and quality.

- Funding: identify approaches and options to address capital, administrative, and operational needs.

- Implementation: formulate a comprehensive plan and achieve results.

To address the regulatory requirements as well as maintain the natural resources in KCMO, the comprehensive program will be based on *"Green Infrastructure"* concepts. Green Infrastructure is an amalgam of non-structural best management practice, low impact development, conservation biology, landscape ecology and smart growth concepts. It involves protecting existing natural resources and waterways by minimizing the adverse impacts from new developments. It also involves enhancing the existing natural resources and waterways by retrofitting existing developments with appropriate green solutions. Emphasis will be placed on avoiding adverse impacts first, then minimizing adverse impacts where unavoidable.

Phase I

The purpose of the work effort is to establish the vision, principles, and priorities for the new stormwater management program in Kansas City, Missouri. These key elements will guide the development of the scope of work, budget, and schedule for the remainder of the project (Phase II). Phase I involves working with the Water Services Department, other City departments, elected officials, regulators, and other stakeholders. Thus, for the first time in Kansas City, all of the stakeholders through an open public process will formulate the scope of work, budget, and schedule for development of the comprehensive city-wide stormwater management program. A series of one-on-one interviews and stakeholder workshops will culminate in a consensus of what the vision, guiding principles, and priorities for stormwater management should be.

Phase I is scheduled for completion in just three months. The first task is the identification of the stakeholders. After significant deliberation, it was agreed that the stakeholders should include:

1. Mayor, City Manager, and City Council Members
2. Director of Water Services Department
3. Directors of City Planning and Development, Environmental Management, Public Works, Codes Administration, and Parks and Recreation Departments.
4. Neighborhood Groups
5. Key Media
6. Chambers of Commerce and Downtown Development Group
7. Home Builders Association and Developers
8. Mid-America Regional Council
9. Kansas City District, Corps of Engineers
10. Missouri Conservation, Natural Resources, and Transportation Departments

One-On-One Interviews

The second step is the formulation and development of a standard questionnaire for the individual stakeholder interviews. The standard questionnaire consists of nineteen questions. Each question is designed specifically to draw out a sense of what the overall vision, guiding principles, and priorities should be from each stakeholder. Also, the questionnaire is intended to develop a dialogue between the stakeholder and the interviewer. The interview team consists of a senior stormwater management professional and a senior public involvement professional.

The stormwater management professional leads the interview by asking the questions. But, more importantly, he listens to the answers and follows up with additional comments and questions in order to elicit as much as possible from each stakeholder. Each interview is scheduled for one hour. The public involvement professional primarily takes notes, but also listens carefully and follows up with additional comments and questions in order to ensure that the notes adequately reflect the interview results. The results from all of the interviews are then compiled and summarized. From the compiled results a preliminary statement of vision and guiding principles is drafted.

Stakeholder Workshops

The individual stakeholder interviews were completed over four to six weeks. The first stakeholder workshop was scheduled at the end of the first six week period. At the facilitated workshop, the results from the individual interviews were shared with all of the stakeholders. The compiled summary results were discussed in detail, with the intent of identifying the areas of broad consensus as well as divergence. Also, the preliminary draft statement of vision and guiding principles were reviewed and discussed. As necessary, follow-up interviews and meetings were held to help reach consensus and identify areas of divergence.

Based upon the results from the first workshop and the follow-up interviews and meetings, preliminary priorities and a preliminary scope of work were drafted. Then a second facilitated stakeholder workshop was held. The primary goals of the second workshop were to resolve the remaining areas of divergence and to reach consensus on the vision and guiding principles. At that point, the draft preliminary priorities and preliminary scope of work were reviewed and discussed. Follow-up meetings were held to resolve the significant issues that remained.

Based upon the results from the second workshop and the follow-up meetings, the preliminary scope of work was revised. Also, a preliminary budget and schedule were drafted. Then, a third facilitated workshop was held at which the draft final scope of work and draft preliminary budget and schedule were reviewed and discussed. All outstanding issues were identified. A fourth facilitated workshop was scheduled at which the last remaining issues were resolved. Based upon the results from the four workshops, the final scope of work, budget, and schedule for Phase II were prepared.

The results from the majority of the one-on-one interviews yielded certain recurring themes. Those recurring themes may be summarized as follows:

1. Focus on reducing flooding, protecting natural resources, and improving water quality.

2. Primary obstacle is funding. Innovative approaches and solutions are needed that combine local state, and federal sources of funding.

3. A big picture view is needed that incorporates green infrastructure into multi-objective and multi-use solutions.

4. Public involvement and education is the key to building support and implementing effective programs.

Phase I Outcomes

Four conclusions may be drawn from the major outcomes of Phase I. The first is that there is wide understanding among the identified stakeholders. Not only do they understand the current situation and its problems, they seem to understand what is needed to solve the problems and improve the situation. Moreover, a majority of the stakeholders have the will to make tough decisions and implement solutions. That will is underpinned by the political support of the local elected officials. The third is a general desire to improve the overall quality of life in the community. That desire is coupled with the notion that the quality of life must be improved in all neighborhoods. The fourth is the realization that no program will be accepted by the community unless it is accountable to the community. Actions or specific solutions will be accepted and funded only if the are effective and economical.

Specific Phase I outcomes include a vision, mission statement, set of guiding principles, stormwater management strategy, and scope of work for Phase II. The KC-One vision is

> *Stormwater management in KC is a holistic regional watershed planning, operating, and growth strategy.*

That vision is supported by the KC-One mission to

- Protect and benefit citizens
- Provide sound and consistent development policy
- Conserve and restore natural resources
- Be regional in scope and implementation

From the foregoing, it follows naturally that in order to provide a stormwater management program that follows the vision and meets the mission a coherent and comprehensive strategy is needed. Such a strategy must provide funding that is adequate to accomplish the stated goals and objectives. The perception that municipal stormwater personnel are lazy, inept, or just ineffective must be overcome. The overall elements of the program must be well executed and cost effective. The public must see action and results at the same time. Moreover, for the overall program to be successful, it is essential that the strategy produce solid results that reduce flooding, improve water quality, and enhance green spaces.

The stakeholders identified eleven essential guiding principles.

1. Maintenance is essential. The overall program must address the never ending responsibility to maintain existing stormwater infrastructure as well as new infrastructure in the future.

2. Focus on reduced flooding and improved green space at the same time. That is, promote multi-purpose solutions.

3. Stormwater must be managed on a watershed basis. Cooperate with neighboring jurisdictions and look for regional approaches and solutions.

4. Provide for adequate funding from the outset.

5. Improve the development process, through streamlining, clearly establish authority, assign responsibility, and measure results.

6. Educate the public because they are stakeholders and recognize that the process never ends.

7. Communication and coordination are essential elements.

8. Elevate stormwater management as a higher City priority.

9. Ensure overall accountability by delineating responsibility, assigning authority, and measuring results.

10. Integrate long term combined sewer overflow control and sanitary sewer overflow control into the stormwater management program.

11. Keep an eye on both the short term and the long term. Plan for short term successes and manage long term trends.

The foregoing work of Phase I culminated in a draft scope of work, schedule, and budget for the proposed work in Phase II. The scope of work consists of ten project elements, including project management and coordination, policy and regulation plan, administration plan, watershed master plan review, City-wide capital improvement plan, FEMA Community Rating System participation, strategic communications plan, strategic funding plan, implementation plan, and executive summary document. The over schedule for Phase II is proposed for two years.

Where Are We Now?

At present, the draft scope of work, schedule, and budget for Phase II of KC-One are in the final stages of negotiation. Full agreement is anticipated soon with work commencing probably in the fall. During the interim, coordination among the various stakeholders is continuing. Lines of communication have been opened and a dialogue has been started between the KC-One Team and the Overflow Control

Program (OCP) Team. The OCP is just beginning the process of developing a long term combined sewer overflow program and a sanitary sewer overflow control plan. In addition, the first steps toward a regional approach to the problem have been taken. The Water Services Department, which has authority for drinking water, wastewater, and stormwater throughout KCMO, invited the largest neighboring jurisdictions to a meeting to discuss issues of mutual concern. Representatives from the Department of Public Works of the Unified Government of Kansas City, Kansas and Wyandotte County, Stormwater Management Program, Johnson County, Kansas, and the Johnson County Wastewater Authority attended. A general discussion took place in which each representative presented a summary of major or significant projects, initiatives, and other issues facing their organization. Two significant results were noted. First, the representatives recognized their obvious inter-connection and inter-dependency from the general discussion. Second, the group agreed to meet again soon and to invite the Mid-America Regional Council to participate. The subject was broached of establishing some sort of regional district or authority with jurisdiction in both states for drinking water, wastewater, and stormwater management.

Next Steps

Future planning, policy development, and project implementation at the local and regional levels continue to evolve at a rapid pace. Continued progress largely depends upon the degree of cooperation and collaboration that actually occurs at all levels to achieve shared goals. Adoption of new trends represents a fundamental shift in the conventional approach to stormwater management. Hence, an adaptive spirit of inquiry is crucial in order to help the community both to celebrate early successes and to learn from failures. The fundamental challenge to the community remains how to establish meaningful goals that help solve ongoing stormwater management problems while, at the same time, improve the quality of life for the overall community.

Chapter 6

Coarse Solids Management

By *Robert Pitt, University of Alabama*

Introduction

Much research has been conducted on the transport, settling, and scour of gross solids in sanitary sewers and in combined sewers, especially in Europe during recent years (Ashley, *et al.* 1999; 2000; 2002; Butler and Karunaratne 1995; Butler, *et al.* 1995; Cigana, *et al.* 1998a, 1998b, 1998c, 1999, 2000, 2001; plus review by Pisano, *et al.* 1998, amongst others). However, relatively little recent research has been conducted concerning the fate of larger particulates that enter separate stormwater drainage systems. Historical design approaches are intended to minimize particulate deposition in sewerage, and usually present a minimum pipe slope or a minimum velocity objective. If followed, these guidelines usually result in minimal maintenance problems associated with particulate accumulations. However, it is still important to **minimize erosion sources in the watershed. In addition, catchbasin sumps have** also been used to trap the larger particulates that enter storm drainage inlets. Therefore, particulate transport in separate storm drainage has not been considered to be a significant problem for public works managers. However, much more is needed to be known about particulate transport in stormwater drainage systems when conducting stormwater quality investigations, especially when examining the effects of source area controls on outfall quality, especially street cleaning. Prior studies have conducted mass balances in urban drainage systems and have found significant accumulations of solids in the drainage systems. These accumulations are mostly of the largest particulates that enter the drainage system, effectively preventing these from being discharged to the receiving waters. Many source area stormwater quality controls, such as street cleaning and the use of catchbasins, also preferentially remove the largest particulates. Modeling of the benefits of these controls therefore typically leads to inaccurate conclusions concerning reduced discharges of these particulates, and associated pollutants. The pollutants being removed would not likely be effectively transported through the drainage system, but would instead accumulate. In addition, source area and inlet samples frequently indicate much larger amounts of large particulates than would be discharged to the receiving water. This also misleads management strategies pertaining to stormwater quality. This paper presents observations from an urban area mass balance investigation along with some traditional particulate transport approaches in an attempt to explain some of these observations.

Interactions between Catchbasin and Street Cleaning in Urban Drainages and Sediment Transport in Storm Drainage Systems

Robert Pitt, University of Alabama, and Shirley Clark, Penn State-Harrisburg

Effects of Catchbasin and Street Cleaning

A study was conducted in Bellevue, Washington (Pitt 1985), in two mixed residential and commercial study areas, as part of the Nationwide Urban Runoff Program (EPA 1983). One task of this research included the monitoring of catchbasins, simple inlets, man-holes, and sewerage sediment accumulations at more than 200 locations for a period of three years. The sediment in the catchbasins and the sewerage was found to be the largest particles that were washed from the streets. The sewerage and catchbasin sediments had a much smaller median particle size than the street dirt and were therefore more potentially polluting than the particulates that can be removed by street cleaning. Cleaning catchbasins twice a year was found to allow the catchbasins to capture particulates most effectively. This cleaning schedule was found to reduce the total residue and lead urban runoff yields at the outfalls by between 10 and 25 percent, and COD, total Kjeldahl nitrogen, total phosphorus, and zinc by between 5 and 10 percent (Pitt and Shawley 1982).

This research examined two study areas, Lake Hills and Surrey Downs, both similar medium density residential areas. Each study area was examined with four separate experimental conditions: no controls, street cleaning alone, catchbasin cleaning alone, and both street cleaning and catchbasin cleaning together. This research was therefore conducted in a replicated complete block design, allowing runoff quality comparisons between periods having these different public works practices. The eight experimental categories were as follows:

Lake Hills, Active CB, No SC (catchbasins were accumulating material, but no street cleaning operations were being conducted during this project period).

Lake Hills, Active CB, SC (catchbasins were accumulating material, and street cleaning operations were being conducted during this project period).

Lake Hills, Full CB, No SC (catchbasins were full and not accumulating material, and no street cleaning operations were being conducted during this project period).

Lake Hills, Full CB, SC (catchbasins were full and not accumulating material, street cleaning operations were being conducted during this project period).

Surrey Downs, Active CB, No SC (catchbasins were accumulating material, but no street cleaning operations were being conducted during this project period).

Surrey Downs, Active CB, SC (catchbasins were accumulating material, and street cleaning operations were being conducted during this project period).

Surrey Downs, Full CB, No SC (catchbasins were full and not accumulating material, and no street cleaning operations were being conducted during this project period).

Surrey Downs, Full CB, SC (catchbasins were full and not accumulating material, street cleaning operations were being conducted during this project period).

Catchbasins were cleaned and surveyed at the beginning of the project. The accumulation of material was then monitored through periodic measurements. The project periods were therefore categorized as "active" or "full." The active periods were when accumulation was taking place in the catchbasins, while the full periods were when the catchbasins were at equilibrium, with no additional accumulation of material.

The following are Student *t*-test results to measure the significance of the difference between selected data groups for outfall total solids concentrations. There would have to be a 50 to 75% difference between the sample means of the two categories to identify a significant difference, with 10 to 15 storms representing each of the two categories for each test site, using a power of 80%, and assuming a typical COV of about 0.75. P values smaller than 0.05 are usually considered as being significantly different (at the 95% confidence level), while larger P values indicate that not enough data are available to distinguish the data groups at the measured differences.

Student's *t*-test results:

2 vs. 6: both street and catchbasin cleaning in both areas, LH vs. SD

P value: 0.71 (not enough data to detect a difference)

3 vs. 7: nothing in both areas, LH vs. SD

P value: 0.031 (significantly different)

2 vs. 3 LH both street and catchbasin cleaning vs. nothing

P value: 0.037 (significantly different)

6 vs. 7 SD both street and catchbasin cleaning vs. nothing

P value: 0.99 (not enough data to detect a difference)

When both street and catchbasin cleaning was being conducted in both areas, the outfall total solids concentrations appeared to be the same (as expected). However, when no controls were in use in either area, the outfall total solids concentrations were significantly different (Lake Hills had lower total solids concentrations compared to Surrey Downs), which was not expected. When both street and catchbasin cleaning was conducted in Lake Hills, the outfall total solids concentrations were significantly larger than when no cleaning was being conducted, which also was not expected. In Surrey Downs, no differences were detected when cleaning was conducted compared to no cleaning.

These results are counter-intuitive. The hypothesis was that the two watersheds would behave in a similar manner when similar activities were being conducted in each, and that the cleaning would reduce the outfall total solids discharges. Over the years, a number of reasons have been given for the observed odd behavior. Older street cleaning equipment was not very efficient in removing the particles that are washed off, and in fact, have been found to actually remove the larger particles that actually armour the finer materials, potentially increasing the solids discharges. However, the catchbasins are removing particles that have washed off the watershed area and have been transported to the drainage system, but this material likely would not have been transported all the way to the outfall. Ashley, *et al.* (1999, 2000, 2002) has extensively researched the transport of solids in combined sewerage. Unfortunately, similar information is currently lacking for separate storm drainages. The initial objective for the use of catchbasin sumps was to reduce the accumulation of coarse debris in the sewerage. These Bellevue tests seem to indicate the substantial benefit of the removal of this material that may otherwise cause potential flow obstruction problems in the drainage system. However, it is quite likely that this large material would rarely flow completely to the outfalls, at least under the relatively mild Bellevue rain conditions and during the time frame of this study.

Particle Size Distributions

The particle size distributions of stormwater at different locations in an urban area greatly affect the ability of different source area and inlet controls in reducing the discharge of stormwater pollutants. A series of U.S. EPA funded research projects has examined the sources and treatability of urban stormwater pollutants (Pitt, *et al.* 1995). This research has included particle size analyses of 121 stormwater inlet samples from three states (southern New Jersey, Birmingham, Alabama; and at several cities in Wisconsin) that were not affected by stormwater controls.

Particle sizes were measured using a Coulter Counter Multi-Sizer IIe and verified with microscopic, sieve, and settling column tests. Figures 1 through 3 are grouped box and whisker plots showing the particle sizes (in μm) corresponding to the 10th, 50th (median) and 90th percentiles of the cumulative distributions for the three areas. If 90 percent control of suspended solids (by mass) was desired, then the particles larger than the 90th percentile would have to be removed, for example. In all cases, the New Jersey samples had the smallest particle sizes (even though they were collected using manual "dipper" samplers of the cascading water at drainage system inlets and not automatic samplers that may miss the largest particles), followed by Wisconsin, and then Birmingham, Alabama, which had the largest particles (which were collected using automatic samplers, but had the largest rain intensities). The New Jersey samples were obtained from gutter flows in a residential neighborhood that was xeroscaped. The Wisconsin samples were obtained from several source areas, including parking areas and gutter flows mostly from residential, but from some commercial areas, and the Birmingham samples were collected from a long-term parking area. The median particle sizes ranged from 0.6 to 38μm and averaged 14μm. The 90th percentile sizes ranged from 0.5 to 11μm and averaged 3μm.

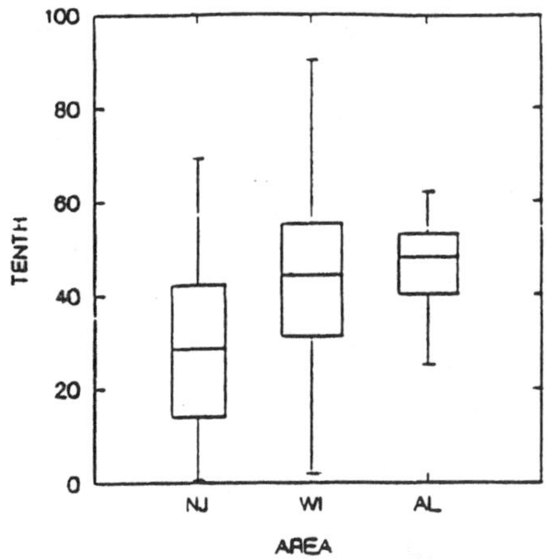

Figure 1. Tenth percentile particle sizes for stormwater inlet flows (Pitt, *et al.* 1997).

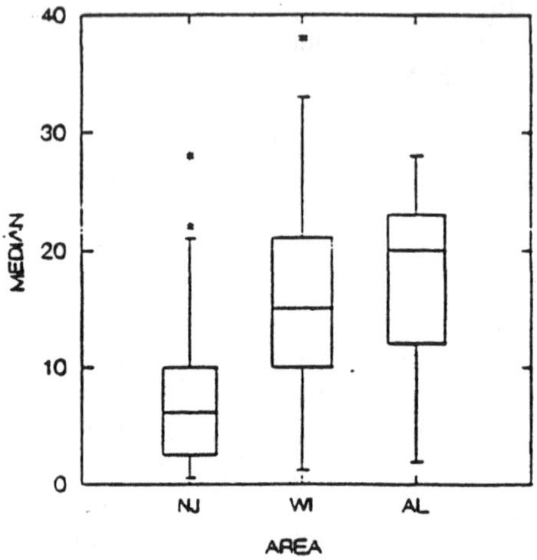

Figure 2. Fiftieth percentile particle sizes for stormwater inlet flows (Pitt, *et al.* 1997).

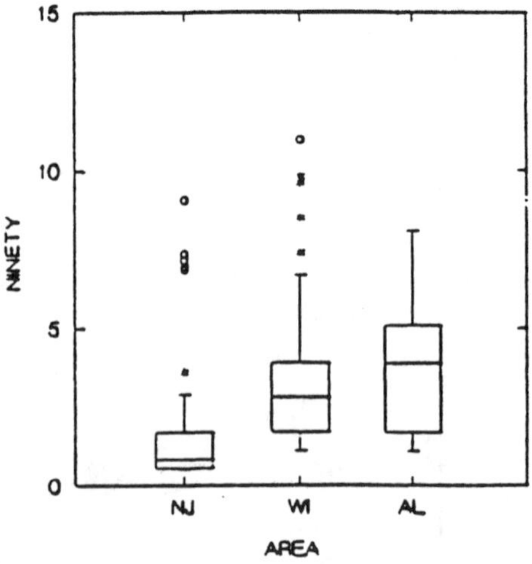

Figure 3. Ninetieth percentile particle sizes for stormwater inlet flows (Pitt, *et al.* 1997).

Stormwater particle size distributions typically do not include bed load components because the sample line velocities in automatic samplers may not be high enough to collect the largest material, plus the line diameter may be smaller than some of the bed load material. During the Monroe St. (Madison, WI) detention pond monitoring, the USGS and WI DNR installed special bed load samplers that trapped the bed load material for analysis. The bedload samplers were liter-sized wide mouth containers which were placed in bored holes in the bottom of the enclosed flat bottomed concrete small box channels right before the pond. Three units were placed in the channel bottom, each having different width slots cut in their lids. The mass of material trapped was directly related to the ratio of the width of the slot to the width of the channel. The material was removed, dried, sieved, and weighed. This particle size distribution was combined with the flow-weighted particle distributions obtained for the runoff events monitored during the same exposure period. Practically all events were monitored, with little flow not represented in these analyses. Figure 4 shows the measured particle size distributions for 16 seasonal samples (each having several runoff events), which also included the bedload particle size distribution data. The bedload samplers were in place for several weeks at a time in order to accumulate sufficient sample for analyses. The bedload material was comprised of the largest material represented on this figure (generally about 300 or 400 μm and larger particulates) and was about 10 percent of the annual total solids loading, but ranged from about 2 to 25 percent for individual periods. The bedload component in Madison was most significant during the early spring rains when much of the traction control sand that could be removed by rains was being washed from the streets. This is not a large fraction of the solids, but it represents the largest particle sizes flowing in the stormwater and it can be easily trapped in most detention ponds or catchbasins.

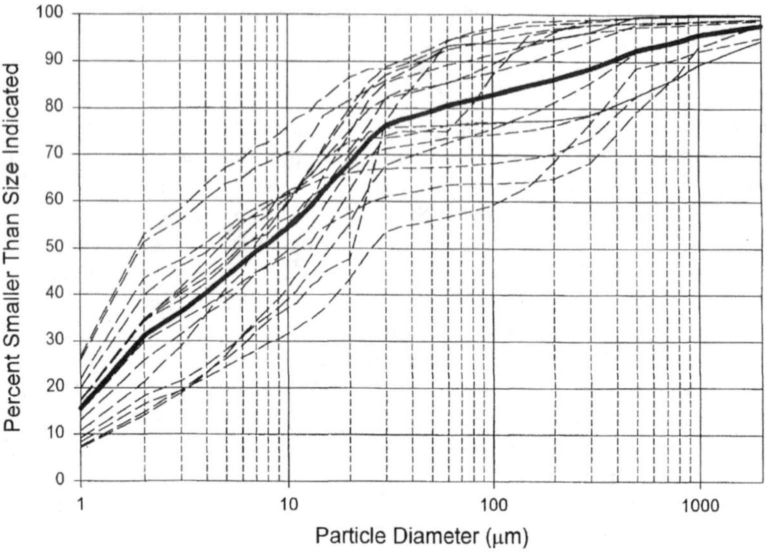

Figure 4. Outfall particle size distributions observed at the Monroe St. wet detention pond (inlet to the pond).

Figure 5 shows a typical "delta" of large material immediately near the inlet to a wet detention pond, along with an accumulation of material along the invert of the corrugated steel drainage pipe. This photo was taken at a pond at a ski resort in Snowmass, CO, in an area of heavy sand applications for traction control in the winter. The bedload sediment material in this photo is quite large, several mm in diameter, and near the upper range of the particle size distribution shown previously. This drainage pipe is relatively short (about 200 ft long), connected to an adjacent parking area. It is rare for this large material to be transported great distances in drainage systems. If it does enter a pond, or any type of sediment device, it is easily trapped near the inlet. However, small to moderate sized particulates can easily be transported quite some distance in the drainage system and be deposited well away from the inlet when discharged into wet detention ponds. The use of forebays, or small pre-settlement ponds, will trap much of the moderate-sized particulates (down to about 25μm) within a relatively small area for easier removal during maintenance operations. The finer particulates (down to just a few μm), containing most of the pollutants, would be trapped in the main pond area. The total sediment depth is relatively small when spread over a large pond area.

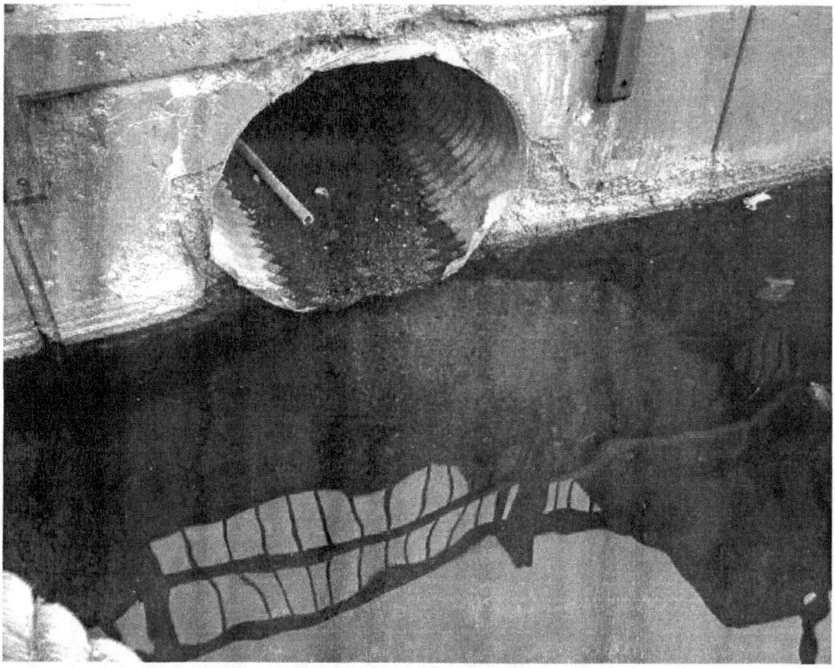

Figure 5. Bedload sediment accumulation in sewerage and near inlet to pond (Snowmass, CO).

Additional data obtained by Pitt, *et al.* (1997) for the USEPA described particle sizes from many different source flows in the Birmingham, Alabama, area. These data did not indicate any significant differences in particle size distributions for different source areas or land uses, except that the roof runoff had substantially smaller particle sizes than the other areas sampled. Also, the source area particle size distributions indicated that larger particles were much more likely to be present at source areas than at outfalls. The larger particles appear to be trapped in the flow paths and drainage system before they reach the outfalls.

After the stormwater particulates enter the storm drainage, they will tend to settle as they flow towards the outfall to the receiving water. If they settle slowly, such as occurs for small particles, they will remain suspended and not become part of the bed load or sediment in the sewerage. However, if they settle to the bottom of the pipe before reaching the outfall, they may become part of the bed load which will bounce along the pipe bottom, or become trapped with other settled debris. When the flow stops, the sediments will tend to dry and become more consolidated. The next runoff event may cause some of this settled material to become resuspended and may move it towards the outfall. Therefore, there are three phases to particulate transport in storm drainage systems: 1) settling of the particulates in the flowing water, 2) movement as bed load during the event, 3) accumulating as sediment and potentially subsequent scour. The following discussion describes these sediment transport phases in storm drainage systems.

Settling of Particulates in Flowing Water in Storm Drainage Systems

The settling velocities of discrete particles are shown in Figure 6, based on Stoke's and Newton's settling relationships. Probably more than 90% of all inlet stormwater particulates are in the 1 to 100 μm range, corresponding to laminar flow conditions, and appropriate for using Stoke's law. This figure also illustrates the effects of different specific gravities on the settling rates. In most cases, stormwater particulates have specific gravities in the range of 1.5 to 2.5 (Pitt 1979). This corresponds to a relatively narrow range of settling rates for a specific particle size. Particle size is much easier to measure than settling rates and it is generally recommended to measure particle sizes using automated particle sizing equipment (such as a Coulter Counter Multi-Sizer) and to conduct periodic settling column tests to determine the corresponding specific gravities and shape effects. If the particle counting equipment is not available, then small-scale settling column tests (using 50 cm diameter Teflon™ columns about 0.7 m long) can be easily used.

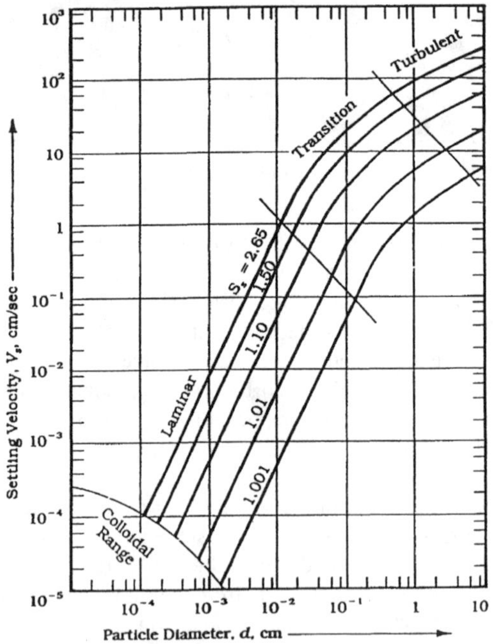

Figure 6. Type 1 (discrete) settling of spheres in water at 10° C (Reynolds 1982).

Pisano and Brombach (1996) obtained solids settling curves for numerous stormwater and CSO samples. They were concerned that many of the samples analyzed for particle size are not representative of the true particle size distribution in the sample. As an example, it is known that automatic samplers do not sample the largest particles that are found in the bedload portion of the flows. Particles having settling velocities in the 1 to 15 cm/sec (100 to 1,000 μm in size) range are found in grit chambers and catchbasins, but are seldom seen in stormwater samples obtained by automatic samplers. It is recommended that bedload samplers be used to supplement automatic water samplers in order to obtain more accurate particle size distributions (Burton and Pitt 2002). Selected US and Canadian settling velocity data are shown in Table 1. The CSO particulates have much greater settling velocities than the other samples, while the stormwater has the smallest settling velocities. The corresponding "Stoke's" particle sizes for the geometric means are about 100 μm for the CSOs, about 50 μm for the sanitary sewage, and about 15 μm for the stormwater.

Table 1. Settling Velocities for Wastewater, Stormwater, and CSO (Pisano and Bromback 1996)

Samples	Geometric Means of Settling Velocities Observed (cm/sec)	Range of Medians of Settling Velocities Observed (cm/sec)
CSO	0.22	0.01 to 5.5
dry weather wastewater (sanitary sewage)	0.045	0.030 to 0.066
stormwater	0.011	0.0015 to 0.15

More than 13,000 CSO control tanks have been built in Germany using the ATV 128 rule (Pisano and Bromback 1996). This rule states that clarifier tanks are to retain all particles having settling velocities greater than 10 m/hr (0.7 cm/sec, or about 100 μm), with a goal of capturing 80% of the settleable solids. Their recent measurements of overflows from some of these tanks indicate that the 80% capture was average for these tanks and that the ATV 128 rule appears to be reasonable for combined sewerage.

In table 2, settling conditions for particulates moving in pipes ranging from 1 to 5 ft in diameter and for flow depths ranging from 10% to 100% of the pipe diameters are presented. Pipe slopes ranging from 0.1 to 2% and particles from 1 to 10,000 μm, all with specific gravities of 2.5, are used in these calculations. This table shows the distances the particles would travel before they would settle to the bottom of the pipe, if starting from the surface of the flow, using Manning's equation to calculate the stormwater velocity, and the combination of Stoke's and Newton's laws for settling rates. A particle settling to the pipe bottom doesn't imply that the particles would be permanently trapped as sediment, but the particles may move (relatively slowly) as part of a mobile bedload. Obviously, the flow distances required for settling of the smallest particles are very long and would remain suspended. Some of the 100 μm particle flow conditions are shown to result in relatively short settling distances, while most of the 1,000 and 10,000 μm particles would most certainly settle to the pipe bottom before reaching an outfall. If the flow and associated bed load movement stops before the particles reach the outfall, they will form a more compact sediment requiring more energy during subsequent rains to scour.

Table 2. Settling Distance (ft) for Particles Flowing in Pipes having Various Diameters and Slopes (n=0.013)

		1 µm particles (2.5 specific gravity)				10 µm particles (2.5 specific gravity)			
Pipe Diameter (ft)	flow depth/pipe diameter ratio	0.001 slope	0.005 slope	0.01 slope	0.02 slope	0.001 slope	0.005 slope	0.01 slope	0.02 slope
1	0.1	>10,000	>10,000	>10,000	>10,000	180	390	550	780
1	0.3	>100,000	>100,000	>100,000	>100,000	1000	2300	3200	4600
1	0.5	>100,000	>100,000	>100,000	>100,000	2200	4895	6900	>10,000
1	0.8	>100,000	>100,000	>100,000	>100,000	4000	8929	>10,000	>10,000
1	1	>100,000	>100,000	>100,000	>100,000	4400	9790	>10,000	>10,000
1.5	0.1	>10,000	>10,000	>100,000	>100,000	350	770	1100	1500
1.5	0.3	>100,000	>100,000	>100,000	>100,000	2000	4500	6400	9000
1.5	0.5	>100,000	>100,000	>100,000	>100,000	4300	9629	>10,000	>10,000
1.5	0.8	>100,000	>100,000	>100,000	>100,000	7900	>10,000	>10,000	>10,000
1.5	1	>100,000	>100,000	>100,000	>100,000	8600	>10,000	>10,000	>10,000
2	0.1	>10,000	>100,000	>100,000	>100,000	560	1300	1800	2500
2	0.3	>100,000	>100,000	>100,000	>100,000	3300	7300	>10,000	>10,000
2	0.5	>100,000	>100,000	>100,000	>100,000	7000	>10,000	>10,000	>10,000
2	0.8	>100,000	>100,000	>100,000	>100,000	>10,000	>10,000	>10,000	>10,000
2	1	>100,000	>100,000	>100,000	>100,000	>10,000	>10,000	>10,000	>10,000
3	0.1	>100,000	>100,000	>100,000	>100,000	1100	2500	3500	4900
3	0.3	>100,000	>100,000	>100,000	>100,000	6400	>10,000	>10,000	>10,000
3	0.5	>100,000	>100,000	>100,000	>100,000	>10,000	>10,000	>10,000	>10,000
3	0.8	>100,000	>100,000	>100,000	>100,000	>10,000	>10,000	>10,000	>100,000
3	1	>100,000	>100,000	>100,000	>100,000	>10,000	>10,000	>10,000	>100,000
5	0.1	>100,000	>100,000	>100,000	>100,000	2600	5800	8100	>10,000
5	0.3	>100,000	>100,000	>100,000	>100,000	>10,000	>10,000	>10,000	>10,000
5	0.5	>100,000	>100,000	>100,000	>100,000	>10,000	>10,000	>100,000	>100,000
5	0.8	>100,000	>100,000	>100,000	>100,000	>10,000	>100,000	>100,000	>100,000
5	1	>100,000	>100,000	>100,000	>100,000	>10,000	>100,000	>100,000	>100,000

Table 2. Settling Distance (ft) for Particles Flowing in Pipes having Various Diameters and Slopes (n=0.013) (cont.)

		100 μm particles (2.5 specific gravity)				1,000 μm particles (2.5 specific gravity)				10,000 μm particles (2.5 specific gravity)			
Pipe Diameter (ft)	flow depth/pipe diameter ratio	0.001 slope	0.005 slope	0.01 slope	0.02 slope	0.001 slope	0.005 slope	0.01 slope	0.02 slope	0.001 slope	0.005 slope	0.01 slope	0.02 slope
1	0.1	1.9	4.4	6.2	8.7	0.1	0.2	0.3	0.4	0.0	0.0	0.1	0.1
1	0.3	11	25	36	51	0.5	1.1	1.6	2.3	0.1	0.2	0.3	0.5
1	0.5	24	54	77	110	1.1	2.4	3.5	4.9	0.2	0.5	0.7	1.0
1	0.8	44	99	140	200	2.0	4.5	6.3	8.9	0.4	0.9	1.3	1.8
1	1	49	110	150	220	2.2	4.9	6.9	10	0.4	1.0	1.4	2.0
1.5	0.1	3.8	8.6	12	17	0.2	0.4	0.5	0.8	0.0	0.1	0.1	0.2
1.5	0.3	22	50	71	100	1.0	2.3	3.2	4.5	0.2	0.5	0.6	0.9
1.5	0.5	48	110	150	210	2.2	4.8	6.8	10	0.4	1.0	1.4	1.9
1.5	0.8	87	200	280	390	3.9	8.8	12	18	0.8	1.8	2.5	3.5
1.5	1	96	210	300	430	4.3	10	14	19	0.9	1.9	2.7	3.9
2	0.1	6.2	14	20	28	0.3	0.6	0.9	1.2	0.1	0.1	0.2	0.2
2	0.3	36	81	120	160	1.6	3.6	5.2	7.3	0.3	0.7	1.0	1.5
2	0.5	77	170	250	350	3.5	7.8	11	16	0.7	1.6	2.2	3.1
2	0.8	140	320	450	630	6.4	14	20	28	1.3	2.8	4.0	5.7
2	1	160	350	490	690	7.0	16	22	31	1.4	3.1	4.4	6.2
3	0.1	12	27	39	55	0.5	1.2	1.7	2.5	0.1	0.2	0.3	0.5
3	0.3	71	160	230	320	3.2	7.2	10	14	0.6	1.4	2.0	2.9
3	0.5	150	340	480	680	6.9	15	22	31	1.4	3.1	4.3	6.1
3	0.8	280	620	880	1200	13	28	40	56	2.5	5.6	7.9	11
3	1	310	680	960	1400	14	31	43	61	2.7	6.1	8.7	12
5	0.1	29	64	90	130	1.3	2.9	4.1	5.8	0.3	0.6	0.8	1.2
5	0.3	170	370	530	750	7.5	17	24	34	1.5	3.4	4.8	6.7
5	0.5	360	800	1130	1600	16	36	51	72	3.2	7.2	10	14
5	0.8	650	1500	2100	2900	29	66	93	130	5.9	13	19	26
5	1	710	1600	2300	3200	32	72	100	140	6.4	14	20	29

Resuspension of Settled Particulates in Storm Drainage

This discussion presents some particulate transport information that can be used to predict if settled particles forming sediment in a pipe may be resuspended or scoured during subsequent events. This information does not allow predictions to be made concerning the accumulation of particulates in the sediment, only the likelihood that previously settled material may scour.

Allowable Velocity and Shear Stress

Allowable Velocity Data

The concept of allowable velocities for various soils and granular materials dates from the early days of hydraulics. Table 3 is an example of allowable velocities from U.S. Bureau of Reclamation research (Fortier and Scobey 1926, reprinted by McCuen 1998), and also shows the corresponding allowable shear stresses and Manning's roughness values. If these velocities are exceeded for an extended period, it is assumed that the channel lining material can become unstable. These values are not directly applicable to pipe flows, but the typically used maximum velocity of about 3 ft/sec for storm drainage design is similar to the values for stiff clays and silts. It is interesting that clays can withstand higher velocities than sands.

Table 3. Maximum Permissible Velocities and Corresponding Unit Tractive Force (Shear Stress) (U.S. Bureau of Reclamation research, Fortier and Scobey 1926)

		Clear Water		Water Transporting Colloidal Silts	
Material	**n**	**V (ft/sec)**	**τ_o (lb/ft^2)**	**V (ft/sec)**	**τ_o (lb/ft^2)**
Fine sand, colloidal	0.020	1.50	0.027	2.50	0.075
Sandy loam, noncolloidal	0.020	1.75	0.037	2.50	0.075
Silt loam, noncolloidal	0.020	2.00	0.048	3.00	0.11
Alluvial silts, noncolloidal	0.020	2.00	0.048	3.50	0.15
Ordinary firm loam	0.020	2.50	0.075	3.50	0.15
Volcanic ash	0.020	2.50	0.075	3.50	0.15
Stiff clay, very colloidal	0.025	3.75	0.26	5.00	0.46
Alluvial silts, colloidal	0.025	3.75	0.26	5.00	0.46
Shales and hardpans	0.025	6.00	0.67	6.00	0.67
Fine gravel	0.020	2.50	0.075	5.00	0.32
Graded loam to cobbles when noncolloidal	0.030	3.75	0.38	5.00	0.66
Graded silts to cobbles when noncolloidal	0.030	4.00	0.43	5.50	0.80
Coarse gravel, noncolloidal	0.025	4.00	0.30	6.00	0.67
Cobbles and shingles	0.035	5.00	0.91	5.50	1.10

Note:

- an increase in velocity of 0.5 ft/sec can be added to these values when the depth of water is greater than 3 ft.
- a decrease in velocity of 0.5 ft/sec should be subtracted when the water contains very coarse suspended sediments.
- for high and infrequent discharges of short duration, up to 30% increase in velocity can be added.

Allowable Shear Stress Data

By the 1930's, boundary shear stress (sometimes called tractive force) was generally accepted as a more appropriate criterion than allowable velocity for channel stability. The average boundary shear stress in uniform flow is calculated by:

$$\tau_o = \gamma RS \quad (\text{lb/ft}^2)$$

where:

γ = specific weight of water (62.4 lbs/ft^3)
R = hydraulic radius (ft)
S = hydraulic slope (ft/ft)

Flow characteristics predicting the initiation of motion of sediment in non-cohesive materials are usually presented in non-dimensional form in the Shield's diagram (Figure 7). This diagram indicates the initial movement, or scour, of non-cohesive uniformly graded sediments on a flat bed. This diagram plots the Shield's number (or mobility number), which combines shear stress with grain size and relative density, against a form of the Reynolds number that uses grain size as the length variable. The ASCE *Sedimentation Manual* (1975) uses a dimensionless parameter, shown on Figure 7, to select the dimensionless stress value. This value is calculated as:

$$\frac{d}{\nu}\left[0.1\left(\frac{\gamma_s}{\gamma}-1\right)gd\right]^{0.5}$$

where:

d = particle diameter (meters)

g = gravitational constant (9.81 m/sec^2)

ν = kinematic viscosity (1.306 x 10^{-6} m^2/sec for 10°C)

γ_s = specific gravity of the solid
γ = specific gravity of water

A series of diagonal parallel lines on Figure 7 represent these calculated values. The dimensionless shear stress value (τ_*) is selected where the appropriate line intersects the Shield's curve. The critical shear stress can then be calculated by:

$$\tau_c = \tau_*(\gamma_s - \gamma)d$$

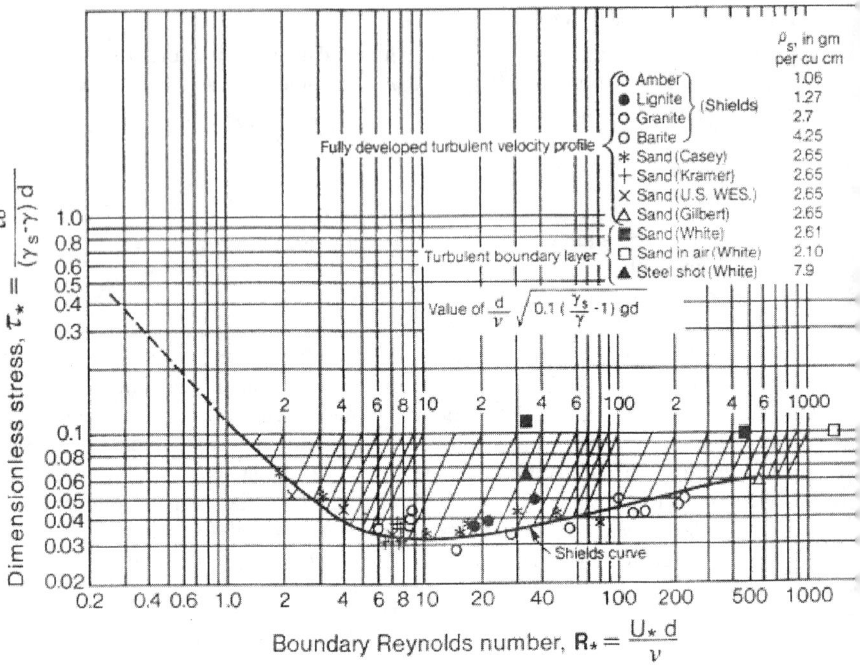

Figure 7. Shield's diagram for dimensionless critical shear stress (COE 1994).

An example evaluation is given by the COE (1994) in their assessment manual. In their example, the use of the Shield's diagram is shown to likely greatly over-predict the erodibility of the channel bottom material. The expected reason they give is that the Shield's diagram assumes a flat bottom channel and the total roughness is determined by the size of the granular bottom material. The actual Manning's roughness value is likely much larger because it is largely determined by bed forms, channel irregularities, and vegetation, and not grain size. They recommend, as a more realistic assessment, that empirical data based on field observations be used. In the absence of local data, they present Figure 8 (from Chow 1959) for applications for channels in granular materials. This figure shows the permissible unit tractive force (shear stress) as a function of the average particle diameter, and the fine sediment content of the flowing water.

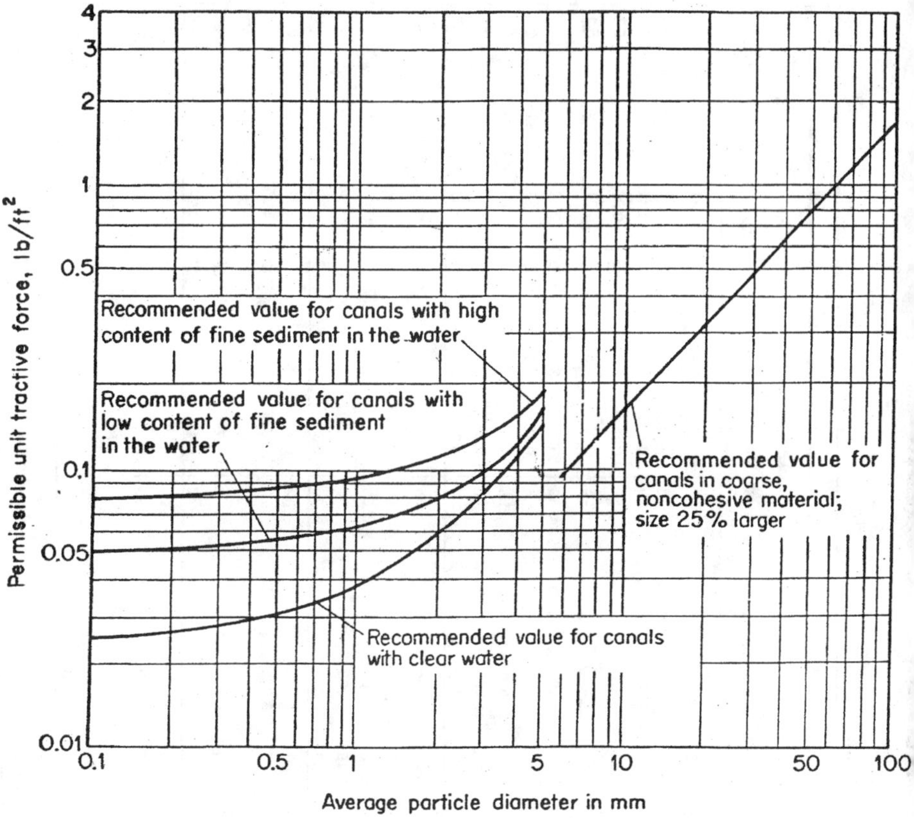

Figure 8. Allowable shear stresses (tractive forces) for canals in granular materials (U.S. Bureau of Reclamation, reprinted in Chow 1959).

Table 4 shows calculated shear stresses, velocities, and discharge quantities for various pipe conditions. Also shown are the estimated maximum particles sizes that would not be scoured during these flow conditions. For the smallest slopes, almost all settled particles would likely remain settled in the sediment and not be scoured, while almost all unconsolidated sediments would be scoured for pipes greater than 1% in slope. This table shows that pipe conditions resulting in at least 3 ft/sec stormwater velocities would also have shear stresses of at least 0.08 lb/ft^2. This shear stress would likely cause scour of particles up to 2,000 µm in size for water having a low content of fine sediment. If the water had a high content of fine sediment, the maximum particles likely to be scoured from the sediment may be only about 100 µm in size. If the sediment was somewhat consolidated (as expected to occur during dry periods between runoff events), then the necessary shear stress to cause sediment scour would be substantially greater, as noted in the following discussion.

Table 4. Calculated Shear Stress and Particle Scour

		shear stress lb/ft²), velocity (ft/sec), and discharge (ft³/sec) for different slopes:													
Pipe Dia. (ft)	flow depth/ pipe dia. ratio	shear at 0.001 slope	Max. particle size not scoured (µm)	velocity at 0.001	discharge at 0.001	shear at 0.005 slope	Max. particle size not scoured (µm)	velocity at 0.005	discharge at 0.005	shear at 0.01 slope	velocit y at 0.01	discharge at 0.01	shear at 0.02 slope	velocity at 0.02	discharge at 0.02
1	0.1	0.0041	<100	0.57	0.028	0.020	<100	1.3	0.063	0.041	1.8	0.089	0.081	2.6	0.13
1	0.3	0.011	<100	1.1	0.21	0.053	250	2.5	0.48	0.11	3.5	0.68	0.21	5.0	0.96
1	0.5	0.016	<100	1.4	0.55	0.078	1,500	3.2	1.29	0.12	4.5	1.7	0.31	6.4	2.5
1	0.8	0.019	<100	1.6	1.2	0.094	2,800	3. 7	2.6	0.20	5.2	3.6	0.38	7.3	5.5
1	1	0.016	<100	1.4	1.1	0.078	1,500	3.2	2.5	0.12	4.5	3.6	0.31	6.4	5.1
1.5	0.1	0.0061	<100	0.75	0.083	0.030	<100	1.7	0.19	0.061	2.4	0.26	0.12	3.4	0.37
1.5	0.3	0.016	<100	1.5	0.63	0.080	1,500	3. 3	1.4	0.16	4.6	2.0	0.32	6.6	2.8
1.5	0.5	0.023	<100	1. 9	1.63	0.12	3,500	4.2	3.6	0.23	6.0	5.2	0.47	8.4	7.3
1.5	0.8	0.028	<100	2.1	3. 4	0.14	5,000	4.8	7.6	0.28	6. 8	11	0.57	9.6	15
1.5	1	0.023	<100	1. 9	3.3	0.12	3,500	4.2	7.4	0.23	6.0	11	0.47	8.4	15
2	0.1	0.0081	<100	0.91	0.18	0.041	<100	2.03	0.40	0.081	2.9	0.57	0.16	4.1	0.80
2	0.3	0.021	<100	1. 8	1.4	0.11	3,300	4.0	3.1	0.21	5.6	4.3	0.42	8.0	6.1
2	0.5	0.031	<100	2.3	3.5	0.16	4,500	5.1	7.98	0.31	7.2	11	0.62	10	16
2	0.8	0.038	<100	2.6	7.3	0.19	>5,000	5.8	16	0.38	8.2	23	0.76	12	33
2	1	0.031	<100	2.3	7.2	0.16	4,500	5.1	16	0.31	7.2	23	0.62	10	32
3	0.1	0.012	<100	1.2	0.53	0.061	700	2. 7	1.2	0.12	3.8	1. 7	0.24	5. 4	2.4
3	0.3	0.032	<100	2.3	4.0	0.16	5,000	5.2	9.0	0.32	7.4	13	0.65	10	18
3	0.5	0.047	<100	3.0	10	0.23	>5,000	6.7	23	0.47	9.5	33	0.94	13	47
3	0.8	0.057	450	3.4	22	0.28	>5,000	7.7	48	0.57	11	68	1.1	15	97
3	1	0.048	<100	3.0	21	0.23	>5,000	6.7	47	0.47	9. 5	67	0.94	13	95
5	0.1	0.020	<100	1.7	2.1	0.10	3,000	3.8	4.6	0.20	5.4	6.5	0.41	7.5	9.3
5	0.3	0.053	400	3.3	16	0.27	>5,000	7.4	35	0.53	10	50	1.1	16	70
5	0.5	0.078	1,500	4.2	41	0.39	>5,000	9.4	92	0.78	13	130	1.6	19	180
5	0.8	0.094	2,800	4.8	84	0.47	>5,000	114	190	0.94	15	270	1.9	22	380
5	1	0.078	1,500	4.2	83	0.39	>5,000	9.4	190	0.78	13	260	1. 6	19	370

assuming n=0.013
water with low content of fine sediment in water (if high content, then require higher shear stress)
conditions with water velocities of at least 3 ft/sec are highlighted

Erodability of Previously Settled Material after Consolidation

Figure 9 is an example of allowable shear stresses for a range of cohesive materials having a range of void ratios corresponding to varying amounts of consolidation. Again, the COE recommends that local field observations or laboratory testing results be given preference. If the void ratio was about 0.4, corresponding to a compact sediment, the shear stress would have to be greater than about 0.3 lb/ft^2 to affect particles larger than clays. This shear stress may only occur for larger and full-flowing pipes having 2% slopes (and about 7 ft/sec velocities).

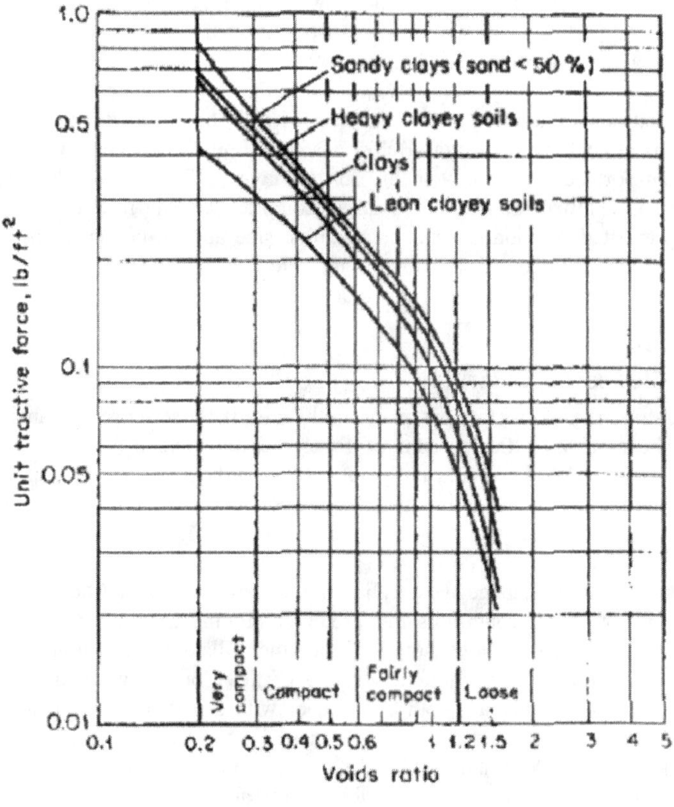

Figure 9. Example of allowable shear stresses (tractive forces) for cohesive materials (COE 1994) (a Leon clayey soil is hardpan. Hardpan is a condition of the soil or subsoil in which the soil grains become cemented together by such bonding agents as iron oxide and calcium carbonate, forming a hard, impervious mass).

Criteria to Ensure Self-Cleaning in Sewerage

Pisano, *et al.* (1998) described the mechanisms associated with deposition and scour of sediment in sewerage. The following paragraphs are summarized from that report.

Not all of the particles of a given size at the flow/sediment boundary dislodge and move at the same time because the flow is turbulent and contains short term fluctuations in velocity. The limiting condition below which sediment movement is negligible and known as the threshold of movement, is usually defined in terms of either the critical bed shear stress, or the critical erosion velocity.

Once sediment is entrained in flowing water, it may travel in the sewer in one of two basic ways: 1) Finer, lighter material tends to travel in suspension, while 2) heavier material travels in a rolling, sliding mode as bedload. In the transport of suspended sediment, there is a continuous exchange between particles settling out and those being eroded and re-entrained back into the flow. Under certain conditions, fine grained and organic particles can form a highly concentrated mobile layer of "fluid mud" near the pipe invert (Ashley and Crabtree 1992).

If the flow velocity or turbulence level decreases, there will be a net reduction in the amount of sediment held in suspension. Below a certain limit, the sediment will form a bed, with transport occurring only in the surface layer. If the flow velocity is further decreased, sediment transport will cease completely. The flow conditions necessary to prevent deposition depend on the pipe size and on properties of the sediment, mostly particle size and specific gravity. Flow velocities needed to entrain sediment tend to be higher than those at which deposition occurs.

Sewer Self Cleansing Criteria

Average shear stress is an important parameter in the criteria for sewer self-cleansing. The average shear stress is the amount of force the fluid exerts on the wetted perimeter of the pipe. Another important parameter is the bed shear stress which is the amount of force the fluid exerts on the bed of sediment in the pipe. Bed shear stress is related to bed load scour and movement.

Historically, the design of drainage systems has included the prevention of deposition of sediments in pipes. These conditions are based on a minimum velocity of flow, or a minimum shear stress that the flow should exert on the walls of the pipe to maintain self-cleansing conditions. The minimum velocity of flow, or the minimum shear stress, corresponds to a particular depth of flow, or with a particular frequency of occurrence. Available design criteria were reported by the Construction Industry Research and Information Association (CIRIA), and a summary of these, outlined by Nalluri and Ab Ghani (1996), are shown in Table 5 and Table 6.

Table 5. Minimum Velocity Criteria

Source	Country	Sewer Type	Minimum Velocity		Pipe Conditions
			(m/s)	(ft/sec)	
American Society of Civil Engineers (1970)	USA	Sanitary	0.6	2.0	Full/Half-full
		Storm	0.9	3.0	Full/Half-full
British Standard (1987)	UK	Sanitary	1.0	3.3	Full
		Combined	1.0	3.3	Full
Bielecki (1982)	Germany	Not noted	1.5	5.0	Full

Table 6. Minimum Shear Stress Criteria

Source	Country	Sewer Type	Minimum Shear Stress		Pipe Conditions
			(N/m^2)	(lb/ft^2)	
Lysne(1969)	USA	Not noted	2-4	0.04 – 0.08	Not noted
ASCE(1970)	USA	Not noted	1.3-12.6	0.027 – 0.26	Not noted
Yao (1974)	USA	Storm	3.0-4.0	0.061 – 0.083	Not noted
		Sanitary	1.0-2.0	0.021 – 0.042	Not noted
Maguire	UK	Not noted	6.2	0.13	Full/Half-full
Bischof (1976)	Germany	Not noted	2.5	0.052	Not noted

These criteria take no account of the characteristics of the sediment, of the suspended sediment concentration, the bed load, or of any cohesion between the sediment particles. Nonetheless, minimum velocity and minimum shear stress levels of 1 m/s (3.3 ft/sec) and 2 N/m^2 (0.04 lb/ft^2) are commonly accepted criteria. However, recent work by many researchers has shown that a single value of minimum velocity or shear stress cannot adequately describe the self-cleansing conditions in all pipes of different size, roughness and gradient for a range of sediment characteristics and flow conditions.

In practice, sewer pipes will not be maintained self-cleansing at all times. The diurnal pattern of the dry weather flow and the temporal distribution and nature of sediments found in sewer flows may result in the deposition of some sediments at times of low flow and the subsequent erosion and transport of these sediments, either as suspended load or bed-load, at times of higher flow. The deposited sediments will exhibit additional strength due to cohesion and, provided that the peak dry weather flow velocity or bed shear stress is of sufficient magnitude to erode these sediments, the sewer will maintain self cleansing operation at times of dry weather. May, *et al.* (1996) presented a definition to describe a self cleansing sewer as “an efficient self-cleansing sewer is one having a sediment-transporting capacity that is sufficient to maintain a balance between the amounts of deposition and erosion, with a time-averaged depth of sediment deposit that minimizes the combined costs of construction, operation and maintenance.” To achieve such self-cleansing performance, the following criteria apply:

Flows equaling or exceeding a limit appropriate to the sewer should have the capacity to transport a minimum concentration of fine-grain particles in suspension (applicable for all types of sewerage systems).

The capacity of flows to transport coarser granular material as bed-load should be sufficient to limit the depth of deposition to a specified proportion of the pipe diameter. This criteria generally relates to combined and stormwater systems. Limit of deposition considerations, i.e., "no deposition" generally applies to sanitary sewer designs. In this context, there must be sufficient shear in sanitary systems to avoid deposition of large particles.
Flows with a specified frequency of occurrence should have the ability to erode bed particles from a deposited granular bed that may have developed a certain degree of cohesive strength (applicable to all systems).

To meet these criteria, new guidelines have been developed by CIRIA (Ackers, *et al.* 1996), and are being adopted throughout Europe for the design of sewers to control sediment problems. Design criteria for the transport of fine grained material in suspension, the transport of coarser sediments as bed load, and the erosion of cohesive sediment deposits and guidelines on the minimum flow velocity and pipe gradient for different types and sizes of sewer are outlined. To account for the effects of cohesion (Criterion 3), the design flow condition should produce a minimum value of bed shear stress of 2.0 N/m2 (0.04 lb/ft2) on a flat bed for a rough concrete pipe (Ackers, *et al.* 1996).

The third criterion is of specific interest to the problem of cleansing accumulated mature sediment beds. Various researchers have studied the flow conditions required to release particles from a deposited bed, which has developed a degree of cohesion. Summaries of investigations forming much of the basis for Criterion 3 are as follows:

Nalluri and Alvarez (1992), whose laboratory studies used synthetic cohesive sediments, concluded that there were two ranges of bed shear stress at which erosion occurred: 2.5 N/m^2 (0.05 lb/ft^2) applying for the weakest material, comprising a surface layer of fluid sediment, and 6 to 7 N/m^2 (0.12 to 0.14 lb/ft^2) for the more granular and consolidated material below. It was found that, after erosion, the synthetic cohesive sediments behaved very much like non-cohesive material.

Ristenpart and Uhl (1993) found in field tests that during dry weather an average bed shear stress of 0.7 N/m^2 (0.015 lb/ft^2) was required to initiate erosion, increasing to an average of about 2.3 N/m^2 (0.05 lb/ft^2) during wet weather, or to 3.3 N/m^2 (0.068 lb/ft^2) after a prolonged period of dry weather and presumably, consolidation of the deposited bed.

Ashley and Crabtree (1992) suggested that the bonds between particles at the surface of a deposited bed are weakened by the presence of the water, so that surface layers can be successively stripped away by the flow. Measurements in the Dundee, Scotland sewers indicated that it began to move at a fluid shear stress of about 1 N/m^2

(0.02 lb/ft^2), with significant erosion of a deposited bed occurring at bed shear of 2 to 3 N/m^2 (0.04 to 0.06 lb/ft^2). Taking account of a review of work by other researchers, Ashley and Crabtree concluded that most deposits should be eroded at a shear stress exceeding 6 to 7 N/m^2 (0.12 to 0.14 lb/ft^2).

Pisano, et al. (1998) also conducted a "desktop" analysis of the sedimentation and scour problems associated with the 3.05 m (10 ft) South Ottawa Tunnel, in Ottawa, Canada. This procedure calculates deposition on a daily basis, determining accumulations of inorganic and organic deposits as well as scour and erosion of prior deposits over a year's period.

The particles in this combined sewerage had settling velocities ranging from 0.1 to 30 cm/sec, corresponding to 75 μm fine sand to 3/8 inch gravel. If the average shear stress is smaller than the critical shear stress for that sieve size and density, deposition will occur. Erosion of the bed load is based on maximum hourly shear stress. If the maximum shear stress is greater than 2 N/m2, then erosion of the bed load occurs. Table 7 shows the full range of deposition and scour that may occur for this large tunnel for a variety of discharge, velocity, and shear stress condtions.

Table 7. Assessment of 3.05 Meter Tunnel (Pisano, *et al.*, 1998)

Discharge		Velocity		Fluid shear		Deposition or erosion potential
(m^3/sec)	(ft^3/sec)	(m/sec)	(ft/sec)	(N/m^2)	(lb/ft^2)	
2.6	93	0.36	1.2	0.27	0.0056	Severe deposition
3.5	130	0.47	1.5	0.48	0.0099	Moderate to severe deposition
4.4	160	0.60	2.0	0.74	0.015	Mild to moderate deposition
5.3	190	0.72	2.4	1.04	0.021	Slight to mild deposition
6.1	220	0.84	2.8	1.43	0.029	Skims juvenile sediments
7.0	250	1.06	3.5	1.86	0.038	None to slight erosion top layer
8.8	320	1.20	4.0	2.85	0.059	Slight to mild erosion of consolidated beds (2-5%)
10.5	380	1.44	4.7	4.15	0.085	Mild erosion of consolidated beds (5-15%)
13.2	470	1.80	5.9	6.47	0.13	Moderate erosion of consolidated beds (15-25%)
15.8	570	2.17	7.2	9.3	0.19	Substantial erosion (25-35%)
17.5	630	2.41	7.9	11.5	0.24	Substantial erosion (35-50%)

Note: K_b = 1.2 millimeters = rough concrete

Accumulation of Sediment in Bellevue Inlet Structures and Storm Drainage

An important part of the Bellevue NURP project was the measurement of the sediment accumulating in the inlet structures (Pitt 1985). The storm drainage system inlets were cleaned and surveyed at the beginning of the project. The 207 inlet structures were then surveyed nine times over two years to determine the depth of accumulating material (from December 1979 through January 1981). The first year rate of accumulation was relatively steady (based on 3 observation periods), while the sediment loading remained almost constant during the second year. During the second year, there was about twice as much contaminated sediment in the separate storm drainage system at any one time as there was on the streets. The scouring of the sewerage sediments out of the drainage systems was not found to be significant during the project period. There was a period of heavy rains in October of 1981 (about 100 mm of rain during a week, very large for Bellevue) during the second year when the accumulated material did not decrease, based on observations made before and after the rain (August 1981 and January 1982). The lack of sediment movement from catchbasin sumps was also observed during earlier tests conducted in San Jose by Pitt (1979). During that study, an idealized catchbasin and sump were constructed based on Lager, *et al.*, (1974) and was filled with clean material having the same particle sizes as typical sump material, along with fluorescent tracer beads. During a year, freezing core samples were obtained and the sediment layers were studied to determine any flushing and new accumulations of material. The sediment material was found to be very stable, except for a very thin surface layer.

The first year accumulation rates (L/month per inlet) ranged from 1.4 in Lake Hills to 4.8 in Surrey Downs, as shown on Table 8. The catchbasins and inlets had sumps (the catchbasin sumps were somewhat larger), while the manholes were much larger, with more volume available for accumulation sediment. The stable volume that occurred during the second year was about 60% of the total storage volumes of the catchbasins and inlets (sump volume below the outlet pipe). If the sumps were very shallow, the maximum sediment depth was only about 12 mm, while the deeper sumps had about 150 mm of accumulated sediment. Individual inlet structures had widely varying depths, but the depth below the outlet appeared to the most significant factor affecting the maximum sump volume available. This "scour" depth generally was about 300 mm. If the sumps were deeper, they generally were able to hold more sediment before their equilibrium depth was reached and would therefore require less frequent maintenance. About 100 L/ha/yr accumulated in Surrey Downs, while only about 2/3 of this value accumulated in Lake Hills. Nine of the most heavily loaded catchbasins in the first summer inventory in Surrey Downs were located very near two streets that did not have curbs and had extensive nearby sediment sources (eroding hillsides). These few catchbasins (about 10% of the total catchbasins) accounted for more than half of the total Surrey Downs sediment observed during that survey. They also represented about 70% of the observed increased loadings between the first winter and summer inventories.

Table 8. Accumulation Rate of Sediment in Inlet Structures in Bellevue, WA (Pitt 1985)

	Number of structures		Sediment accumulation (L/month)		Approx. months to stable volume	Stable volume (L)	
	total	per ha	per ha	per unit		per ha	per unit
Surey Downs (38.0 ha)							
Catchbasins	43	1.1	5.3	4.8	13	68	62
Inlets	27	0.7	2.0	2.8	20	40	57
Manholes	6	0.2	0.8	4.0	19	15	76
Average	76 total	2.0 total	8.1	4.2	15	123 total	62
Lake Hills (40.7 ha)							
Catchbasins	71	1.7	2.4	1.4	18	43	25
Inlets	45	1.1	1.5	1.4	14	22	20
Manholes	15	0.4	1.6	4.0	23	36	90
Average	131 total	3.2 total	5.5	1.7	18	101 total	31

Besides inlet sediment surveys, pipe surveys were also conducted during the study. Very few storm drain pipes in either test area had slopes less than one percent, the assumed critical slope for sediment accumulation. n Lake Hills, the average slope of the 118 pipes surveyed was about 4 percent. Only 7 percent of the Lake Hills pipes had slopes less than 1 percent. The 75 pipes surveyed in Surrey Downs had an average slope of 5 percent, and 12 percent had slopes less than 1 percent. A pipe sediment survey was conducted in October of 1980. Very little sediment was found in the storm drains in either study area. The pipes that had significant sediment were either sloped less than 1-1/2 percent, or located close to a source of sediment. The characteristics of the pipe sediments were similar to the characteristics of the sediment from close by inlets and catchbasins, indicating a common source, and the eventual movement of the inlet sediments. The volume of sediment found in the Lake Hills pipes was about 1-1/2 m^3, or about 0.04 m^3 per ha, or about 40% of the total sediment in the inlet structures (about 0.1 m^3 per ha stable volume). This was equivalent to about 70 kg of sediment/ha. In Surrey Downs, much more sediment was found in the storm drainage: more than 20 m^3 of sediment was found in the pipes, or about 0.5 m^3/ha or 1,000 kg/ha. Most of this sediment was located in silted-up pipes along 108th St. and Westwood Homes Rd., which were not swept and were close to major sediment sources.

The chemical quality of the captured sediment was also monitored. Tables 9 and 10 show the sediment quality for Surrey Downs inlet structures sampled between January 13 and June 17, 1981. The sediment quality shown on this table is very similar to the street dirt chemical quality that was simultaneously sampled and analyzed. It is interesting to note that the COD values increased with increasing particle sizes, likely corresponding to increasing amounts of organic material in the larger material. The nutrients are generally constant with size, while the metal concentrations are much higher for the smaller particles, as expected for street dirt. As indicated on the table, the lead values were likely much higher when these samples were taken compared to current conditions. Current outfall lead concentrations are now about 1/10 of the values they were in the early 1980s.

Table 9. Chemical Quality of Bellevue, WA, Inlet Structure Sediment (mg constituent/kg total solids) (Pitt 1985)

Particle Size (µm)	COD	TKN	TP	Pb*	Zn
<63	160,000	2,900	880	1,200	400
61-125	130,000	2,100	690	870	320
125-250	92,000	1,500	630	620	200
250-500	100,000	1,600	610	560	200
500-1,000	140,000	1,600	550	540	200
1,000-2,000	250,000	2,600	930	540	230
2,000-6,350	270,000	2,500	1,100	480	190
>6,350	240,000	2,100	760	290	150

* these lead values are much higher than would be found for current samples due to the decreased use of leaded gasoline since 1981.

Table 10. Annual Calculated Accumulation of Pollutants in Bellevue, WA, Inlet Structures (Pitt 1985)

	Solids						
	L/ha/yr	kg/ha/yr	COD kg/ha/yr	TKN kg/ha/yr	TP kg/ha/yr	Pb kg/ha/yr	Zn kg/ha/yr
Surrey Downs	96	147	37	0.17	0.25	0.49	0.10
Lake Hills	66	100	7.5	0.07	0.07	0.07	0.02

These accumulations are substantial. As an example, the total suspended solids discharged at the outfalls was about 100 to 200 kg/ha/yr, with the amount accumulating in the catchbasin inlets about the same quantity.

Conclusions and Recommendations

To achieve self-cleaning drainage systems in an urban areas, Pisano, *et al.* (1998) listed the following criteria:

Flows equaling or exceeding a limit appropriate to the sewer should have the capacity to transport a minimum concentration of fine-grain particles in suspension (applicable for all types of sewerage systems).

The capacity of flows to transport coarser granular material as bedload should be sufficient to limit the depth of deposition to a specified proportion of the pipe diameter. This criteria generally relates to combined and stormwater systems. Limit of deposition considerations, i.e., "no deposition" generally applies to sanitary sewer designs. In this context, there must be sufficient shear in sanitary systems to avoid deposition of large particles.

Flows with a specified frequency of occurrence should have the ability to erode bed particles from a deposited granular bed that may have developed a certain degree of cohesive strength (applicable to all systems).

Ashley and Crabtree (1992) made sediment movement measurements in the Dundee, Scotland, sewers and found that sediment began to move at a fluid shear stress of about 1 N/m^2 (0.02 lb/ft^2), with significant erosion of a deposited bed occurring at bed shear of 2 to 3 N/m^2 (0.04 to 0.06 lb/ft^2). Taking account of a review of work by other researchers, they concluded that most deposits should be eroded at a shear stress exceeding 6 to 7 N/m^2 (0.12 to 0.14 lb/ft^2). This is likely to occur for most flow conditions in pipes at least 2 ft in diameter at a slope greater than 0.5%.

For the smallest slopes (<0.5% slopes), almost all settled particles would likely remain and not be scoured, while almost all unconsolidated sediments would be scoured for pipes at greater than 1% in slope. Pipe conditions resulting in at least 3 ft/sec stormwater velocities would also have shear stresses of at least 4 N/m^2 (0.08 lb/ft^2). This shear stress would likely cause scour of particles up to 2,000 μm in size for water having a low content of fine sediment. If the water had a high content of fine sediment, the maximum particles likely to be scoured from the sediment may be only about 100 μm in size.

Required Sample Line Velocities to Minimize Particle Sampling Errors

The collection of stormwater samples representing the complete range of likely particle sizes can be challenging. Manual samples can be obtained as the flowing water drops into an inlet, or as the water cascades out of an outfall. These samples, if frequent during an event, should adequately represent the complete particle size distribution. However, if an automatic sampler using a pump system is used, there may be some losses of particles during the sampling operation. Typical sample lines are Teflon™ lined polyethylene and are 10 mm in diameter. The water velocity in the sample line is about 100 cm per second, enabling almost complete sampling of a wide range of particle sizes. However, bedload sampling equipment is usually needed to adequately collect samples of bedload at outfalls. Table 11 shows the particle sizes that would be lost in vertical sampling lines at a pumping rate of 30 and 100 cm/second (Burton and Pitt 2001).

Table 11. Losses of Particles in Sampling Lines (Burton and Pitt 2001)

	30 cm/sec flow rate		**100 cm/sec flow rate**	
	Critical settling rate (cm/sec)	**Size range (μm, for ρ = 1.5 to 2.65 g/cm³)**	**Critical settling rate (cm/sec)**	**Size range (μm, for ρ = 1.5 to 2.65 g/cm³)**
100% loss	30	2,000 - 5,000	100	8,000 - 25,000
50% loss	15	800 - 1,500	50	3,000 - 10,000
25% loss	7.5	300 - 800	25	1,500 - 3,000
10% loss	3.7	200 - 300	10	350 - 900
1% loss	0.37	50 - 150	1	100 - 200

A water velocity of 100 cm/sec (about 3 ft/sec) would result in very little loss of stormwater particles. Particles of 8 to 25 mm would not be lifted in the sample line at all at this velocity, but these sized particles would not fit through the openings of the intake or even fit in most sample lines. They are also not likely to be present in stormwater, but may be a component of bedload in a stream, or gravel in the bottom of a stormdrain pipe, requiring special sampling. Very few particles larger than several hundred micrometers occur in stormwater and these should only have a loss rate of 10% at the most. Most particles in stormwater outfall discharges are between 1 and 100 μm in diameter and have a density of between 1.5 and 2.65 g/cm^3. Even at 30 cm/sec, these particles should experience insignificant losses. A pumping rate of about 100 cm/sec would add extra confidence in minimizing particle losses. ASTM (1995) in method D 4411 recommends that the sample velocity in the sampler line be at least 17 times the fall rate of the largest particle of interest. As an example, for the 100 cm/sec example above, the ASTM recommended critical fall rate would be about 6 cm/sec, enabling a particle of several hundred micrometers in diameter to be sampled with a loss rate of less than 10%. This is certainly adequate for most stormwater sampling needs. Bedloads comprised of particles of up to several mm can account for about 5 to 10% of the sediment load under some conditions, and special sampling would be needed to collect these larger particulates. In many cases, this larger material would not move through the pipe, but accumulate as a semi-permanent sediment deposit.

Figure 10 contains photographs of bed load samplers at a residential site in Madison, WI, being studied by the WI DNR and the USGS. Three samplers are placed in a row to ensure that all of the bedload material is captured. They are able to remove the samples after each monitored event (also being sampled using conventional automatic samplers) to add this solids component to the total sediment discharge amount, and to adjust the particle size distribution. Their preliminary data has indicated that this bedload (not being captured by the automatic samplers) comprises about 4% of the total sediment loading for the events. This sampling is being conducted in conjunction with street cleaner tests investigating high-performance street cleaning equipment.

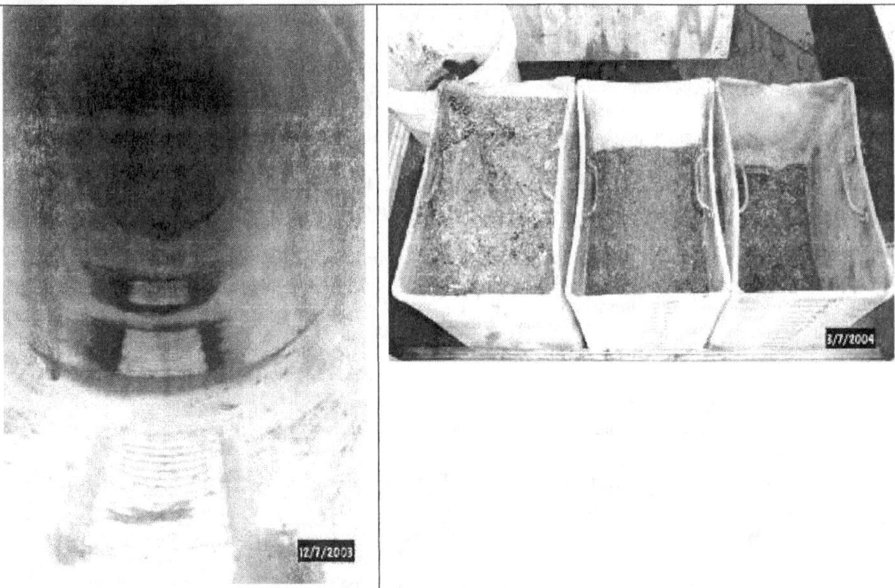

Figure 10. Recent bed load sampling in a residential area, Madison, WI (R. Bannerman, personal communication).

References

Ackers, J.C., D. Butler, and R.W.P. May. *Design of Sewers to Control Sediment Problems* Construction Industry Industry Research and Information Association (CIRIA), Report 141, 1996.

Allison, R.A.; Walter, K.A.; Marx, D.; Lippner, G.; and Churchwell, R. A Method for Monitoring and Analyzing Litter in Freeway Runoff as Part of the Caltrans Litter Management Pilot Study. *2000 Joint Conference on Water Resources Engineering and Water Resources Planning and Management,* July 2000, Minneapolis, MN. American Society of Civil Engineers, CD-ROM. 2000.

American Society of Civil Engineers. *Design and Construction of Sanitary and Storm Sewers*, ASCE Manuals and Reports on Engineering Practice, No. 37, 1970.

Armitage, N.P., and Rooseboom, A. The Removal of Litter from Stormwater Conduits in the Developing World. *Water Sci. Technol.* (G.B.). 39, 9, 277. 1999.

Armitage, N., and Rooseboom, A. The Removal of Urban Litter from Stormwater Conduits and Streams: Paper 1 - The Quantities Involved and Catchment Litter Management Options. *Water SA*. **26**, 181. 2000a.

Armitage, N., and Rooseboom, A. The Removal of Urban Litter from Stormwater Conduits and Streams: Paper 2 - Model Studies of Potential Trapping Structures. *Water SA*. **26**, 189. 2000b.

Armitage, N.; Marais, M.; and Pithey, S. Reducing Urban Litter in South Africa through Catchment Based Litter Management Plans. *Models and Applications to Urban Water Systems, Monograph 9*. 37. 2001.

Aronson, G., D. Watson, and W. Pisano. Evaluation of Catchbasin Performance for Urban Stormwater Pollution Control. U.S. EPA. Grant No. R-804578. EPA-600/2-83-043. 78 pages. Cincinnati, June 1983.

ASCE. *Sedimentation Engineering*. Edited by Vito A. Vanoni. Manuals and Reports on Engineering Practice, No. 54. 1975.

Ashley, R.M. and R.W. Crabtree. "Sediment origins, deposition and build-up in combined sewer systems." *Water Science Technology*, Vol. 25, No. 8, 1992.

Ashley R.M.; Hvitved-Jacobsen, T.; Vollertsen, J.; McIlhatton, T.; and Arthur S. Sewer Solids Erosion, Washout, and a New Paradigm to Control Solids Impacts on Receiving Waters. *Proc. the Eighth International Conference on Urban Storm Drainage*. August 30 – September 3, 1999, Sydney, Australia. Edited by IB Joliffe and JE Ball. The Institution of Engineers Australia, The International Association for Hydraulic Research, and The International Association on Water Quality, 171. 1999.

Ashley, R.; Crabtree, B.; and Fraser, A. Recent European Research into the Behavior of Sewer Sediments and Associated Pollutants and Processes. *2000 Joint Conference on Water Resources Engineering and Water Resources Planning and Management,* July 2000, Minneapolis, MN. American Society of Civil Engineers, CD-ROM. 2000.

Ashley, R.M.; Dudley, J.; Vollertsen, J.; Saul, A.J.; Blanksby, J.R.; Jack, A. The effect of extended in-sewer storage on wastewater treatment plant performance. *Water Science and Technology*, **45**(3), 239-246. 2002.

British Standard Institute. Sewerage – Guide to New Sewerage Construction, BS 8005, Part 1. 1987

Burton, G.A. and R. Pitt. *Stormwater Effects Handbook*. CRC Press. Boca Raton, FL. 2001.

Butler, D. and S.H.P.G. Karunaratne. "The suspended solids trap efficiency of the roadside gully pot." Wat. Res. Vol. 29, No. 2. pp. 719-729. 1995.

Butler, D., Y. Xiao, S.H.P.G. Karunaratne, and S. Thedchanamoorthy. "The gully pot as a physical and biological reactor." Wat. Sci. Tech. Vol. 31, No. 7, pp. 219-228. 1995.

Chang, H.H. *Fluvial Processes in River Engineering*. John Wiley & Sons. 1988.

Chow, Ven Te. *Open Channel Hydraulics*. McGraw-Hill. 1959.

Cigana, J.; Couture, M.; Lefebvre, G.; and Marche, C. Evidence of a Critical Velocity in Underflow Baffle Design for Floatables Control in Combined Sewer Overflows (CSOs). Proc. Adv. in Urban Wet Weather Pollut. Reduction, Cleveland, Ohio, WEF (CP3805), 275. 1998a.

Cigana, J.; Lefebrve, G.; Marche, C.; and Couture, M. Design Criteria of Underflow Baffles for Control of Floatables. *IAWQ 19th Biennial Int. Conf.*, Vancouver, Can., 8, 58. 1998b.

Cigana, J.; Lefebrve, G.; Marche, C.; and Couture, M. Design Criteria of Underflow Baffles for Control of Floatables. *Water Sci. Technol.* (GB), 38, 10, 57. 1998c.

Cigana, J.; Couture, M.; Meunier, C.; and Comeau, Y. Determination of the Vertical Velocity Distribution of Floatables in CSOs. *Water Sci. Technol.* (G.B.). 39, 2, 69. 1999.

Cigana, J.F.; Lefebvre, G.; and Marche, C. Experimental Capture Efficiency of Floatables using Underflow Baffles. *Collection Systems Wet Weather Pollution Control: Looking into Public, Private, and Industrial Issues,* May 2000, Rochester, NY. Water Environment Federation, CD-ROM. 2000.

Cigana, J.; Lefebvre, G.; and Marche, C. Critical Velocity of Floatables in Combined Sewer Overflow (CSO) Chambers. *Water Sci. Tech.* **44**:287. 2001.

Croke, T. "Reliable channel design using turf reinforcement mats." *Erosion Discussion*. Vol. 5, No. 1. North American Green (availabe from their web page: www.nagreen.com). June 2000.

Croke, T. "Accounting for bends in channel design." *Erosion Discussion*. Vol. 6, No. 1. North American Green (availabe from their web page: www.nagreen.com). Summer 2001.

Dammel, E.E.; Berger, B.J.; Regenmorter, L.C.; and Lippner, G.S. Evaluating Drain Inlet Cleaning as a Stormwater Best Management Practice. *5th International Conf.: Diffuse/Nonpoint Pollution and Watershed Management*. CD-ROM. 2001.

Davis, D.P.; MacArthur, D.; Martin, S. CSO floatables abatement: How Onondaga County is screening, booming and bagging its way to floatables control. *WEF/CWEA Collection Systems 2002 Conf. Proc., May 2002*. Water Environment Federation. CD-ROM. 2002.

EDP (Environmental Design and Planning, Inc.). Evaluation of Catchbasin Monitoring. Contract No. R804578010, U.S. Environmental Protection Agency, Cincinnati, Ohio, March 1980.

Fischer, R.; Turner, R. Operation and maintenance experience with CSO Floatables control facilities. *WEF/CWEA Collection Systems 2002 Conf. Proc., May 2002*. Water Environment Federation. CD-ROM. 2002.

Fortier, S. and F.C. Scobey. "Permissible canal velocities." *Trans. ASCE*, Vol. 89, paper No. 1588, pp. 940-984. 1926.

Grey, G. and Oliveri, F. Catchbasins - Effective Floatables Control Devices. Proc. Adv. in Urban Wet Weather Pollut. Reduction , Cleveland, Ohio, WEF (CP3805), 267. 1998.

Grey, G.M.; Oliveri, F.; and Rozelman, S. The Role of Catchbasins in a CSO Floatables Control Program. *Proc. Water Environ. Fed. 72nd Annu.Conf. Exposition*, [CD-ROM], New Orleans, LA. 1999.

Grottker, M. "Pollutant removal by gully pots in different catchment areas." The Science of the Total Environment. Vol. 93. pp. 515-522. 1990.

Irgang, L.M.; Atasi, K.Z.; and Scholl, J.E. Effects of a Catchbasin Cleaning on Stormwater Quality: a BMP Demonstration Project. *WEFTEC 2001 Conf. Proc.* CD-ROM. 2001.

Irvine, K.N. Chapter 10: Buffalo River floatables control and continuous water quality monitoring. *Best Modeling Practices for Urban Water Systems.* Vol. 10 in the Monograph Series. W. James, Ed. 151. 2002.

Lager, J.A., W.G. Smith, W.G. Lynard, R.M. Finn, and E.J. Finnemore. Urban Stormwater Management and Technology: Update and Users' Guide. U.S. Environ. Protection Agency, EPA-600/8-77-014, 313 p. Cincinnati, Ohio, September 1977.

Lau, S.L., and Stenstrom, M.K. Solid-phase Extraction for Oil and Grease Analysis. *Water Environ. Res.*, **69**, 3, 368. 1997.

Lau, S.L.; Khan, E.; and Stenstrom, M.K. Catchbasin Inserts to Reduce Pollution from Stormwater. *Water Sci. Tech.* **44**:23. 2001.

Lippner, G.; Johnston, J.; Combs, S.; Walter, K.; and Marx, D. Results of California Department of Transportation Litter Management Pilot Study. *Transportation Res. Record.* **1743**:10. 2001.

Lysne, D.K. "Hydraulic design of self-cleaning sewage tunnels." *Journal of Sanitary Engineering Division of ASCE*, Vol. 95, 1969.

Marais, M.; Armitage, N.; and Pithey, S. A Study of the Litter Loadings in Urban Drainage Systems - Methodology and Objectives. *Water Sci. Tech.* **44**:99. 2001.

May, R.W.P *Sediment Transport in Pipes and Sewers with Deposited Beds*. HR Wallingford, LTD., Wallingford, Report SR320, 1993.

McCuen, R.H. *Hydrologic Analysis and Design*, 2nd Edition. Prentice Hall. 1998.

Memon, F.A.; Butler, D. Assessment of gully pot management strategies for runoff quality control using a dynamic model. Science of the Total Environment, 295(1-3), 115-129. 2002.

Nalluri, C. and E.M. Alvarez. " The influence of cohesion on sediment behavior." Water Science Technology 1992

Nalluri, C. and A. Ab-Ghani. "Bed load transport with deposition in channels of circular cross section." *Proceedings of the Sixth International Conference on Urban Storm Drainage*. Niagra Falls, Canada. 1993.

Newhouse, W.R.; Maisch, F.E.; Bizzarri, R.E. Hydraulic Design Provides Low Maintenance CSO Floatables Control. *Proc. Water Environ. Fed. 72nd Annu.Conf. Exposition*, [CD-ROM], New Orleans, LA. 1999.

Newman, T.L.; Leo, W.M.; and Gaffoglio, R. A No Cost, Best Management Practice for Floatables Control in New York City. *Proc. Water Environ. Fed. 72nd Annu.Conf. Exposition*, [CD-ROM], New Orleans, LA. 1999.

Newman, T.L., II; Leo, W.M.; and Gaffoglio, R. Characterization of Urban-Source Floatables. *Collection Systems Wet Weather Pollution Control: Looking into Public, Private, and Industrial Issues,* May 2000, Rochester, NY. Water Environment Federation, CD-ROM. 2000.

Newman, T.L. A Methodology to Design and/or Assess Baffles for Floatables Control. *Models and Applications to Urban Water Systems, Monograph 9*. 51. 2001.

Othmer Jr., E.F.; Friedman, G.; Borroum, J.S.; and Currier, B.K. Performance Evaluation of Structural BMPs: Drain Inlet Inserts (Fossil Filter and StreamGuard) and Oil/Water Separator.

Proc. ASCE EWRI Conf. - Bridging the Gap: Meeting the World's Water and Environmental Resources Challenges. CD-ROM. 2001.

Phillips, D.I. A New Litter Trap for Urban Drainage Systems. *Water Sci. Technol.* (G.B.), 39, 2, 85. 1999.

Pisano, W.C., J. Barsanti, J. Jopyce, and H. Sorensen. *Sewer and Tank Sediment Flushing: Case Studies*. EPA/600/R-98/157. Water Supply and Water Resources Division, U.S. Environmental Protection Agency, Cincinnati, OH Dec. 1998.

Pitt, R. Demonstration of Nonpoint Pollution Abatement Through Improved Street Cleaning Practices. U.S. EPA. Grant No. S-804432. EPA-600/2-79-161. 270 pages. Cincinnati, August 1979.

Pitt, R. and G. Shawley. A Demonstration of Non-Point Source Pollution Management on Castro Valley Creek. Alameda County Flood Control and Water Conservation District (Hayward, CA) for the Nationwide Urban Runoff Program, U.S. Environmental Protection Agency, Water Planning Division, Washington, D.C., June 1982.

Pitt, R. Characterizing and Controlling Urban Runoff through Street and Sewerage Cleaning. U.S. EPA. Contract No. R-805929012. EPA/2-85/038. PB 85-186500/AS. 467 pages. Cincinnati, June 1985.

Pitt, R., S. Clark, and K. Parmer. *Protection of Groundwater from Intentional and Nonintentional Stormwater Infiltration*. U.S. Environmental Protection Agency, EPA/600/SR-94/051. PB94-165354AS, Storm and Combined Sewer Program, Cincinnati, Ohio. 187 pgs. May 1994.

Pitt, R., B. Robertson, P. Barron, A. Ayyoubi, and S. Clark. *Stormwater Treatment at Critical Areas: The Multi-Chambered Treatment Train (MCTT)*. U.S. Environmental Protection Agency, Wet Weather Flow Management Program, National Risk Management Research Laboratory. EPA/600/R-99/017. Cincinnati, Ohio. 505 pgs. March 1999.

Quasebarth, T.; Schroeder, D.; Chappell, R.; Churchwell, R.; and Lippner, G. An Investigation of Factors Influencing Solids Transport and Deposition into Highway Drain Inlets. *Proc. ASCE EWRI Conf. - Bridging the Gap: Meeting the World's Water and Environmental Resources Challenges*. CD-ROM. 2001.

Raudkivi, A. J., and Tan, S. K. "Erosion of cohesive soils," *Journal of Hydraulic Research*, Vol 22, No. 4, pp 217-233. 1984.

Ristenpart, E. and M. Uhl. "Behavior and pollution of sewer sediments." Proceedings of the Sixth International Conference on Urban Storm Drainage. Niagra Falls, Canada. 1993.

Sartor, J. and G. Boyd. Water Pollution Aspects of Street Surface Contaminants. U.S. EPA. Contract No. 14-12-921. EPA-R2-72-081. 236 pages. Washington, D.C., November 1972.

SCS (Soil Conservation Service). Handbook for Channel Design for Soil and Water Conservation. SCS-TP-61, rev. 1954.

SCS (Soil Conservation Service). *Design of Open Channels*. TR-25. 1977.

Siegel, J.P., and Novak, R.J. Duration of Activity of the Microbial Larvicide VectoLex CG (R) (*Bacillus sphaericus*) in Illinois catchbasins and waste tires. Journal of the American Mosquito Control Association. 15, 3, 366. 1999.

Sutherland, R.C. and Jelen, S.L. Quantiyfing the optimum urban runoff pollutant load reductions associated with various street and catchbasin cleaning practices. Proc. 9th Int. Conf. Urban Drainage – Global Solutions for Urban Drainage. CD-ROM. 2002.

U.S. Army Corps of Engineers (COE). *Hydraulic Design of Flood Control Channels*. EM 1110-2-1601. U.S. Army Engineer Waterways Experiment Station, Vicksburg, MS. undated.

U.S. Department of Agriculture (USDA). Design of Open Channels. Technical Release No. 25, Soil Conservation Service, Washington, DC. 1977.

U.S. Army Corps of Engineers (COE). Engineering and Design: Channel Stability Assessment for Flood Control Projects. EM 1110-2-1418. U.S. Army Engineer Waterways Experiment Station, Vicksburg, MS. 1994.

USDA. Stability Design of Grass-Lined Open Channels. Agricultural Handbook 667. 1987:

Walker, D.E.; Heath, G.R.; and Kubiak, D.A. CSO Floatables Control Using Underflow Baffles. *Proc. Water Environ. Fed. 70th Annu. Conf. Exposition*, Chicago, Ill., **2**, 665. 1997.

Walker, D. E.; Heath, G. R.; and Kubiak, D. A. (1998) Floatables Control. *Water Environ. Technol.*, 10, 2, 45. 1998.

Williams, A.T., and Simmons, S.L. Estuaries Litter at the River/beach Interface in the Bristol Channel, United Kingdom. *J. Coastal Res.* (U.K.), **13**, 4, 1159. 1997a.

Williams, A.T., and Simmons, S.L. Movement Patterns of Riverine Litter. *Water, Air, Soil Pollut.* (Neth.), **98**, 1-2, 119. 1997b.

Williams, A.T., and Simmons, S.L. Sources of Riverine Litter: The River Taff, South Wales, UK. *Water, Air, & Soil Pollution*, 112, 1-2, 197. 1999.

Williams, D. T., and Julien, P. Y. "Examination of stage-discharge relationships of compound/composite channels," *Proceedings of the International Conference on Channel Flow and Catchment Runoff.* B.C. Yen, ed., University of Virginia, Charlottesville, 22-26 May 1989, pp 478-488. 1989.

Yao, K.M. " Sewer line design based on critical shear stress." *Journal Environmental Engineering Division of ASCE*, Vol. 100. 1974.

Chapter 7

Coarse Solids Management 2

Introduction

By *Betty Rushton and Shirley Clark*

This chapter emphasizes the value of considering the particle size distribution of solids and their associated pollutants when designing and analyzing stormwater treatment projects. Studies show particle size affects the type of treatment techniques best suited for removing pollutants and indicates a series of practices may be necessary to meet the more stringent goals required by Total Maximum Daily Loads (TMDL) and the National Pollutant Discharge Elimination System (NPDES) regulations. Also emphasized in this chapter are methods to calculate particle transport, calculations for including gross solids in efficiency calculations, and impact of the size distribution in designing filter systems to remove particles. The problems associated with collecting representative samples that include the variety of stormwater particle sizes, including the largest sizes, in studies conducted using automatic water quality samplers are also discussed. The contribution of gross solids (>100 um) has been overlooked in most stormwater monitoring efforts.

Section 1 presents observations from an urban area mass balance investigation along with the application of several traditional particulate transport approaches to the problem. It describes the particle sizes found in the studies and provides tables and figures to calculate their movement in sewer pipes. Studies documenting the accumulation of sediments in storm drains and recommendations to achieve a self-cleaning drainage system are presented.

Section 2 includes methods to incorporate the entire particle size distribution into the material captured by gross pollutant traps in the calculation for removal efficiencies, which historically only measured the particles in suspension. It provides several formulas suitable for including the gross solids in efficiency calculations and emphasizes pitfalls associated with laboratory techniques.

Section 3 presents data from a study designed to investigate the use of media filters for their ability to remove stormwater solids and their associated heavy metals, and also evaluated sorption media for their ability to remove dissolved solids (passes through a 0.45-um filter). The results indicate the importance of influent metal concentration on pollutant removal including the lower limit of metals removal by commonly-available and inexpensive filter media. In addition, the ability of media to retain pollutants during extreme conditions of pH and anaerobiosis was evaluated.

The Florida Study

Betty Rushton, Ph.D. Southwest Florida Water Management District, 2379 Broad Street, Brooksville, Florida. Ph: (352) 796-7211, FAX: (352) 797-5806, Email: betty.Rushton@swfwmd.state.fl.us/

Introduction

This retrofit project was designed to reduce the amount of pollution discharged to the Hillsborough River and ultimately Tampa Bay by installing a MODEL PSW100_60 (32 CFS capacity) CDS unit and a constructed linear marsh. Only the CDS unit is discussed here, but the first year progress report that includes all the data is available (Rushton 2004). The drainage basin is approximately 132.4 acres in size and includes a 30.6-acre high intensity strip commercial district immediately upstream. The remainder of the watershed includes multi-family and residential land uses as well as a golf course and major urban thoroughfares. The drainage basin receives no sewage overflow and the streets are periodically cleaned with a street sweeper. The purpose of this paper is to present a possible method for calculating the pollution removal ability of the CDS unit by incorporating the material collected in the CDS sump as well as the pollutants in the storm flow. Comments and suggestions about the method would be appreciated. Although the human derived trash and polycyclic aromatic hydrocarbons (PAHs) were not able to be included when using this method, they were effectively collected by the CDS unit and are also discussed.

Methods

Measuring Hydrology and Concentrations

Hydrology. Hydrology measurements were calculated using Star velocity meters, ISCO water level sensors, and tipping bucket rain gauges. See Figure 1b for location of sensors. Levels were converted to flow using appropriate weir and pipe formulas. This information was stored in data loggers until retrieved and downloaded into spreadsheets to be processed into tables and figures. Bypass flow over a diversion weir was also measured. The flow was measured continuously and this flow was divided into storm flow, base flow and bypass flow. The flow calculations were checked against an estimated water budget for the pond (Rushton 2004).

Water quality. Flow weighted samples were collected to measure water quality for both storm flow and base flow using automated samplers. Samples were taken in front of and after the CDS unit. It was not possible to sample the overflow concentrations because these became almost immediately mixed with the CDS outflow downstream of the diversion weir. For bypassed flow, it was assumed that the same concentrations would be present as were measured at the inflow of the CDS unit. The system discharged continuously, and the base flow samples were collected several times during the month with each composite sample representing about a two-day period.

Gross Solids. Representative samples of the material in the CDS unit were analyzed each time the unit was vacuumed out. The litter was removed from the mass of material and was quantified separately. About two liters of material collected at one-foot depth intervals with an Ekman Dredge while the unit was being vacuumed out were taken from separate sides of the CDS unit. These were divided into sample A and sample B and each sample was well mixed before removing sub-samples to be sent to the laboratory. Standard methods for soil analysis were used to quantify the material. The amount collected in the CDS unit was estimated by using the volume of the sump basin and the depth of the material in the sump.

Particle Size Analysis. The material captured in the sump of the CDS unit was analyzed by particle size for two of the clean out periods. The laboratory separated the samples into particle size by the dry sieving method and samples were analyzed for constituents by particle size. Also a portion of each sample was not sieved and analyzed separately for one clean-out period.

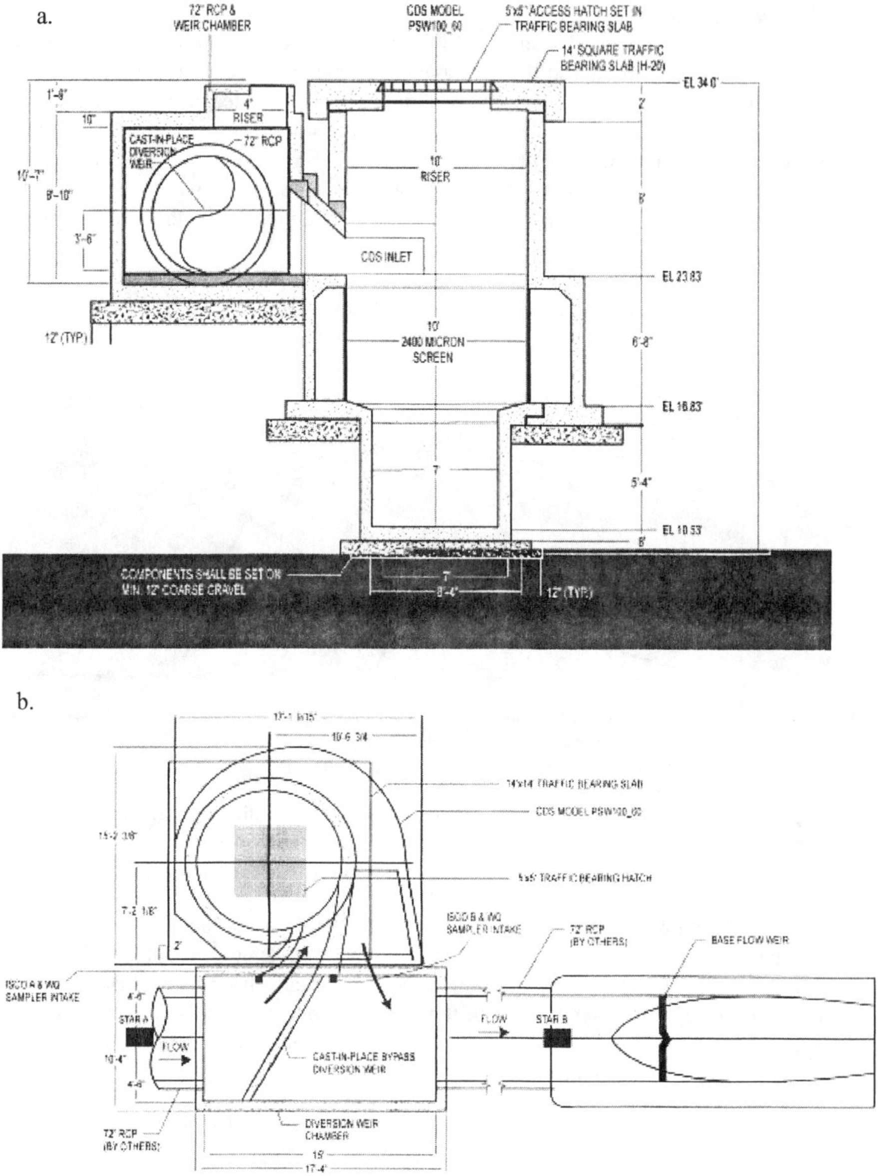

Figure 1. Vertical section (a) and plan view (b) of the Continuous Deflective Separation System (CDS).

Calculating pollution removal

During the first year of the study the CDS unit collected almost all of its gross solids during a six-month period in the spring and summer (Figure 2). The efficiency data in this report represent this six-month period, which extended from February 1st to July 14th, 2003. During this time period, 18 storm samples and 26 base flow samples were analyzed for water quality. This included 57 percent of all the storm event rainfall and many of the smaller events that were not sampled were included in the base flow samples. All of the flow was measured and included in the calculations for mass loading. Both average and median values for water quality were used in the calculations to estimate load reductions. The CDS unit collected 336 cubic feet of material and most of the collected mass was leaves (55 to 75%).

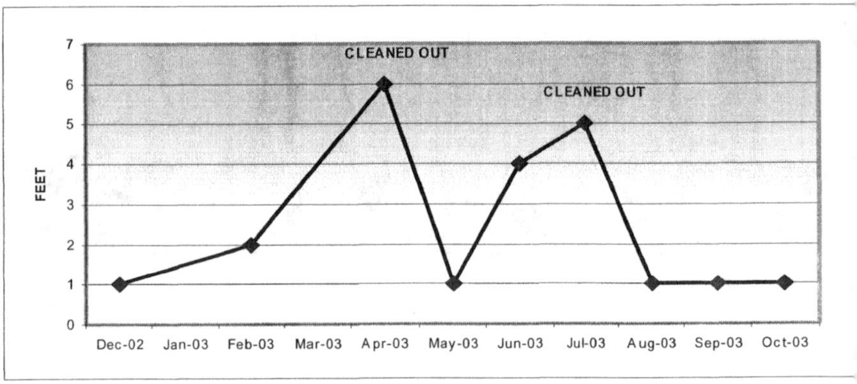

Figure 2. Measurements taken inside the CDS unit to determine how much material had been collected and to schedule clean-outs. The period from February 1 to July 14, 2003 includes the data analyzed for this report.

One of the purposes of the study was to calculate load reductions that could be compared to other stormwater studies and that are appropriate for determining TMDLs. The automatic water quality samplers usually used to evaluate pollution reduction do not adequately sample the material collected by CDS units and other manufactured devices. The following section explains the method used to include gross solids in the calculation for load reductions.

A summary of all the flow, water quality and gross solids data collected during the six-month period used to calculate pollutant load reduction is shown in Table 1. Several assumptions were made including:

- The same volume of water left the CDS unit as entered at the inflow, except for the flow that was discharged over the bypass weir.
- The water that discharged over the bypass weir had the same water quality as the water that entered the CDS unit.

- The average and median values for water quality analyzed for the sampled storms (57%) are representative of all storms.
- The average and median values for water quality analyzed for the base flow sampled (about 30%) are representative of all base flow.
- The gross solids that were collected in the CDS unit represent all the gross solids that entered the system (This is not entirely true because floating litter and leaves did bypass the CDS unit during high flow and considerable amounts of leaves and pavement material were noted in the downstream pond).
- The calculations for solids use total suspended solids for the water quality data and total solids collected by the CDS unit for the gross solids.

The standard method (referred to as the Scoping Method in the Caltran study) for measuring BMP system efficiency (the amount of pollution reduced or increased) is calculated with the following formula:

$$\%Efficiency = \frac{(Load_{in} - Load_{Out})}{(Load_{in})} \times 100$$

To use the formula all the loads must be in comparable units. For our analysis we used grams and kilograms.

Step 1.

To convert the gross solids loads, the volume of material collected in the CDS unit (minus the litter) had to be converted from cubic feet to kilograms. The bulk density measurement was used for this calculation. .

(Mass of gross solids)*(bulk density)*(conversion factors)

Where:

Mass of gross solids = 336 cubic feet (measured in the CDS unit)

Bulk Density = 0.54 g/cm^3 form (laboratory analysis on an as received sample)

Using these calculations our site had 5137 kilograms of debris and coarse sediments measured during the two cleanout periods.

Step 2.

To convert the individual constituents measured by the laboratory for the gross solids material. The constituents were reported in mg/kg. The calculation took the form: (Constituent mg/kg) X (5137 kg of gross solids) X (conversion factors)

The results from these calculations are shown as the loads captured in CDS unit in Table 1.

Step 3.

To convert water quality data to loads (average and median values were calculated separately but they both use the same method).

(constituent mg/l) X (flow for time period ft^3) X (conversion factors)

Where:

Storm flow for period = 101,794 cu meter (3,594,432 cubic ft)

Bypass flow for period= 46,876 cu meter (1,655,237 cubic ft)

Base flow for period= 37,351 cu meter (1,318,881 cubic ft)

Constituent concentrations from Table 1

The results are shown as mass loading in Table 1.

Step 4.

The efficiency of the CDS unit for removing pollutants is calculated by the standard formula and uses either the average or median loads as calculated above.

%efficiency = (storm in+bypass in+base in+sump) – (storm out + bypass in +base out)

(storm in + bypass in + base in + sump)

Where:

Storm in = storm loads in front of CDS unit calculated by Step 3 above

Bypass in=bypass loads in front of CDS unit calculated by step 3 above.

Base in=base flow loads in front of CDS unit calculated by step 3 above

Sump=loads calculated in step 2 above.

Storm out =storm loads after CDS unit calculated by step 3 above

Base out=base flow loads after CDS unit calculated by step 3 above

Table 1. Summary data for water quality and gross solid concentrations and loads for the period Feb. 1 to July 12, 2004.

SAMPLE TYPE	FLOW AMOUNT	TOTAL NITROGEN		TOTAL PHOSPHORUS		RECOVERABLE COPPER		RECOVERABLE ZINC		TOTAL* SOLIDS		ORTHO PHOSPHORUS		AMMONIA		NITRATE+ NITRITE		TOTAL KJELDAHL-N	
		BEFORE CDS	AFTER CDS	BEFORE CDS	AFTER CDS	BEFORE CDS	AFTER CDS	BEFORE CDS	AFTER CDS	BEFORE CDS	AFTER CDS	BEFORE CDS	AFTER CDS	BEFORE CDS	AFTER CDS	BEFORE CDS	AFTER CDS	BEFORE CDS	AFTER CDS
WATER QUALITY		(mg/l)	(mg/l)	(mg/l)	(mg/l)	(ug/l)	(ug/l)	(ug/l)	(ug/l)	(mg/l)	(mg/l)	(mg/l)	(mg/l)	(mg/l)	(mg/l)	(mg/l)	(mg/l)	(mg/l)	(mg/l)
STORM SAMPLES																			
# Samples		18	18	18	18	18	18	18	18	17	17	18	18	18	18	18	18	18	18
Mean		1.141	1.217	0.104	0.108	9.5	11.8	51.8	58.7	14.2	14.4	0.027	0.023	0.057	0.036	0.440	0.373	0.701	0.843
Median		1.150	1.115	0.092	0.092	9.4	9.8	46.8	49.2	11.7	12.8	0.024	0.019	0.018	0.018	0.441	0.378	0.631	0.730
St. Dev.		0.291	0.351	0.048	0.049	3.3	5.3	23.7	34.7	11.7	10.0	0.020	0.016	0.071	0.047	0.228	0.186	0.271	0.370
Max		1.740	1.900	0.235	0.205	18.6	25.5	112.0	147.0	50.1	49.4	0.081	0.063	0.205	0.160	1.010	0.819	1.198	1.519
MASS LOADING																			
FLOW***	(cu meters)	(kg)	(kg)	(kg)	(kg)	(grams)	(grams)	(grams)	(grams)	(kg)	(kg)	(kg)	(kg)	(kg)	(kg)	(kg)	(kg)	(kg)	(kg)
MEANS																			
STORM	101,794	116.2	123.8	10.6	11.0	965.1	1198.6	5276.9	5973.6	1443.6	1461.3	2.7	2.3	5.8	3.7	44.8	38.0	71.4	85.9
BYPASS**	46,876	53.5	53.5	4.9	4.9	444.4	444.4	2430.0	2430.0	664.8	664.8	1.3	1.3	2.7	2.7	20.6	20.6	32.9	32.9
BASE	37,351	55.6	51.4	2.6	2.8	436.6	429.7	861.6	968.1	218.2	253.4	0.9	0.7	0.8	1.4	34.2	26.3	21.4	25.0
MEDIANS																			
STORM	101,794	117.1	113.5	9.4	9.4	959.4	1000.6	4764.0	5003.2	1191.0	1303.0	2.4	1.9	4.6	2.6	23.2	19.0	64.2	74.3
BYPASS	46,876	53.9	53.9	4.3	4.3	441.8	441.8	2193.8	2193.8	548.5	548.5	1.1	1.1	0.8	0.8	20.6	20.6	29.6	29.6

* Total solids are total suspended solids for water quality and total solid for CDS material.

Results

Pollution Removal Efficiency

Including the sump material in the calculations improves the accuracy of the removal efficiency calculation and gives a higher efficiency rating for the CDS unit. This method shows that the CDS unit is quite good at reducing some pollutants and is not effective for removing others. The percent efficiencies calculated for loads both with and without the sump material are shown in Table 2.

Total Solids – The unit is quite effective at removing the larger solid material (>75 microns) found in bed loads such as street dirt, leaves and other large size particles. Total solids were removed by 68 to 71 percent. This is not surprising since the units were designed to capture this type of material. If the 32 percent of flow that bypassed the CDS unit is considered by assuming that the bypassed flow contained the same proportion of material as was collected by the CDS unit then the percent reduction is between 56 and 58 percent. The increase in suspended loads in the water column indicates that particles are being broken down into smaller particles as they move around in the unit.

Table 2. Load Efficiencies including loads for water quality samples only (the Scoping Method in the Caltran study) and loads that also include the amount retained in the CDS unit. Negative percentages indicate higher loads were discharged from the CDS unit than entered.

Constituent	Water Quality Only		Includes CDS Loads	
	Sample Means	Sample Medians	Sample Means	Sample Medians
Total Solid *	-2%	-8%	68%	71%
Total Kjeldahl Nitrogen	-15%	-15%	2%	3%
Total Phosphorus	-4%	-3%	15%	18%
Recoverable Copper	-12%	-1%	-4%	6%
Recoverable Zinc	-9%	-2%	-1%	6%
Total Nitrogen	-2%	4%	na	na
Ammonia **	15%	-6%	na	na
Nitrate + Nitrite	15%	20%	na	na
Ortho Phosphorus	13%	13%	na	na

* Water Quality samples include only the suspended solids loads and the CDS loads are total loads.

** The median calculations for ammonia were skewed by six rain events where ammonia was measured below the laboratory limit of detection. When these events are eliminated the efficiency is +29 percent

na = not analyzed in the material collected by the CDS unit.

Total Phosphorus – The efficiency measured for total phosphorus (15% to 18%) probably reflects the fact that phosphorus easily attaches to soil particles and organic material and it is expected it would be reduced by attaching to solids. Ortho phosphorus, the inorganic portion of total phosphorus is reduced by 13% in the water samples indicating it is being transformed to organic nitrogen or attaching to particles. Better efficiency would probably have been measured except that leaf litter releases phosphorus when soaked in water (Strynchuk et al. 1999). In bench studies, concentrations of phosphorus increased by 89 percent in water after it was soaked with grass clippings and tree leaves for only one day. The authors also reported a corresponding decrease in the phosphorus measured in the solids of 54 percent.

Total Kjeldahl Nitrogen (TKN) – When the sump material is included in the calculations, TKN improves from an increase of 15 percent to a small amount of removal (2% to 3%). Some of this can be explained by nitrogen transformations. TKN is the combination of organic nitrogen and ammonia. Much of the ammonia and nitrate is converted to organic nitrogen and these soluble nutrients are reduced in the CDS unit as shown for the water quality sample means. The negative efficiency for water quality samples for ammonia was caused by six rain events where concentrations for both the inflow and outflow were below the laboratory detection limit. When these values were removed, the ammonia in the CDS was reduced by 29 percent instead of the 6 percent increase calculated in the table. It might be expected that nitrogen would increase in the water column as it passed through the CDS unit. TKN was increased by 31 percent in water that was used to soak grass clippings and tree leaves in bench experiments when measured after one day (Strynchuk et al. 1999). The authors also reported a corresponding decrease in the TKN concentrations in the litter of 11 percent.

Recoverable Metals – Copper, lead, and zinc were measured at low concentrations at the site and probably exhibited no reduction by the CDS unit. Besides the low concentrations, the low pollution removal can possibly be explained by the tendency of metals to attach to the smaller sized particles such as clay, which were not collected by the CDS unit. Particle size analysis revealed only a low percentage of particles were retained that were less than 75 microns. Although organic material has been found to be an effective sink for metals and over 50 percent of the material collected by the CDS unit was tree leaves, the zinc, lead, and copper concentrations retained by the unit were quite low and averaged values were below levels considered toxic to sensitive organisms. (Lead was not evaluated for this paper because most concentrations were below the laboratory detection limit).

Additional Captured Pollutants

The CDS unit was quite effective at removing two types of pollutants that are not usually measured in storm water BMP monitoring studies – PAHs and trash. Although these pollutants could not be analyzed for removal efficiency using the method described above, they are undoubtedly reduced and the amount retained by the CDS unit was quantified.

Polycyclic Aromatic Hydrocarbons (PAHs) – Toxic levels of PAHs were measured in the material in the CDS unit as well as in soils where the unit discharges into the pond. The effect that gross solid deposition has upon polluted sediments has not always been emphasized in storm water studies that base their interpretation and pollution reduction on water quality data for individual storm events. Sediments can accumulate pollutants through mechanisms of direct disposition of solids, or through various processes where soluble pollutants precipitate and contaminate the sediments. Scouring of storm conveyance systems and ultimately streams and rivers takes a long time and is difficult to relate to storm events. These polluted sediments probably have a greater toxic effect on runoff biota than the dissolved toxicants in the water column. Many studies have shown the severe detrimental effects of urban runoff on receiving water organisms (Pitt 1995). Other studies have documented that even tolerant species are eliminated when toxic levels of metals and PAHs are measured in the sediments (Rushton *et al.* 2004).

PAHs enter the environment mostly from volcanoes, forest fires, residential wood burning, pavement, and exhaust or leakage from automobiles and trucks (ATSDR 2001). PAHs do not easily dissolve in water, but stick to solid particles and settle to the bottoms of rivers or lakes. Breakdown of PAHs in soil and water generally takes weeks to months and is caused primarily by the actions of microorganisms. The sediment data at the Broadway Outfall site suggest that the CDS unit is reducing the amount of PAHs being transported to the Hillsborough River. The 2001 data was collected in a ditch downstream from the Broadway Outfall pipe before the retrofit and the 2002 data was collected in a shallow pond that had replaced the ditch as part of the retrofit. The data for both time periods indicate that PAHs in the soils were confined close to where they entered the sunlit open aerobic ditch or pond (Figure 3).

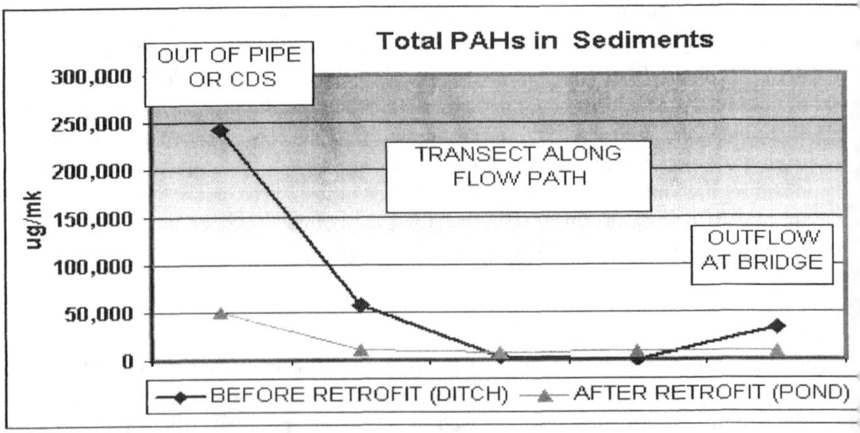

Figure 3. Concentrations of PAHs measured in the sediments of the ditch before the retrofit and in the pond after construction of the CDS unit. The X-axis represents distances downstream of the CDS unit.

The PAH concentrations were measured at much higher levels in the soils before the retrofit in 2001, but are still measured at levels where they are considered probably toxic (>23,000 ug/kg) at the inflow to the pond in 2002. Concentrations were also much higher near the storm sewer where it flowed into the open water ditch. Concentrations tapered off to non-detectible concentrations as water flowed down the ditch until it reached the bridge, where additional storm runoff entered the flow stream. Although PAH concentrations were much lower in 2002, they were detected at two stations where low or no concentrations had been detected in 2001. This was likely due to disturbance of the soil when the pond was constructed, depositing some of the contaminated soils into the pond area.

When the pre-construction PAH soil samples collected in July 2001 near where the storm sewer pipe entered the ditch are compared to the samples collected in the CDS unit in April 2003, concentrations are similar (Table 3). Both sediments exceed the concentrations considered toxic to organisms in fresh water. The material collected by the CDS unit is reducing these toxic levels of hydrocarbons that were previously released to the environment. Most of the PAHs are now removed from the flow stream and disposed of in a location where they can no longer contaminate a water body. The CDS unit has reduced a serious long-term impact related to the re-suspension of previously deposited polluted material that has been reported as even more detrimental to the environment than the short-term "first flush" effect associated with urban runoff (Heaney 1978). However, the total PAHs measured in the sediments after the construction of the CDS unit, while reduced, still do not meet state standards. This may be attributed to the 32% of flow that bypasses the unit.

Table 3. Concentrations of PAHs in soil samples collected at the Broadway Outfall retrofit site compared to concentrations in the CDS unit during cleanout and to State Standards (FDEP 2003).

	Soil conc before retrofit July-01	CDS Conc. for Each Clean-Out		Total for Year	Sediment Standards*	
		April-03	July-03			
RATIOS =====>		0.56	0.44	1	TEC**	PEC**
CONCENTRATIONS FOR PAHs (ug/kg)						
Acenaphthene	2,000	740	520	643	7	89
Anthracene	5,900	1,818	1,360	1,616	57	850
Benzo (a) anthracene	24,000	14,496	8,200	11,726	110	1,100
Benzo (a) pyrene	19,000	17,465	9,400	13,916	150	1,500
Benzo (b) fluoranthene	30,000	24,225	15,200	20,254	NG	NG
Benzo (g,h,I) perylene	7,000	11,146	5,400	8,617	NG	NG
Benzo (k) fluoranthene	10,000	21,310	5,400	14,310	NG	NG
Chrysene	26,000	26,150	11,800	19,836	170	1,300
Fluoranthene	41,000	39,770	24,000	32,831	420	2,200
Fluorene	2,800	1,018	740	896	NG	NG
Indeno (d,2,3I,-cd) pyrene	12,000	14,434	6,600	10,987	NG	NG
Phenanthrene	26,000	18,870	10,800	15,319	200	1,200
Pyrene	35,000	38,175	18,400	29,474	200	1,500
TOTAL PAH (ug/kg) >>	240,700	229,616	117,820	180,426	1,600	23,000

* Numerical sediment quality assessment guidelines for Florida Inland Waters (FDEP 2003)

** TEC=Threshold Effect Level, PEC=Probable Effect Level, NG=no guidelines

Trash (Litter)

The human derived trash was collected, weighed and sorted for each cleanout period and represents a year's worth of litter extending from July 2002 to July 2003 (Table 4). The sample included the litter that had been skimmed off each month as well as the litter retrieved from the mass of material removed by the vactor truck at time of clean out. Although the amount of litter is small (31.62 ft^3) when compared to the other material collected by the CDS unit (413 ft^3), it is an eye sore and has the potential to impact wildlife as well as leach pollutants. Plastics were measured more often than any other litter category, but aluminum and Styrofoam were also found in significant quantities. Slightly more trash in fewer categories was collected during the July clean out.

Table 4. Amount of litter collected in the CDS unit during the first year. It included over 100 pounds of assorted material, which measured a total volume of 31.62 cubic feet.

	APRIL 2003 CLEAN OUT				JULY 2003 CLEAN OUT				TOTAL FOR YEAR			
CATEGORY	KG	LB	M^3	FT^3	KG	LB	M^3	FT^3	KG	LB	M^3	FT^3
Plastic	13.98	30.66	0.25	8.79	19.31	42.34	0.33	11.76	33.29	73.00	0.58	20.55
Aluminum	1.65	3.63	0.04	1.52	2.68	5.88	0.05	1.75	4.34	9.51	0.09	3.27
Styrofoam	0.39	0.85	0.08	2.91	0.40	0.89	0.03	0.99	0.79	1.73	0.11	3.90
Miscellaneous	0.94	2.05	0.00	0.11	2.64	5.78	0.01	0.48	3.57	7.83	0.02	0.58
Wood	1.65	3.63	0.04	1.52	2.68	5.88	0.05	1.75	4.34	9.51	0.09	3.27
Paper	0.42	0.91	0.00	0.04	0.00	0.00	0.00	0.00	0.42	0.91	0.00	0.04
Glass	0.05	0.12	0.00	0.00	0.00	0.00	0.00	0.00	0.05	0.12	0.00	0.00
Cig. Butts	0.02	0.04	0.00	0.01	0.00	0.00	0.00	0.00	0.02	0.04	0.00	0.01
TOTALS	**19.09**	**41.87**	**0.42**	**14.90**	**27.71**	**60.77**	**0.47**	**16.72**	**46.81**	**102.64**	**0.89**	**31.62**

Particle Size Analysis

For one clean out period the gross solids samples were analyzed by particle size and the results compared to a whole sample that was not sieved and the concentrations were quite different (Figure 4). The whole samples that were not sieved are designated as "TOTAL" in Figure 4. These data represent material collected from July 2003 to March 2004. Most of the largest size particles (65%) were leaves that didn't pass through the largest sieve size. Twenty percent of the samples represent sediments between 75um and 850um and ten percent of the samples were less than 75um. Total nitrogen concentrations were about equally divided between the two particle sizes, but phosphorus, metals and to some extent polycyclic aromatic hydrocarbons (PAHs) had higher concentrations in the sediments represented by the smaller particle sizes. Of some concern is the fact that sieving the samples seems to increase the concentrations for all constituents except PAHs. This type of result has been seen in other studies of samples sorted by particle size in stormwater (Harper, personal communication). This phenomenon needs additional studies to determine its cause.

Conclusions

The CDS unit is effective for removing gross solids including large size particles (> 75 microns), trash and toxic levels of PAHs. It does not remove significant amounts of suspended pollutants measured in the water column in traditional storm water studies. Studies on other hydrodynamic devices have shown similar results. This should be no surprise since storm water is not retained in the devices long enough for much biological treatment to take place. A report that includes all the data collected for year one is available if requested (betty.rushton@swfwmd.state.fl.us).

Hydrodynamic devices could be an important first step in a treatment train of several storm water treatment techniques if large sized particles, trash or PAHs are a concern for the basin. When calculating treatment efficiency for hydrodynamic devices, the sump material should be included in the calculation to give a total solids removal, rather than only TSS removal calculated from water samples. In addition, if a bypass is used in the BMP, the volume of water and associated pollutants, which bypass the device should be included in the calculation. Accounting for a total mass balance of the flows and pollutants through and around a hydrodynamic device gives a much more accurate picture of the pollutant removal efficiency, as opposed to a TSS calculation of the liquid phase only. These types of devices are not designed to provide significant removal of particles small enough to be defined as TSS. Rather, they are designed to remove large particles considered to be gross solids.

Using gross solids removal efficiency is an effective way to compare the performance of hydrodynamic devices. However, it should not be used to compare them to other BMPs since gross solids removal is not measured with other BMPs. When comparing hydrodynamic devices to other BMPs, only the pollutant removal efficiency in the water column passing through the device should be considered. This assures that the different BMPs are being compared against the same standards. In general, hydrodynamic devices perform poorly for pollutants suspended in the water column when compared to traditional BMPs, and they will rarely meet most design criteria that require a specified TSS removal efficiency. Changing the definition of the targeted pollutants for TMDL compliance could make the devices more acceptable to regulatory agencies.

Acknowledgements

This project was funded in part by a Section 319(h) Nonpoint Source Management Program grant from the U.S. Environmental Protection Agency through a contract with the Stormwater/nonpoint Source Management Section of the Florida Department of Environmental Protection. The total cost of construction and monitoring of the project is $540,934 and $342,560 of this is being provided by US EPA. David Worley and Eric Livingston of FDEP are invaluable in providing support and guidance. Rebecca Hastings, Kara Teague and Stan Kane have been indispensable in collecting and processing data. Gordon England has reviewed the draft report and his insight gained from studying similar devices added important information.

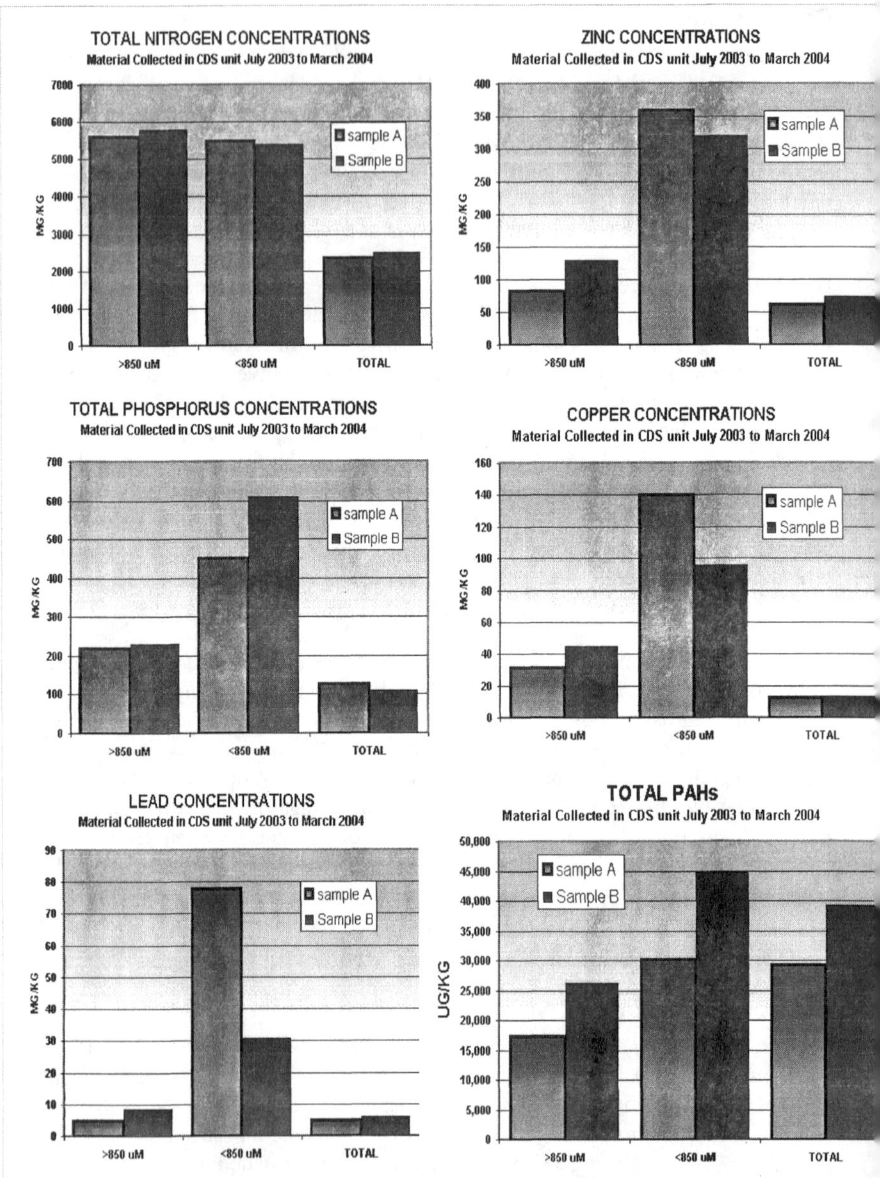

Figure 4. Comparisons of constituent concentrations for two particles sizes with a sample analyzed without being sieved (TOTAL). Data from March 2004 clean out. Most of the largest particle sizes were leaves that did not pass through the sieve

References

Agency for Toxic Substances and Disease Registry (ATSDR). 2001. *Polycyclic Aromatic Hydrocarbons* (PAHs). <http://www.atsdr.cdc.gov/ToxProfiles/phs8806.html > (March 2001)

Florida Department of Environmental Protection (FDEP) 2003. *Development and Evaluation of Numerical Sediment Quality Assessment Guidelines for Florida Inland Waters*. Technical Report. <http://www.dep.state.fl.us/water/monitoring/docs/seds/SQAGs> April 2003.

Heaney, J. P. 1978. *Nationwide Assessment of Receiving Water Impacts from Urban Storm Water Pollution.* Department of Environmental Engineering Sciences, University of Florida, Gainesville , Florida.

Pitt, R. E. 1995. "Biological Effects of Urban Runoff Discharges". *In Stormwater Runoff and Receiving Systems: Impact, Monitoring, and Assessment.* E.E.Herricks (ed). Lewis Publishers, New York.

Rushton, B. T. 2004. *Broadway Outfall Stormwater Retrofit Project. Progress Report for Year One.* Southwest Florida Water Management District, 2379 Broad Street, Brooksville, Florida 34604.

Rushton, B.T., D. Huneycutt, and K. Teague. 2004. *Characterization of Three Stormwater Ponds*. Southwest Florida Water Management District, 2379 Broad Street, Brooksville, Florida 34604

Syrynchuk, J., J. Royal and G. England. 1999. "Grass and Leaf Decomposition and Nutrient Release Study Under Wet Conditions". In Proc. *6th Biennial Stormwater Research & Watershed Management Conference.* Southwest Florida Water Management District, 2379 Broad Street, Brooksville, Florida 34604.

Media Filtration to Remove Solids and Associated Pollutants from Stormwater Runoff

Shirley E. Clark, Ph.D., P.E.[1], Pauline D. Johnson, Ph.D.[2], Robert Pitt, Ph.D., P.E.[2], Sarah Gill, Ph.D.[2], and Mukesh Pratap, M.S.C.E.[1]

[1]Environmental Engineering Program, The Pennsylvania State University at Harrisburg

[2]Department of Civil and Environmental Engineering, The University of Alabama

Introduction

Stormwater runoff from critical source areas, or 'hotspots' (vehicle service areas, heavy equipment storage and maintenance areas, public works yards, auto recyclers/junkyards, large parking lots, bus or truck (fleet) storage areas, and vehicle and equipment washing/steam cleaning facilities) may contain substantially higher pollutant loadings of hydrocarbons, toxic trace metals, nutrients, pathogens, and/or other toxicants and pollutants as compared to 'normal' runoff from most other areas (Bannerman et al. 1993; Pitt et al. 1995; Claytor and Schueler 1996). Opportunities for the control of these pollutants can be grouped into three basic strategies: source avoidance/reduction; passive systems; and installed treatment technologies. While the concept of pollution avoidance and minimization at critical sources is the most preferred option, it is unlikely that any one method will provide either a realistic or effective solution to the problem. The scope of this project was to address emerging stormwater control technologies, and focusing on media filtration systems, for the capture of solids and the associated heavy metals from urban runoff.

Heavy metals in urban stormwater primarily originate from automobile-related activities and the exposure of building materials to rain. Heavy metals may occur as dissolved, colloidal or particulate bound species. Particle associations and speciation critically affect the toxicity and bioavailability of metals and are greatly dependent upon chemical and physical parameters. The general rule is that most metals are predominantly associated with particulates, but with several exceptions. One, particular source areas such as scrapyards (Table 1) may have runoff where a significant fraction of some metals remains in the dissolved phase. Second, if the pH of the runoff (such as in acid rain conditions) is decreased below neutral, the portion of dissolved trace metals may increase (Dempsey et al. 1993). Third, dissolved solids, and therefore dissolved metals, in urban runoff are also associated with the use of de-ice chemicals on roads during the winter and from fertilizer and pesticide use on residential lawns, parks, golf courses, and roadsides (Merkel et al. 1988). Therefore, control of solids is expected to significantly reduce the metals concentration reaching a receiving water.

Filtration has been shown by the water and wastewater treatment industry to be an excellent mechanism for removing solids. Sand filters made of a fine sand (d_{10} = 0.2 to 0.3 mm) have been shown to remove most solids greater than 20 – 50 μm (Clark 2000). Prior research from the Birmingham, Alabama, area has shown that the range of median particle diameters in urban runoff is between 5 and 50 μm (Clark 1996), indicating that filtration is a potentially viable treatment solution. Removals of particles less than 20 μm may occur if the contact time is extended. If inadequate filtration occurs, the suspended particles will exit the filter and potentially increase the turbidity and pollutant content of the receiving water.

Table 1. Concentrations of Metals in Scrap Yard Runoff (Data from Clark, 2000)

	Fe (mg/L)	Al (mg/L)	Zn (mg/L)
Total	20 – 810	15 – 70	1.6 – 8
< 63 μm diameter	22 – 767	15 – 58	1.5 – 7.6
< 38 μm diameter	21 – 705	15 – 58	1.4 – 7.4
< 20 μm diameter	15 – 534	12 – 50	1.1 – 7.2
< 0.45 μm diameter (filterable fraction)	0.1 - 38	0.1 – 5	0.1 – 6.7
	Cu (mg/L)	Pb (mg/L)	Cr (mg/L)
Total	1.1 – 3.8	0.6 – 1.7	0.1 – 1.9
< 63 μm diameter	1.1 – 3.6	0.1 – 1.6	0.1 – 1.6
< 38 μm diameter	1.1 – 3.3	0.1 – 1.6	0.1 – 1.4
< 20 μm diameter	1.0 – 2.8	0.1 – 1.6	0.1 – 1.2
< 0.45 μm diameter (filterable fraction)	0.1 – 0.3	0.1 – 0.3	0.1 – 0.3
	Ni (mg/L)	Ca (mg/L)	Na (mg/L)
Total	0.1 – 3.3	31 – 230	22 – 1200
< 63 μm diameter	0.1 – 3.1	31 – 220	3 – 1175
< 38 μm diameter	0.1 – 2.9	31 – 210	3 – 1155
< 20 μm diameter	0.1 – 2.2	31 – 210	3 – 1155
< 0.45 μm diameter (filterable fraction)	0.1 – 0.2	8 – 210	3 – 1051
	K (mg/L)	Mg (mg/L)	Mn (mg/L)
Total	8.4 – 54	4.5 – 50	0.4 – 7.7
< 63 μm diameter	7.9 – 50	4.5 – 48	0.3 – 7.3
< 38 μm diameter	7.9 – 49	4.5 – 42	0.3 – 6.7
< 20 μm diameter	7.7 – 47	4.2 – 31	0.3 – 5.0
< 0.45 μm diameter (filterable fraction)	5.9 – 41	1.8 – 12	0.1 – 1.2

This project investigated the use of media filters for their ability to remove solids and the associated heavy metals, and included the investigation of sorption media to remove a portion of the metals that pass through a 0.45-μm filter. It also investigated filters operated in both a downflow and an upflow mode, because it was felt that upflow operation had the potential to delay filter clogging.

Methodology

Actual stormwater runoff was used for many of the metals-removal laboratory-scale tests conducted in this study. The runoff source was a UA campus parking lot in Tuscaloosa. A series of runoff samples was collected from this parking lot in order to characterize the runoff's quality. Iron, zinc and copper were the most prevalent metals detected in the collected runoff along with small amounts of particulate bound lead. Prior to using this parking lot for collecting runoff for the tests, a series of samples was collected from this lot in order to characterize the runoff's quality (Tables 2 and 3).

Table 2. Results of 'First Flush' Samples from Parking Lot at the Univ. of Alabama.

Date		10/5/2001	11/20/2001	11/24/2001	11/27/2001	6/22/2002	6/27/2002
Cu (mg/L)	total	0.036	0.051	0.018	0.030	0.181	0.075
	dissolved	0.017	0.030	0.012	0.030	0.038	0.013
Cr (mg/L)	total	0.015	< 0.010	< 0.010	*0.013*	0.016	0.011
	dissolved	< 0.003	< 0.010	< 0.010	< 0.010	0.002	< 0.002
Pb (mg/L)	total	0.024	0.014	*0.004*	*0.009*	*0.038*	0.037
	dissolved	< 0.004	< 0.003	< 0.003	< 0.003	< 0.008	< 0.008
Zn (mg/L)	total	0.203	0.136	0.034	0.079	0.256	0.197
	dissolved	*0.009*	0.051	*0.001*	0.006	0.028	< 0.003
Cd (mg/L)	total	< 0.001	< 0.001	< 0.001	< 0.001	< 0.0004	< 0.002
	dissolved	< 0.001	< 0.001	< 0.001	< 0.001	*0.001*	< 0.002
Fe (mg/L)	total	5.872	4.469	1.373	2.823	5.976	6.924
	dissolved	0.020	0.033	0.016	0.035	0.027	0.013
Na (mg/L)	total	0.855	2.996	0.755	1.618	2.651	0.337
	dissolved	0.878	2.929	0.828	2.396	0.906	0.241
Mg (mg/L)	total	8.618	8.673	1.970	3.309	24.934	8.665
	dissolved	2.048	5.422	1.130	1.203	5.680	0.501
Ca (mg/L)	total	35.760	46.060	10.189	15.702	139.323	20.949
	dissolved	18.959	38.692	8.956	10.054	41.593	4.454
K mg/l	total	2.784	7.224	1.450	2.653	10.119	2.194
	dissolved	1.567	5.881	1.227	2.004	3.203	0.684
pH		7.42			7.84	7.65	7.85
Color						285.00	55.00
TSS mg/l		271.33	260.00	33.00	82.00	104.00	348.67
TDS mg/l		106.70	309.33	59.80	75.40	396.55	34.10

Table 2 has the results of the first flush samples collected from the parking lot while Table 3 contains the results for composite samples collected of the runoff. Iron, zinc and copper were the most prevalent metals detected in the collected runoff along with small amounts of particulate bound lead. These results showed that the metals' concentration in the runoff was similar to that reported in the NURP and MS4 (NSQD) studies, and therefore, it would be acceptable for these filtration studies.

Twelve media were initially evaluated through batch equilibrium and kinetic tests compare the rate and extent of metals capture. The three best performing media: peat-sa mix, compost, and zeolite were then selected for an in-depth study column study usi parallel upflow columns in packed media beds. Metal removal efficiency was examined different rates of flow and influent conditions. TCLP tests were performed on the spe media to investigate disposal options. A summary of methods employed follows.

Table 3. Results of Runoff Characterization Tests on Composite Samples from Parking Lot at the University of Alabama.

Date	10/5/2001	11/20/2001	11/24/2001		11/27/2001	6/22/2002	6/27/2002
Cu (mg/L)	total	0.059	0.034	0.027	0.046	0.072	0.064
	dissolved	0.041	0.022	0.021	0.011	0.031	0.015
Cr (mg/L)	total	*0.005*	< 0.010	< 0.010	< 0.010	0.007	*0.005*
	dissolved	< 0.003	< 0.010	< 0.010	< 0.010	*0.001*	< 0.002
Pb (mg/L)	total	*0.009*	*0.007*	< 0.003	*0.004*	*0.014*	0.027
	dissolved	< 0.004	< 0.003	< 0.003	< 0.003	< 0.008	< 0.008
Zn (mg/L)	total	0.089	0.080	0.026	0.045	0.116	0.137
	dissolved	0.019	0.039	0.009	*0.001*	0.022	*0.003*
Cd (mg/L)	total	< 0.001	< 0.001	< 0.001	< 0.001	< 0.0004	< 0.002
	dissolved	< 0.001	< 0.001	< 0.001	< 0.001	< 0.0004	< 0.002
Fe (mg/L)	total	1.898	1.967	0.641	1.274	2.755	4.006
	dissolved	0.025	0.023	0.019	0.058	0.024	0.012
Na (mg/L)	total	0.545	1.817	0.605	2.447	1.230	0.367
	dissolved	0.532	1.806	0.620	1.517	0.862	0.286
Mg (mg/L)	total	4.049	5.662	1.395	2.181	10.946	4.992
	dissolved	1.617	4.359	1.001	1.182	5.333	0.831
Ca (mg/L)	total	20.089	34.712	8.705	12.447	61.485	16.197
	dissolved	13.781	31.146	7.690	10.557	39.950	7.302
K (mg/L)	total	2.461	4.623	1.508	2.264	4.166	1.837
	dissolved	2.032	4.187	1.370	1.820	2.677	1.025
pH	7.33				7.46	7.71	7.92
Color						260.00	45.00
TSS mg/l	70.00	91.00	11.00		38.00	103.33	204.00
TDS mg/l	74.80	216.20	46.80		65.00	348.70	41.80

Note: Italicized numbers were above ICP-OES LOD (limits of detection) but below LOQ (limits of quantification). Concentrations below LOD are reported as less than the LOD value for the ICP run in which that sample was tested.

Laboratory Studies of Equilibrium and Kinetics

Equilibrium adsorption isotherm studies were performed to determine the relationship between metal sorption/desorption from the media and equilibrium metal ion concentration in solution. Fifty milliliters of solution was placed with a known mass of media in 60 mL plastic powder jars then sealed with parafilm and screw top lids. These jars were shaken for 4 hr at 250 rpm on an Innova 2100 Platform Shaker from NewBrunswick Scientific. Kinetic testing determined that 4 hrs was sufficient for equilibrium to be reached. Then 5-10 mL samples of solute were filtered through 0.45 μm metricel membrane filters from Gelman Sciences, preserved with 0.1-0.2 mL of ultra pure nitric acid to create a 2% nitric acid solution, and refrigerated until metal analysis. This procedure was used to examine the equilibrium partitioning of metals between the solution and media, the 'ultimate' uptake capacity of the media for zinc (ultimate given the listed conditions), the uptake ability of the different media for low concentrations of metals from actual stormwater, the leaching of metals from the media, and the sorption/desorption of the major ions Ca, Mg, K, and Na from the media during metal uptake. The kinetics tests were performed using baffled batch adsorbers constructed according to standardized tank design for Continuously Stirred Tank Reactors (CSTRs). All components of the batch system design were constructed from perspex and PVC to minimize metal sorption to the fixtures. A known mass of media was added to a stirred tank with a known metals concentration. At increasing time intervals over a four-hour period, aliquots of sample were withdrawn. The kinetic test were conducted to compare the rate of metals uptake between the media. All metals analysis was performed using a Perkin Elmer Inductively Coupled Plasma – Optical Emission Spectrophotometer (ICP-OES) DV3000.

Laboratory-scale filter columns were then constructed for fixed-bed column studies so as to evaluate the dynamic characteristics of the metal capture process. Stormwater runoff was treated using parallel lab-scale upflow filter columns to investigate parameters affecting the breakthrough curve and performance of the column. Stormwater was used in a parametric study to investigate the effect of variables such as media and flowrate on the breakthrough curve. A sump area was provided below the media to allow space for particulates to settle. The depth of media in the columns was approximately 30 cm. The runoff was collected from the parking lot prior to testing. The stormwater entered at the side of the column below the media and above the sump and flowed upward through the media. The effluent exited the column at the top through a section of vinyl tubing. Samples of the effluent were collected at regular intervals. Samples of the influent were also collected periodically from the holding tanks throughout the column run to check for changes in the influent concentrations. The design is shown in Figure 1. This is also the design that was used to investigate the ability of upflow filters to remove solids by particle size.

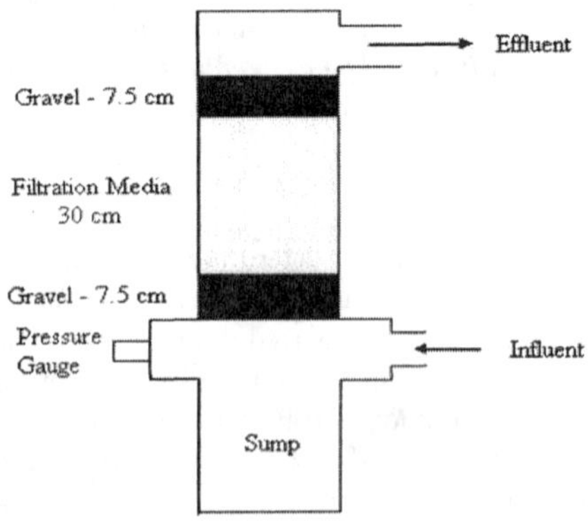

Figure 1. Upflow filter design.

Use of Upflow Filters to Reduce Solids Concentration in Runoff

For the upflow testing, where solids removal was the only parameter evaluated, simulated stormwater was used. The first series of tests used a sieved local soil which had particles whose nominal diameter was less than 200 μm. The second test solid was a mixture of 45% Sil-Co-Sil 106, 45% Sil-Co-Sil 250 and 10% fine sand. The solution concentration in all upflow tests was approximately 500 mg/L.

Four different filtration media were used in this phase of the research: fine (sandblast) grade sand, compost-sand (50/50 v/v) mixed media, peat moss-sand (50/50 v/v) mixed media and pool sand. The test water for the two series of tests is described above. The concentration of 500 mg/L was selected to represent the high end of typical, non-construction erosion, stormwater solids concentrations (500-1000 mg/L) and the size was selected to represent the range commonly found in urban runoff. Both the influent and effluent were collected and analyzed for turbidity, total solids and particle size distribution. The purpose of doing particle size distribution analysis was to determine the filtration efficiency for various particle sizes. The test methods used in this research were in accordance with the EPA and other standard test methods. The individual filter runs were conducted until breakthrough was achieved, i.e., the influent and effluent concentrations were the same, or until the filter clogged. The second series of tests ran the filters on an intermittent basis, similar to what would be seen in actual operation.

The Role of Biomass in Metals Uptake

Engineered filter systems that have been developed to treat metal-contaminated waters in field situations most frequently develop microbial populations (Pitt, personal communication). The effect of microbial growth within filter systems in stormwater metal retention was studied in laboratory column experiments using identical columns – one column microbially colonized and the other designed to inhibit microbial growth. The growth-inhibited column was constructed with sterilized sorbent matrix, isolated from light (to prevent growth of phototrophic microorganisms) and fed with UV treated stormwater. The microbially colonized column contained non-sterilized sorbent matrix, was exposed to light and fed with untreated stormwater. Microbial biomass (as protein) was quantified using the BCA protein determination procedure (Urrutia and Roden, submitted). Metal retention, head loss and pH in both columns were compared to determine the influence of biomass on the capture of metals.

Testing of the Pilot-Scale Device

Testing on the pilot-scale filters was performed in a manner expected to simulate the intermittent use that they would receive in a full-scale application. Eight filtering events were performed over a 4-week period, using settled stormwater runoff from two medium-density residential areas in Hoover, AL. Periodically during a sampling event, the effluent flow rate and depth of water on top of the filter were noted for each media. Sample collection and handling was in accordance with *Standard*

Methods for the Examination of Water and Wastewater (American Public Health Association [APHA], *et al.* 1995). The collected samples were analyzed for pH, conductivity, turbidity, color, hardness, toxicity, heavy metals and major cations (Pb, Cd, Zn, Cu, Cr, Fe, Ca and Mg), and solids (total, dissolved, suspended and volatile).

Prior to performing the pilot-scale testing, several potential endpoints for the filtering were noted: (1) physical clogging, (2) chemical breakthrough for several pollutants, or (3) end of the project. By testing these filters intermittently for several weeks, the effects of intermittent drying (by comparing the results with the laboratory experiments where the filters were run continuously until chemical breakthrough occurred) on filter performance could be determined.

The filtration columns used during the pilot-scale tests were constructed in large (55 gallon [0.21 m^3]) polyethylene tanks purchased from Aquatic Eco-Systems, Inc. in Apopka, Florida. The inner diameters of the tanks were 0.53 m and the surface area of the tank at any cross-section was 0.217 m^2. The media selected for study in the pilot-scale filters included those previously used in the laboratory experiments (sand, activated carbon, peat moss, compost, cotton textile waste, and agrofiber), as well as a loamy topsoil and a lightweight "sand." The two additional filter containers included a Jacuzzi filter set (sold by Aquatic Eco-Systems, Inc., Apopka, Florida) and the StormFilter cartridge test tank (for use with the compost) supplied by Stormwater Management, Inc., of Portland, Oregon. The filtration columns were constructed using the same guidelines as were used in the bench-scale tests (see Johnson [2003] for additional information). Gravel was placed in the bottom of the tank to a depth above the top of the spigot (approximately 5 cm). Above the gravel was a 0.15 m layer of sand (the same medium as used in the sand filter). A 0.3-m layer of the medium of choice (usually a mixed media, 50/50 v/v with sand) was placed above the sand. The filters were then rinsed thoroughly with tap water, and the initial flow rates through the media were measured. The filters then were mounted on a trailer for easy transport to the sampling locations. Nonwoven, synthetic fabrics were also placed on the surface of the filters to improve run times and make cleaning of the filters easier. The porosity of the fabric was such that significant interception of particles in the pre-settled runoff by the filter was not anticipated. The primary reason for using the filter fabric was that, during the design of the filter chamber of the MCTT (Pitt, *et al.* 1999), bypassing of large amounts of filter area was noted.

Test water to the pilot-scale filters was supplied through two submersible pumps. The pumps were attached to a distribution manifold. The purpose of the manifold was twofold: (1) to ensure that the water entering any specific filter plus the sample collected as the influent was a random subsample of the water received from the submersible pump, and (2) the flow to any filter could be regulated through the use of gate valves to maintain a flow rate no greater than that dictated as the optimum contact time (approximately 10 to 15 minutes for most pollutants for the media tested).

Results and Discussion

Laboratory-Scale Batch Studies

Results of the batch equilibrium and kinetic studies emphasized the importance of characterizing the stormwater before selecting a treatment media since the type and quantity of metals, pH, and other runoff characteristics can vary a great deal between sites. For example, determining the range of metal concentrations to be treated is crucial to selecting the best media, since the removal efficiencies of the media relative to each other changed with varying metal concentration. Media that were effective at high metals concentrations were outperformed by some media at the low metals concentrations typically found in stormwater. In addition, some media that offer potential as sorbents of metals may create water quality problems of their own. In equilibrium tests for all media, correlations were evident between the metals sorbed and the Ca, Mg, K, and Na ions desorbed. The ions desorbing and their quantity varied depending on the media and exchange metals present. The three best performing media (peat-sand mix, compost, and zeolite) were then selected for in-depth study in filters.

Maximum zinc sorption capacity of media – comparison of equilibrium isotherm curves

To examine the metal partitioning relationship between the media and the solution and to compare the maximum uptake capacity of the media under similar conditions, batch equilibrium capacity tests were performed using zinc as the target metal because it is one of the most common metals found in runoff. Also, while most metals in runoff are particulate bound, up to 50% of the Zn in runoff is often in the dissolved phase. Metals in the dissolved phase are more available to aquatic life and therefore present more potential for harm. Sorption is therefore likely to be the technique of choice for Zn removal from runoff.

The results of the equilibrium capacity tests were used to generate equilibrium uptake capacity graphs plotting metal concentration at equilibrium versus mass of metal removed from solution per mass of media. An example of these graphs is shown in Figure 2. Each test was run in triplicate and the resulting error bars are included on the graph. From these graphs, an estimate of the ultimate uptake capacity of the media under the test conditions was made. Table 4 lists the approximate maximum zinc uptake capacity for the given conditions for each of the media.

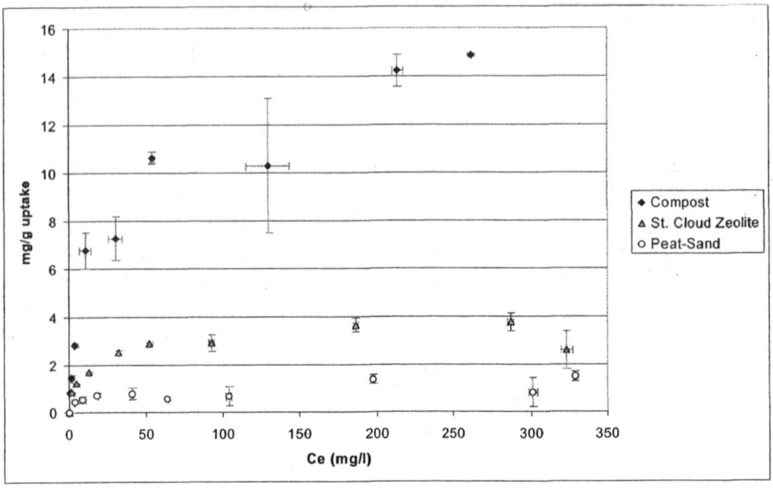

Figure 2. Batch equilibrium for zinc onto compost, St. Cloud zeolite, and peat-sand.

An additional batch equilibrium test was performed using zinc at low concentrations. The objective was to examine whether the media with the best metal sorption at high concentrations also removed metals better than the other media at low metal concentrations. For this test the TP 207 resin, compost and St. Cloud zeolite were used since these three media had very different 'ultimate' uptake capacities for zinc. The results of the low concentration isotherm test (Figure 3) give a very different view of the media than the initial batch equilibrium studies.

Table 4. Approximate Uptake Capacity of Media for Zinc.

Media	Maximum Uptake of Zinc (mg/g)
TP 207*	90
Compost	15
Kudzu	8.5
Peanut Hulls	3.5
St. Cloud Zeolite	3.5
Ammonia Zeolite	3.5
Activated Carbon	3.0
Agrofiber	2.0
Iron-Oxide Coated Sand	2.0
Peat-Sand Mix	1.25
Cotton Waste	< 1
Sand	< 1

*Metals concentrations necessary to achieve saturation on the resin exceeded the range of the current study but a study by Parrish (1998) using an identical procedure found a maximum uptake of 90 mg/g.

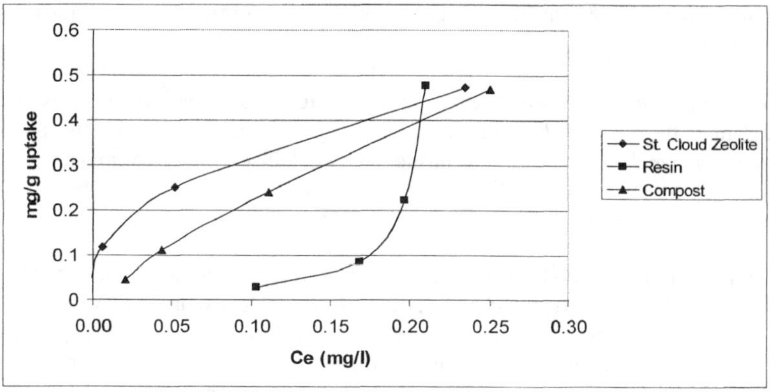

Figure 3. Batch equilibrium for zinc onto media at low initial concentrations.

Unlike the high concentration batch tests, this test did not include replicates, so the reproducibility of the data is uncertain. However, the data indicated that metal concentration could have a significant impact on which media was the best at metal removal. Therefore, a second set of batch equilibrium tests (with replicate samples) using actual runoff with low concentrations of metals were conducted to examine leaching and uptake of metals at these low metal concentrations. In this new batch study, the ability of the media to remove metals from spiked runoff varied depending on the metal being examined and its concentration. The peat-sand mix, the St. Cloud zeolite, compost, peanut hulls, and iron-oxide coated sand showed the best overall performance. The results confirmed that the media that had the best metal uptake at high metal concentrations were not necessarily the media that had the best metal uptake at low metal concentrations, which are more typical of stormwater. Therefore, while understanding the ultimate capacity of the media is important, for applications to stormwater filters, it is more important to examine how the media will perform at the low metal concentrations actually found in stormwater.

Finally, the kinetic test examining rate of uptake of metals by the media was performed. For the kinetic tests, 2 L of a solution containing 0.25 mg/L of each of the six metals of interest in de-ionized water was mixed with 10 g of media for four hours and samples were withdrawn at intervals. St. Cloud zeolite, peanut hulls, cotton mill waste, and peat-sand were found to have the fastest rates of uptake.

Based on the kinetic test data, the order of preference for removal on a mass basis was Pb=Cr > Cu > Cd > Fe > Zn for the peat-sand, Pb=Cr > Cu=Fe > Zn=Cd for the St. Cloud zeolite, and Pb=Cd>Zn>Fe=Cr>Cu for the compost. The order of preference for removal based equilibrium conditions on the low concentration metal uptake tests, was Cd=Pb > Zn=Cu > Cr > Fe for peat-sand, Zn=Cd > Pb > Cu > Cr > Fe for St. Cloud zeolite, and Cd > Zn > Pb > Cu > Cr > Fe for compost. The differences between the results from the kinetic and the low concentration uptake tests were probably due to the different solutions used for the tests. In extensive comparisons of the three media, the peat-sand mix performed best at removing dissolved metals.

Based on the results of the two batch equilibrium studies and the kinetic study the peat-sand mix, the St. Cloud zeolite, and the compost were found to be the most promising media and were selected for further study in lab-scale filter columns.

Laboratory-Scale Filter Studies

All three media removed total suspended solids well over all laboratory column runs, but only peat showed any removal of total dissolved solids. Compost increased the total dissolved solids concentrations. As expected, the behavior of the total suspended solids and particulate-bound metals paralleled each other. Copper and chromium typically had lower removal efficiencies than the other metals; one possible explanation is that these metals had a larger fraction of their particulate-bound metal concentrations associated with smaller particulates which passed more easily through the columns. Peat had the best removal efficiencies for particulate-bound metals. Removal efficiencies of compost and zeolite were approximately the same.

Peat-sand had the best dissolved metal removal efficiencies followed by compost and then zeolite. Peat-sand had the greatest headloss and the greatest change in headloss over the course of each run. Zeolite and compost beds offered considerably less headloss. The degree of change in headloss increased as the total suspended solids concentration of the influent increased for all media. The advantages and disadvantages of the three media are summarized in Table 5 below.

Table 5. Overview of Metals Removal in Laboratory-Column Studies

Media	Metal Uptake	Other
Peat-Sand mix	Best metal capture capability.	Most detrimental impact on pH, the greatest headloss, and showed the most potential for clogging
Compost	Second best metal capture capability.	Added color to the effluent. less impact on the pH of the effluent, less headloss, and exhibited less potential for clogging
Zeolite	Lowest metal capture capability	Less impact on the pH of the effluent, less headloss, and exhibited less potential for clogging

Overall, there was a significant improvement in dissolved metal removal for all three media when residence time was extended from 0.5 minutes to 5 minutes, but extending the residence time from 5 to 10 minutes only slightly increased overall dissolved metal removal. However, this conclusions is based on the total removal of all the dissolved metals, there was a great deal of variation depending on the metal, the media, and the series of runs being examined. Total suspended solids and particulate bound metal removal in the peat-sand and zeolite columns were not impacted by residence time. Total suspended solids and particulate bound metal removal in the compost column improved slightly as residence time increased from 0.5 to 3 minutes, but extending residence time past 3 minutes did not result in additional improvement in removal. Particulate metal removal in the media columns was slightly higher than in the control column even at low flow rates, but at flow rates less than 50 m/day most of the particulates and particulate bound metals were primarily being removed by settling. However, as the flow rate was increased to 150 m/day and above, straining of the particulates by the media began to play a larger role in removal.

TCLP analysis was performed on samples of the three filter column media (peat-sand mix, St. Cloud zeolite, and compost) that had been loaded with different concentrations of metals in order to establish the regulatory classification of used media with different degrees of metal saturation. Cadmium leached the easiest from all three media, which could be especially problematic since it has the lowest regulatory level. The St. Cloud zeolite had the worst metal retention capability of the three media. Overall, the peat-sand and compost retained metals far better than zeolite. At similar levels of media saturation (mg/g metal on media) the compost and peat-sand leached about 2 % of loaded lead and chromium and 5-20% of loaded cadmium, while the zeolite leached 20-25% of loaded lead, 10-20% of loaded chromium and 60% of loaded cadmium. In addition, given the media/solution metal partitioning seen in the equilibrium tests and the generally low dissolved concentrations of chromium and lead in runoff (median <0.010 mg/L) it is doubtful that lead concentrations on the peat-sand or chromium and lead concentrations on the compost would reach high enough levels to result in the spent media being hazardous waste based on leaching of these metals unless the media were being used to treat runoff with unusually high concentrations of chromium or lead.

Upflow vs. Downflow Filtration. Upflow columns proved more effective than downflow columns in the control of detention time and a reduction in clogging of the media by solids and associated head loss in the columns. At residence times of three to ten minutes, most of the suspended solids settled out in the sump area of the columns. Compost and zeolite columns showed little sign of increased headloss during these runs, even when influent suspended solids concentrations were around 400 mg/L.

Solids removal in upflow filters was shown to be effective at a laboratory-scale when flow rates were varied between 190 and 250 m/day (Pratap, 2004, submitted). Four media (compost-sand, peat-sand, sandblast-grade sand, and pool filter sand) were compared. The results for turbidity show that the mixed media filters, compost-sand and peat-sand, have the greatest percent removals (consistently > 60% until breakthrough begins) (Figure 4; error bars ± 1 standard deviation based on 3 runs for each media at the same flow rate). The error bars for the peat-sand are small, indicating good reproducibility between the independent runs. A statistical comparison of the filter media showed that the compost-sand and peat-sand had significantly greater removals of turbidity during the early and intermediate parts of the filter run. However, when near breakthrough, only the compost-sand could be statistically differentiated from the other media. This was due to the washout of fines from the compost media, resulting in significantly poorer performance. All four filter media were capable of achieving effluent turbidities in the range of 5 – 20 NTU during the early and intermediate sections of the filter run while operating at flow rates in the range of rapid filtration. A similar analysis of the results was performed for the monitoring parameters of total solids and particle size counts (1 – 175 μm particle size range). The media were indistinguishable from each other when evaluated in terms of total solids and particle volume removal. The filters had average percent removals of total solids between 50-80% until breakthrough occurred and all filters had effluent total solids concentration in the range of 70-160 mg/L during the early and intermediate parts of the filter run.

Similar results were seen for intermittent operation with the exception of an initially highly-turbid slug of water that exited the filter immediately upon restart. It is thought that this is likely due to particles in the water on top of the filter settling onto the filter surface and being flushed out quickly after restart (Figure 5). A quiescent time near the end of a filter's life was seen to improve performance slightly and extend the run longer than would be anticipated based on the shape of the breakthrough curve.

Heavy metals removal. Figure 6 depicts the ability of the filters to remove copper. Copper, lead, and zinc removals were not found to be statistically significant for any of the filter media, even though the media loading was small. It has been assumed in the modeling equations that adsorption is irreversible, assuming that the influent water characteristics, especially pH and conductivity, do not change significantly. However, at a certain low concentration, it would be expected that an equilibrium would be established where removal below that equilibrium concentration would not occur. Concentrations below the equilibrium concentration, whether due to low removal efficiencies or to low influent concentrations, would encourage the reversal of the driving force for the sorption reaction, resulting in desorption. The locations of this equilibrium for many of the media are indicated on the figures for these three metals. For copper and lead, the minimum effluent water concentration attainable appears to be approximately 5 to 10 μg/L, and approximately 10 to 20 μg/L for zinc, for all media.

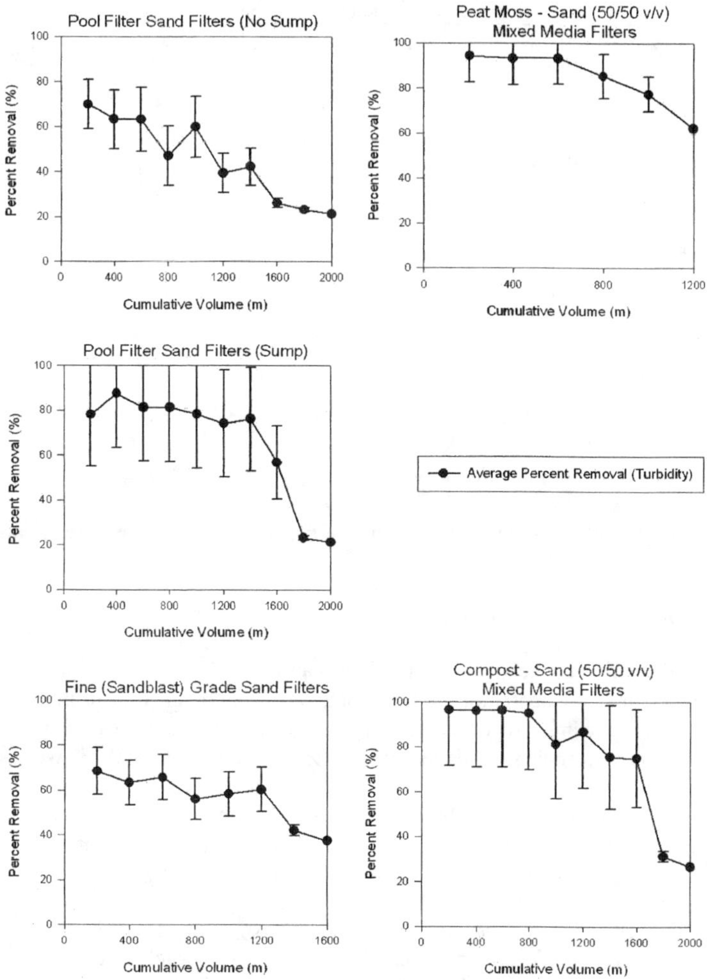

Figure 4. Average percent removal of turbidity versus cumulative loading on media.

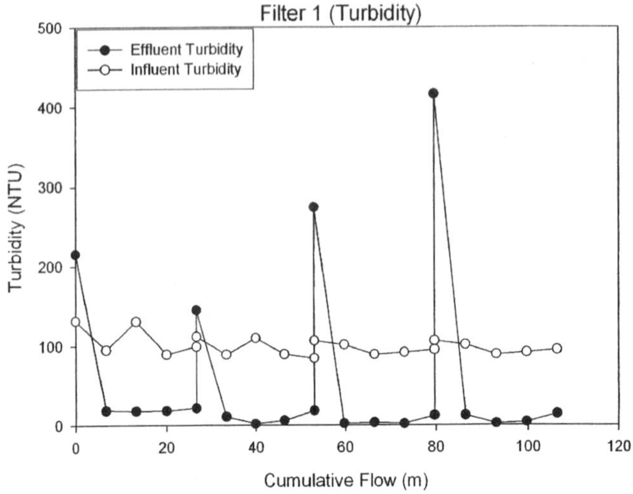

Figure 5. Turbidity vs. cumulative loading for a representative upflow sand filter.

Results of the Pilot-Scale Testing

The influent iron concentrations are significantly greater than the influent copper, lead and zinc concentrations. Therefore, adsorption is the primary removal mechanism occurring. It is overwhelming the desorption and makes it appear that the adsorption is irreversible. Statistically significant removals ($\alpha \leq 0.05$) were seen for the sand, cotton-sand, agrofiber-sand and carbon-sand filters. The peat-sand and compost-sand filters had a probability of 0.054 that the effluent iron concentrations were less than the influent concentrations – not statistically significant at $\alpha \leq 0.05$ but with more testing, the results would likely show statistically significant removals.

Major ion removal (calcium, magnesium, hardness). The removal ability for these ions is not shown. The loam, peat-sand, and cotton-sand filters were able to consistently remove calcium from the influent water. The behavior of the peat-sand filter was not unexpected based upon the literature review and the behavior of the laboratory filter column. The potential of the cotton to remove a small amount of calcium was also seen in the laboratory columns. Only the peat had the ability to almost completely remove the calcium from the influent water. The compost-sand filter consistently added calcium to the runoff water. This also is in agreement with the laboratory column results and with the ability of compost-sand filters to neutralize its influent when the influent pH is significantly different from 7.0 to 7.5.

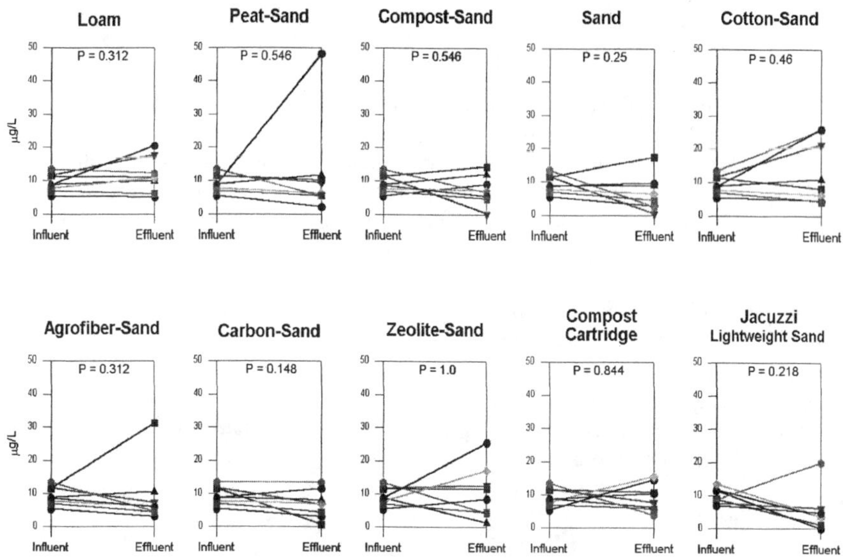

Figure 6. Copper removal during pilot-scale filtration experiments.

No filter had the ability to remove magnesium consistently. Hardness removal was only statistically significant with the peat-sand and loam filters, with only the peat-sand filter able to remove large percentages of hardness. This is not unexpected given the ability of the peat filter to remove calcium. Mimicking its behavior with calcium, the compost-sand filter contributed hardness to the runoff water.

pH. The peat-sand filter is the only filter that had a measurable effect on pH. Hydrogen as the hydronium ion, H_3O^+, is one of peat's easily exchangeable ions, and in general, as peat adsorbs other cations, it releases the hydronium ion. pH has been proposed as a method of monitoring remaining life in a peat-filter. The other filter media, when treating an influent at a neutral or near-neutral pH, tend not to affect the pH measurably. This is in contrast to the previous research in Clark (1996) where when the influent water to a compost-sand column was not near a neutral pH, compost-sand attempted to neutralize it.

In real-world applications for stormwater treatment, filters that contain adsorption media typically clog before the media can experience chemical break-though. It is not yet clear if depth filtering media will be a cost-effective stormwater control, considering the pre-treatment needed to prevent this clogging. The necessary pretreatment alone may provide adequate control, without the additional filtration cost. Large-scale filtration installations (especially sand) have been shown to perform well for extended periods of time with minimal problems. The use of supplemental materials (such as organic compounds) should increase their performance for soluble compounds. The use of upflow filtration is also expected to increase the life of filters before clogging, for some media types (not for peat/sand combinations).

The confirmation of the modeling equations for a few pollutants for some media indicate that this modeling approach has the potential to provide an estimate of the life of the filter, i.e., the time until media replacement is needed, for applications where the influent concentration is not quite as low as it was in the two detention ponds used in this pilot-scale study. The best application for these filters may then be to further treat the effluent from a detention pond in critical source areas, such as scrap metal recyclers, rather than as a polisher for effluent from a detention pond in a residential area. During these field pilot-scale tests and related full-scale tests, the minimum attainable concentration for these metals using media filtration appears to be approximately 5 to 10 μg/L for copper and lead, and approximately 10 to 20 μg/L for zinc, irrespective of media. These tests do confirm the utility of different filtration arrangements, especially concerning the problems associated with clogging and reduced filtration performance.

Role of Biomass in Metals Removal

The effect of microbial growth within filter systems on stormwater metal retention was studied in laboratory column experiments. The filter media used in the study was sand. Because the ability of unmodified sand to remove dissolved metals was more limited than many of the other media examined in this study, it was a good media to use to examine the impact of microbial growth on dissolved metal uptake. Microbial growth in filters did appear to increase metal uptake and retention. The mass of dissolved metal sorbed per gram of media was higher in the microbially colonized columns than in the growth-inhibited columns for all the metals examined except chromium. Also, statistically significant differences were found between dissolved metal removal efficiency (at equivalent levels of media saturation) in the microbially colonized and growth-inhibited columns for all of the metals examined except chromium. Microbially colonized columns also retained more particulate metals than the growth-inhibited columns, possibly due to decreased void space caused by microbial growth. Leaching experiments performed with the columns found that at equivalent effluent pHs the microbially colonized columns leached a smaller percentage of their initial metal content than the growth-inhibited columns. Comparison of the metal content and the protein content (representing microbial biomass) in media from different column sections showed that there was a fairly linear relationship between increasing metal content and increasing protein content. Based on the metal contents determined from the media digestions the microbially colonized columns retained 11-19% more Cu, Cr, Pb, and Zn and 39% more Cd than the growth inhibited columns. For the lab-scale sand media filters examined in this study, the difference between the two types of columns when comparing percent removal at the same meq/g or mg/g media saturation was statistically significant - with the microbial columns having the higher removal efficiencies - but small. However, the situation examined was one where microbial growth was limited. The sand media provided no nutritional source to encourage microbial growth. In media that would provide a better environment for microbial growth, such as compost, the impact could be more significant.

Related tests were conducted to measure the disassociation potential of heavy metals and nutrients under aerobic and anaerobic conditions having extreme Eh values. Studies on the effect of anaerobiosis on metal retention by filter systems indicated that heavy metals were not mobilized from filter systems under anaerobic conditions. It was found that metal retention within the filters, with the exception of iron on compost, was not different from what was observed in oxygenated environments (Figures 7, 8, and 9 for copper, iron and zinc, respectively). However, it is plausible that under certain specific environmental conditions, co-precipitation of metals by iron- and sulfate- reducing bacteria may take place in stormwater treatment systems.

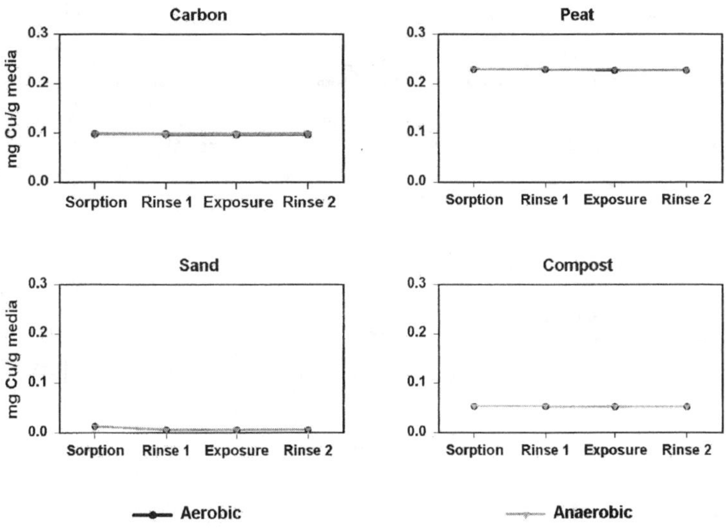

Figure 7. Behavior of copper under aerobic and anaerobic conditions.

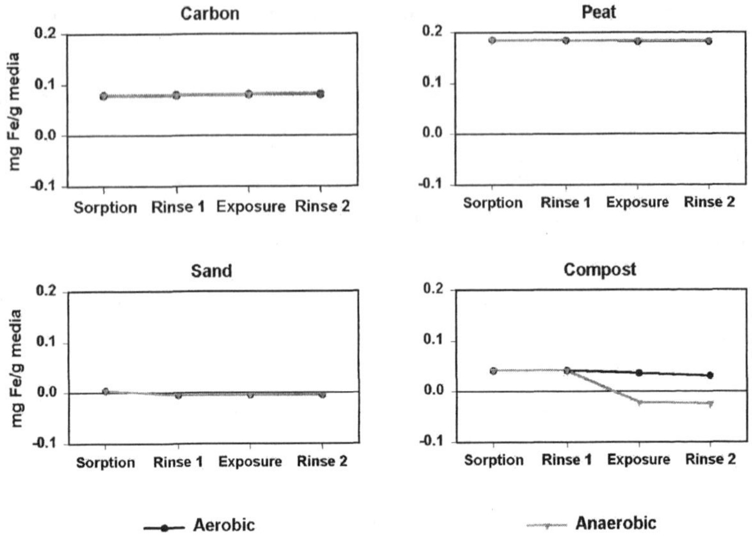

Figure 8. Behavior of iron under aerobic and anaerobic conditions.

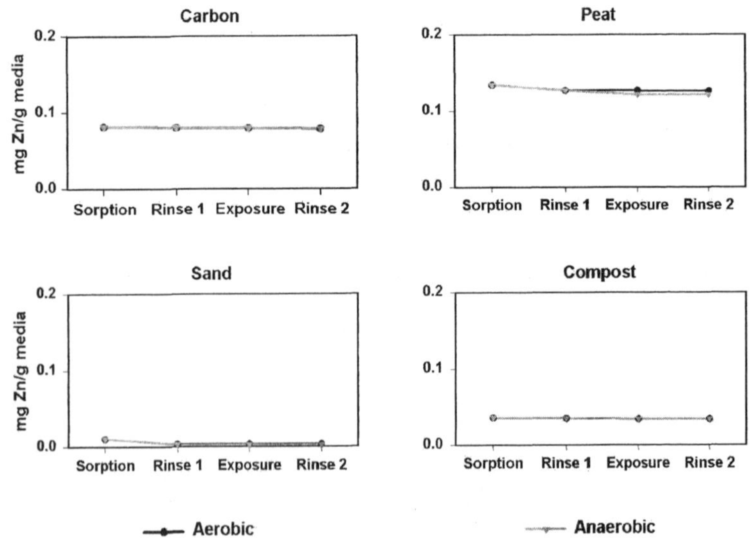

Figure 9. Behavior of zinc under aerobic and anaerobic conditions.

Treatability Testing

Treatability tests were performed to assess the effectiveness of different treatment trains and processes quantifying improvements in stormwater toxicity and metals capture. The treatability tests included intensive analyses of samples from twelve sampling locations in the Birmingham, AL, area that all had elevated toxicant concentrations, compared to the other urban source areas initially examined. The treatability tests conducted were: settling column, floatation, screening and filtering, photo-degradation, aeration, combined photo-degradation and aeration and an undisturbed control sample. More than 900 toxicity tests were performed using the Microtox™ procedure. Turbidity was also analyzed on all samples. Results indicated a reduction in toxicity as the level of treatment increased and in particular, as the solids were increasingly removed from the original stormwater (Figure 10). All samples, with one exception, showed dramatic reductions in toxicity with increasing settling times.

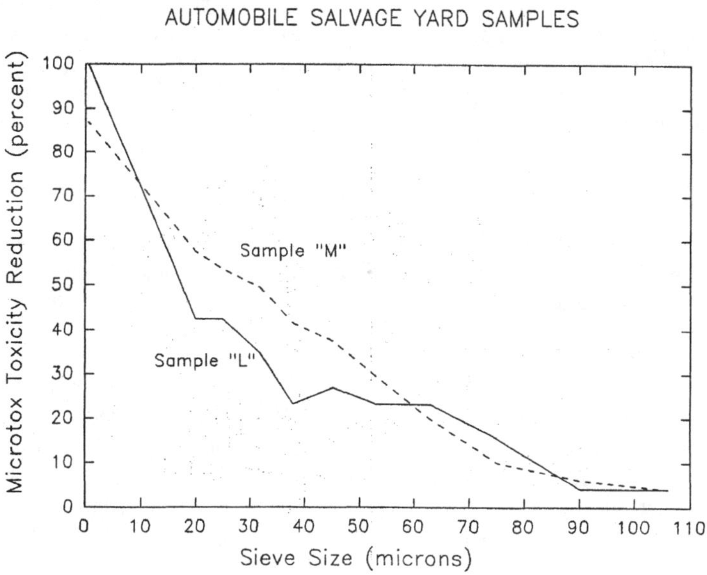

Figure 10. Toxicity reduction from sieve treatment - automobile salvage yards (Pitt, et al. 1999).

Metals Associations

Metal-particulate association tests on the parking lot runoff revealed that more than 90% of the filterable forms of Ca, Mg, K, Fe, and Zn were in ionic forms, with very little colloidal, or other bound forms. Also, more than 80% of the filterable Cr and Pb were also ionic, while only about 50% of the filterable Cu and 30% of the filterable Cd were ionic. This data can be used to estimate the level of control that may be associated with different designs of particle trapping devices. Some pollutants can be significantly reduced by a reduction in particulates, such as suspended solids, total phosphorus and most heavy metals. Other pollutants, such as nitrates, are reduced much less, even after filtration down to 0.45 μm.

Experiments were also conducted to examine the likelihood of the metals disassociating from the particulates under pH conditions ranging from about 4 to 11 with both weak and strong acids. These tests indicated that the heavy metals of concern remain strongly bound to the particulates during long exposures at the extreme pH conditions likely to occur in receiving water sediments and in control device sumps and stormwater pond sediments. Additional testing is ongoing at The University of Alabama.

Related tests were conducted as part of the filter media evaluation task of this research to measure the disassociation potential of heavy metals and nutrients under aerobic and anaerobic conditions having extreme Eh values. Studies on the effect of anaerobiosis on metal retention by filter systems indicated that heavy metals were not mobilized from filter systems under anaerobic conditions. It was found that metal retention within the filters was not different from what was observed in oxygenated environments. However, it is plausible that under certain specific environmental conditions, co-precipitation of metals by iron- and sulfate- reducing bacteria may take place in stormwater treatment systems.

Conclusions

The results of this investigation emphasize the importance of characterizing the stormwater before selecting a treatment media since the type and quantity of metals, pH, and other runoff characteristics can vary a great deal between sites. For example, determining the range of metal concentrations to be treated is crucial to selecting the best media, since the removal efficiencies of the media relative to each other changed with varying metal concentration. Media that were effective at high metals concentrations were outperformed by some media at the low metals concentrations typically found in stormwater. Upflow columns proved more effective than downflow columns in the control of detention time and a reduction in clogging of the media by solids and associated head loss in the column. Studies on the effect of anaerobiosis on metal retention by filter systems indicated that heavy metals were not mobilized from filter systems under anaerobic conditions. It was found that metal retention within the filters was not different from what was observed in oxygenated environments. However, it is plausible that under certain specific environmental conditions, co-precipitation of metals by iron- and sulfate- reducing bacteria may take place in stormwater treatment systems. Tests also indicate that the heavy metals of concern remain strongly bound to the particulates during long exposures at the extreme pH conditions likely to occur in receiving water sediments. They will also likely remain strongly bound to the particulates in stormwater control device sumps or detention pond sediments where particulate-bound metals are captured. The pilot-scale testing showed that there is a lower limit, however, to the effluent quality. Removals to concentrations below 10 – 20 μg/L for most metals are not feasible.

Acknowledgements

We wish to extend our sincerest gratitude to WERF for funding this project (Project No. 97-IRM-2). This project also could not have been completed without the assistance of many graduate students at UA, at the University of Alabama at Birmingham, and at Penn State Harrisburg.

References

American Public Health Association. (1995). *Standard Methods for the Examination of Water and Wastewater, 18th edition*. American Public Health Association, American Water Works Association, Water Environment Federation.

Bannerman, R.T.; Owens, D. W.; Dodds, R. B.; Hornewer. N. J. (1993) Sources of pollutants in Wisconsin stormwater. *Water Sci. & Technol.*, 28 (3-5), 241-259.

Clark, S.E. (1996). *Evaluation of Filtration Media for Stormwater Runoff Treatment.* M.S.C.E. Thesis, University of Alabama at Birmingham, Birmingham, Alabama, 442 pages. 1996.

Claytor, R.; Schueler, T. (1996) *Design of stormwater filtering systems (DRAFT manual)*. U.S. Environmental Protection Agency Region V, Chesapeake Bay Research Consortium, Center for Watershed Protection, Endicott, MD.

Collins, M.R., T.T. Eighmy, J.M. Fenstermacher, Jr. and S.K. Spanos (1992). Removal of natural organic matter by conventional slow sand filtration. *J. Am. Water Works Assoc.* 84 (5): 80-90.

Crites, R.W. "Micropollutant removal in rapid infiltration." In *Artificial Recharge of Groundwater*, ed. Takashi Asano, Boston: Butterworth Publishers, 1985, 579-608.

Johnson, P., R. Pitt, S.R. Durrans, M. Urrutia, and S. Clark. (2003). *Innovative Metals Removal Technologies for Urban Stormwater*. Report No. WERF97-IRM-2. Water Environment Research Foundation, Alexandria, VA. 200 pages.

Pitt, R., B. Robertson, P. Barron, A. Ayyoubi, and S. Clark March 1999. *Stormwater Treatment at Critical Areas: The Multi-Chambered Treatment Train (MCTT).* U.S. Environmental Protection Agency, Wet Weather Flow Management Program, National Risk Management Research Laboratory. EPA/600/R-99/017. Cincinnati, Ohio. 505 pgs.

Pitt, R.E.; Field, R.; Lalor, M.; Brown, M. (1995) Urban Stormwater Toxic Pollutants: Assessment, Sources, and Treatability. *Water Environ. Res.,* 67 (3), 260-275.

Pratap, M.R.; S.E. Clark; R. Pitt. (2004) Upflow Filters For Stormwater Treatment at Critical Source Areas. II: Media Comparison. Submitted to *Water Environment Research*, August 2004.

The Caltran Study

Roger James, P.E. Senior consultant, Water Resources Management, 63 Ivy Drive, Orinda, CA 94563

Introduction

The California Department of Transportation (Caltrans) as part of the Best Management Practice (BMP) Retrofit Pilot Program monitored the performance of two structural end-of pipe BMPs. Each had a 31.2 l/s (1.1 cfs) capacity and was installed to treat runoff at the Interstate freeway-210 Orcas Avenue 0.44 ha (1.09 acres) and Filmore Street 1.0 ha (2.52 acres) sites located in the City of Lakeview, Los Angeles County. The expected flows from these sites are 7.1 l/s (0.25 cfs) (1-year event) and 89.8 l/s (3.17 cfs) (25-year event) and 17 l/s (0.60 cfs) (1-year event) and 210.1 l/s (7.42 cfs) (25-year event) respectively.

Flow and rainfall data for the 2000-01 and 2001-02 seasons were collected at each site. Gross pollutant and trace element data was collected during 4 storm events during 2000-01 and 6 events during 2001-02 at the Orcas Avenue site and 7 events during 2000-01 and 10 events during 2001-02 at the Filmore Street. Material trapped in the sumps of the BMP was characterized only during the 2001-02 season.

The storm water monitoring program was implemented pursuant to the BMP Operation, Maintenance, and Monitoring Plan (OMMP, Caltrans 1999) to evaluate the performance of the pilot BMPs to capture and retain floatables, trash and debris greater than 6.35 mm (0.25 inch) and fine sand and larger particles of sediment. Pollutant removal efficiency analysis using a Scoping Study Methodology (Scoping Method) derived from a Federal Highway Administration 1989 study (Woodward Clyde 1989) and the Mass Balance Approach (MBA) developed by CDS™ Technologies Inc.(2000 and 2001) produced significantly different results.

BMP Efficiency Evaluation

Researchers have used a number of different methods to evaluate the performance of storm water BMPs that primarily rely on reductions in pollutant concentrations or loads. ASCE and USEPA (April 2002) describe ten different methods for BMP water quality monitoring data analysis. Caltrans used the Scoping Method and a modification of the MBA to determine the efficiency of the BMPs. The significantly different results produced by these two methods is attributed to the sensitivity of the Scoping Method when small data sets are available for analysis, differences in the Scoping Method and MBA, incorrect application of the MBA analysis and the limitations of automatic samplers to collect representative samples of storm water runoff.

Although a number of parameters were monitored during the Caltrans studies Total Suspended Solids (TSS) is used in the following analysis of the differences in the results between the two methods.

Caltrans Scoping Study Methodology

BMP efficiencies were evaluated from a comparison of effluent and influent loadings (over a period of time) from the following relationship:

Efficiency (%) = [(Loading in – Loading out)/Loading in] x 100

The Scoping Method estimates loadings by:

- Determining influent and effluent Event Mean Concentrations (EMCs) from at least five representative storms
- Validating that the EMCs are log-normally distributed
- Computing the expected seasonal mean EMCs using log-normal distribution
- Measuring the wet season runoff volume or calculating that volume using runoff coefficients, watershed area and rainfall depth
- Calculating the expected wet season influent and effluent loads by multiplying the expected mean EMC by the measured seasonal runoff volumes

The Scoping Method was obtained from the Federal Highway Administration (FHWA) report that characterized storm water runoff pollutant loads from highways and predicted water quality impacts (Woodward-Clyde Consultants, May 1989). The FHWA report based on 993 separate storm events collected at 31 highway runoff sites distributed among 11 states notes that there is a lack of confidence in data when the number of samples is small and that a single sample can distort results. Caltrans acknowledged these limitations of the procedure when applied to small data sets and indicated that other techniques would be employed as needed to make effective use of data.

Mass Balance Analysis Methodology

The MBA was developed to overcome the inherent difficulties using automatic samplers to obtain representative samples of the wide range of solids suspended in storm water runoff including sediments larger than very fine sand and gross pollutants. The MBA was also developed to address the erroneous data that results from the TSS method of analysis of suspended sediments that has been identified by the United States Geological Survey (USGS 2000 and Gray 2000).

The MBA requires that material captured in the sump or storage area be removed, sampled and characterized for pollutants of concern. This information combined with the mass of pollutants discharged and bypassed from the unit and mass of pollutants contained in the sump and the separation chamber provides the most accurate method of determining influent loading (Figure 1). The efficiency is determined by:

$$\text{Efficiency} = \frac{\text{Sump Load} + \text{Separation Chamber Load}}{\text{Outlet Load} + \text{Sump Load} + \text{Separation Chamber Load} + \text{Bypass Load}}$$

Critical to the MBA are the accurate measurement of the runoff volumes over the period of evaluation, quantification of each of the loads and analysis of samples of each load with laboratory methods that use comparable temperatures.

Monitoring Results

2000-2001 Efficiencies

Caltrans (2001) reported TSS removal efficiencies of 29% for the Orcas Avenue unit and 3% for the Filmore Street unit as determined by the Scoping Method.

A MBA analysis was not possible during 2000-2001 because material captured in the CDS sump was not fully quantified or characterized as specified in the OMMP and MBA protocols.

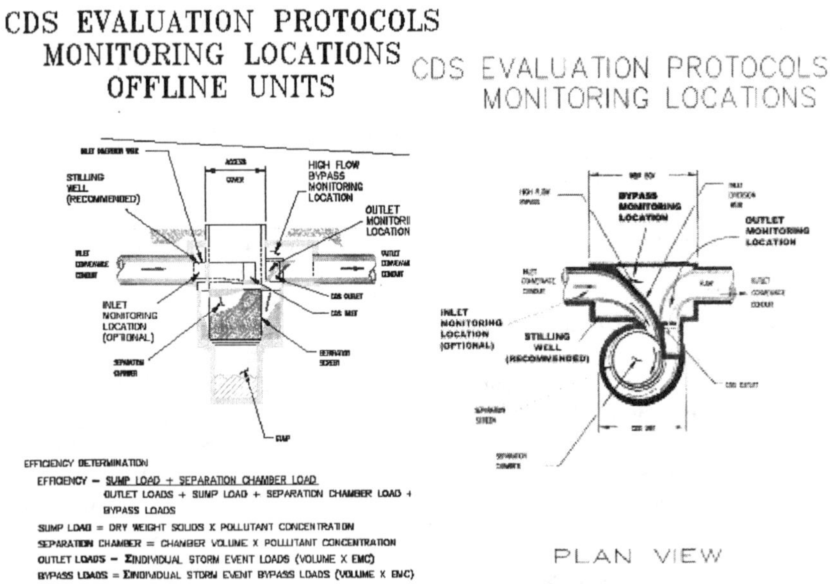

Figure 1. CDS Mass Balance Analysis Evaluation Protocols and Monitoring Locations

The following observations were made from an analysis of the gross pollutant data:

- The gross pollutants at both sites are dominated by vegetative material with vegetative material 76 to 91% of the gross pollutant volume.

- The units removed over 78% of the gross pollutants by volume and 76% by weight at the Orcas Avenue site; however, these results were skewed by 48 and 56% removals when the unit was bypassed in January 2001 during a major storm event (11 cms (4.33 inches) of rainfall) with intensity of 4.9 cm/hr (1.92 in/hr). During this event the unit lacked capacity due to inadequate maintenance and hydraulic capacity.

- The unit at the Filmore Street site removed over 93% of the gross pollutants by volume and 90% by weight.

2001-02 Efficiencies

Caltrans (2002) reported 2001-02 TSS removal efficiencies of 11% for the Orcas Avenue unit and negative 81% for the Filmore Street unit as determined by the Scoping Method. Capture of over 32.3 kg of dry solids at the Orcas Avenue site and over 63.7 kg at the Filmore Street site strongly suggest that the Scoping Method does not provide an accurate method of determining efficiency of the units to capture solids suspended in storm water runoff. It is highly unlikely that the CDS unit generated solids as indicated by the negative 81% efficiency determined by the Scoping Method for the Filmore Street site. This unlikely negative annual removal was highly influenced by the first storm event of the year, which reported a negative 142 % removal. This is impossible as the unit was cleaned prior to this storm event. With no material in the unit scour could not have occurred, thus the result would mean the unit created material. The results also confirm the FHWA's and Caltrans' concern in applying the Scoping Method when there are only a limited number of events monitored and corresponding data points.

Caltrans reported the MBA efficiencies for both of the sites as 33%; however, there were several significant inconsistencies in the methods used to calculate the MBA efficiencies:

- Outlet loads were calculated using the Scoping Method instead of available measured values for monitored events.

- The sump loads used in the MBA were based on incinerated samples even though the MBA calls for determination of all loads using comparable temperatures i.e., TSS is performed at 105°C while TVS (Inorganic Mass) are performed at 550±50°C that drives off organic matter, intrinsic water and volatile minerals that are measured in the TSS method of analysis.

Use of Scoping Method in Mass Balance Approach

Caltrans used the Scoping Method for determination of the outlet loads in the MBA analysis rather than using measured values during monitored events and estimated values for non-monitored events.

Pollutant loadings were determined by multiplying the annual lognormal mean EMCs determined from the monitored events and the total volume of runoff during the wet season. This approach assumes that the EMCs are the same for both monitored and non-monitored events. The approach also significantly overestimates the loadings for non-monitored events because many of those events had very small intensities where TSS values are expected to be less than the annual means derived from monitored events where storm events had higher intensities and volumes.

During the 2001-02 season at the Orcas Avenue site 64.4% of the total rainfall and 87.6% of the runoff were monitored and at the Filmore Street site 76.7% of the rainfall and 83.3% of the total runoff were monitored. This represents a significant capture or sampling of rainfall events and measured flow and provides a sound basis for using a combination of actual data for monitored events and estimated data for non-monitored events.

An analysis of the monitored loadings and calculated loadings for non-monitored and measured runoff revealed a significant discrepancy with the Scoping Method. The Filmore Street site in 2001-02 had monitored loadings of 20.59 kg of TSS and estimated loadings of 10.37 kg during non-monitored events. Total annual rainfall was 133.98 mm and 102.74 mm occurred during sampled events. Many of the non-monitored storm events were 3 mm (0.12 inches) or less that would likely produce little runoff and low runoff coefficients (C). Calculated "C" values for measured rainfall events ranged from 0.117 to 0.258 when rainfall ranged from 3.25 to 28.14 mm. A highly unlikely "C' of 0.413 was calculated for the non-monitored events where rainfall ranged from 0.25 to 5.76 mm. The EMC applied for the non-monitored events was 78.73 mg/l while the average TSS measured in 2001-02 was 50.7 mg/l. It seems improbable that the non-monitored event loads would be 50% of the monitored loads when the non-monitored events rainfall was only 23% of the total rainfall. Similar results were also observed at the Orcas site
Efficiency of TSS removal as calculated using different methods is summarized in Table1. A comparison of the efficiencies for both the Orcas Avenue and Filmore Street units determined by the Scoping Method using the Annual EMC for monitored and non-monitored events and using a combination of measured values for monitored events and the Annual EMC for non-monitored events produces significant differences in efficiency demonstrating the sensitivity of the Scoping Method when there is a small data set.

Sensitivity of Scoping Study Method to Small Data Sets

To demonstrate the sensitivity of the Scoping Method when limited data is available an analysis of the Filmore Street site is presented. 10 storm events were monitored in 2001-02 season at the Filmore Street CDS unit. Using the Scoping Study method, expected seasonal influent and effluent EMCs were calculated using a lognormal distribution. The expected removal efficiency at this site was calculated to be a negative 81%.

A closer look at the effluent EMC data indicates that effluent EMCs for two events seem to be extreme measurements when compared to the rest of the data. To demonstrate the sensitivity of the Scoping Study method on these extreme events for small data sets, three sets of calculations were performed on the EMC data for 2001-02 wet season.

First, the seasonal average calculated using the Scoping Method and ignoring an event that had an effluent concentration of 186 mg/L indicates that the efficiency increased to negative 42%. In other words, a single event had an almost 50% change in removal efficiency.

For the second set of calculations, an event that had an effluent concentration of 1 mg/L was discarded. The seasonal removal efficiency using the Scoping Method was found to be negative 33%. In this case, the seasonal removal efficiency increased by 60% when this event was discarded. This event had a very low effluent EMC and one would expect that deleting this event would lower the seasonal removal efficiency. To the contrary, there was about 60% increase in the efficiency when this event was discarded.

Lastly, the seasonal removal efficiency was calculated to be negative 6% when both events 1 and 7 were discarded. This amounts to approximately 93% change from the original calculated value.

These calculations clearly demonstrate that use of the Scoping Method in the project, where sample size is small, is not appropriate and provides misleading results. Removal efficiencies at the Filmore site for the 2001-02 season changed by approximately 93% when calculations were performed using the Scoping Method and simply deleting two monitoring events. One event deleted had an extremely high EMC (186 mg/L) compared to the rest of the data while the other event had a very small EMC (1 mg/L). This re-iterates limitations of this procedure when applied to small data sets as acknowledged by FHWA and Caltrans.

Temperature Correction

To correctly perform the MBA it is essential that all analysis of the monitored effluent and captured solids (sump load) be performed on a comparable basis. The MBA calls for determination of all loads using comparable temperatures i.e. TSS is performed at 103-105°C and captured solids (sump load) are measured at a comparable temperature. The TVS (Inorganic Mass) method of analysis should not be used on the captured solids (sump load) because it is performed at a temperature of 550±50°C that drives off organic matter, intrinsic water and volatile minerals that are measured by the TSS method of analysis.

Captured sediment data presented (Caltrans 2002) is a combination of both the dry mass where analysis were performed at 200°F (93.3°C) and inorganic mass where analysis were performed at a temperature >500°C. The post-burned samples lost from 26–49% of their mass during this incineration process confirming that the TVS method removes a significant portion of the solids captured and removed by the units. Caltrans in its MBA calculations used the post-burned load that significantly underestimates the efficiency of the units as the analysis compared inorganic captured material to inorganic and organic material in the effluent load. As shown in Table 1, using comparable analytical temperatures (TSS at 103-105°C and "dry mass" at 93.3°C) the calculated TSS removal efficiencies were much higher than those reported.

Table 1 – TSS Removal Efficiencies Using Different Methods

Method of Evaluation	Orcas Avenue	Filmore Street
Scoping Study	11%	-81%
Mass Balance – A	33%	33%
Mass Balance - B	65%	67%

Mass Balance - A – Outlet Loads determined using Scoping Method
Mass Balance - B - Outlet Loads determined by applying Annual EMC only to non-monitored events and applying individual EMCs to individual volumes and using Pre-Burned (dry mass) data

Discussion of Alternative Methods

The Scoping Method determines efficiency of storm water BMPs comparing influent and effluent pollutant loadings determined by EMCs applied to flows measured or estimated over a period of time. There are several fundamental deficiencies in the method when applied to BMPs designed to capture sediments larger than 63 micron (very fine sand) and when there are limited data sets. An analysis of data collected during the project to evaluate the units confirms the FHWA's and Caltrans' concern and clearly demonstrates that wide variations in BMP efficiencies are reported when limited data is available for analysis. T he Caltrans OMMP further relies on use of automatic samplers to collect samples of runoff and that have been found to narrowly define TSS when larger size sediments and gross pollutants are present. The limitations of automatic samplers to collect representative particle size distributions greater than 63-128 μm (fine sand) and the technical limitations of automatic samplers to meet site-specific conditions and runoff-quality characteristics have been identified by USGS in studies conducted with the cooperation of the FHWA. The report on these studies found that commercially available automatic samplers used in the Caltrans study are not designed for collection of bed load found in highway and urban runoff. The USGS recommended research is needed to develop and test adaptation of automatic pumping samplers to collect representative samples of bed load material for highway runoff studies (Brent 2000).

The MBA approach overcomes the Scoping Method and OMMP shortcomings by measuring material captured within the unit and sampling the effluent or discharge from the unit after larger sediment particles are removed. The MBA has an additional advantage of capturing trash and debris allowing the characterization of gross pollutants that are not sampled by traditional sampling techniques. The MBA also allows the full characterization of pollutant load reductions to document compliance with Total Maximum Daily Loads (TMDLs).

Summary and Conclusions

The Scoping Study Method and Mass Balance Approach produce significantly different results that are attributed to:

- The results from a single event can significantly distort the analytical results when the Scoping Method is applied to a small data set.

- Application of mean EMCs to the entire annual runoff to determine influent and effluent loads rather than using actual measured values for monitored events.

- Limitations of automatic samplers to collect representative samples of solids in storm water runoff.

The true efficiency of the CDS units to capture solids suspended in storm water runoff is likely to be closer to the methods used to determine Mass Balance - B than the Scoping Method (Table 1).

New techniques and protocols are needed to obtain representative samples of all solids found in storm water runoff. Improved sample management and analytical methods are needed to better characterize the wide range of solids found in storm water runoff. These improved methods are needed to better characterize all solids in storm water runoff, in the design and selection of sediment capture BMPs and in the development and determining compliance with TMDLs where solids and pollutants associated with those solids are of concern.

References

Brent, Gardner C., Gray, John R., Smith, Kirk P. and Glysson, Douglas G. (2000), *A Synopsis of Technical Issues for Monitoring Sediment in Highway and Urban Runoff*, Open-File Report 00-497, U.S. Department of the Interior, U.S. Geological Survey, prepared in cooperation with the Federal Highway Administration

Caltrans BMP Retrofit Pilot Program (1999), District 7, BMP Operation, Maintenance, and Monitoring Plan, Continuous Deflective Separation Units, Volume II, Appendix I-H

CDS Technologies, Inc. (2000) Evaluation Protocols for Continuous Deflective Separation Performance

CDS Technologies, Inc. (2001) Evaluation Protocols for Continuous Deflective Separation Performance Addendum # 1

Caltrans BMP Retrofit Pilot Program (2001), District 7, 2000/2001 Report, Best Management Practices, Operations, Maintenance & Monitoring, Continuous Deflective Separation Units

Caltrans BMP Retrofit Pilot Program (2002), District 7, Los Angeles, Fourth Year 2001/2002 Report, Continuous Deflective Separation Units, Operations, Maintenance & Monitoring, CTSW-RT-02-037, 2002

Gray, John R., Glysson Douglas, Turcios, Lisa M. and Schwarz, Gragory E. (2000), *Comparability of Suspended-Sediment Concentration and Total Suspended Solids Data*, Water-Resources Investigations Report 00-4191, U.S. Department of the Interior, U.S. Geological Survey

Woodward Clyde Consultants (1989), Pollutant Loadings and Impacts from Highway Stormwater Runoff, FHWA

USGS (2000), *USGS Policy on Collection and Use of Total Suspended Solids Data*, Office of Water Quality and Office of Surface Water Technical Memorandum No. 2001.03

Chapter 8

Performance of BMPs

Introduction

By *Shirley E. Clark*

Stormwater managers are increasingly faced with the need to address many potentially-conflicting issues as part of their activities. First, they need to control the potential flooding waters generated after urbanization. This has led to the use of detention facilities and control of peak flow rates to pre-development levels. However, these designs that focused on flow control for larger storms have been providing inconsistent performance for the control of other pollutants. Many other techniques, in addition to detention/sedimentation, have been proposed to alleviate these pollutant concerns. Stormwater ordinances/regulations and the development of TMDLs (with its subsequent need to control loadings to a stream) has driven the improvement of existing treatment technologies (often referred to as structural best management practices [BMPs]) and the development of new ones. Assumptions have made regarding the range of performance of these devices, based on the field data that currently exists, and as new data is submitted for publication, these new insights are added into the body of knowledge on BMP design and performance, specifically on the design needed to provide a specific performance.

This chapter focuses on the performance of several best management practices (BMPs) for urban runoff treatment. In addition, it also raises the question of how should devices be evaluated, which is a vital question for designers and vendors as many states are requiring that stormwater treatment technologies have a verifiable performance. The first paper, by Lenhart and de Ridder, reviews the issue of defining acceptable performance (percent removal vs. actual effluent concentrations) as faced by the authors while participating in a Washington state BMP performance-verification evaluation. The second paper, by Michael Barrett and Joseph Malina, investigates the ability of silt fences to provide adequate removal of solids in construction runoff. Vegetated buffers and their performance are the focus of the third paper (Michael Barrett), while the performance of bioinfiltration areas are reviewed in the last paper, by William Hunt and Al Jarrett.

Sampling Issues and Meeting Sampling Protocol Requirements for BMP Performance Verification in the State of Washington

James H. Lenhart, P.E. and Scott A. de Ridder
Stormwater Management, Inc. 12021-B NE Airport Way, Portland, OR 97220

Background and Introduction

In 2003, the Washington State Department of Ecology (Ecology) made substantial revisions to its 1992 Stormwater Management Manual for the Puget Sound. The new manual, *Stormwater Management for Western Washington (Manual)*, provides criteria defining a design storm and the expected treatment-device removal performance for various stormwater contaminants. Through the revised Manual, Ecology also established a process for the application and evaluation of new Best Management Practices (BMPs).

To oversee this process, Ecology also formed a Technical Review Committee (TRC) comprised of water quality program managers from municipalities throughout Washington and Oregon. The TRC was given the task of establishing a protocol for accepting and evaluating BMPs, which became known as the "Technology Acceptance Protocol—Ecology" (TAPE), and was first released in 2002. (Ecology, 2002). The TAPE defines both the treatment and acceptance classification of a BMP under treatment categories such as Pre-Treatment, Basic (Suspended Solids <500-um [TSS-WA], Enhanced treatment (dissolved Cu and Zn), and Total Phosphorus treatment.

BMP acceptance classifications are listed as Pilot Level (PLD), Conditional Short Term Use Level (CUD), or General Use Level Designation (GULD). Pilot level is used for BMP's which have little or no data and are accepted for the purpose of monitoring only. Conditional Short Term Use implies that the existing performance data available for the technology is sufficient to suggest probable attainment of performance requirements and thus can be widely implemented during the monitoring phase, while GULD means that monitoring is no longer required and the technology is unconditionally accepted.

To begin the approval process, a BMP manufacturer or designer of a new non-proprietary BMP must present an application that includes a technical description of the technology, maintenance and cost information, an analysis of laboratory and field data, and design information specific to Ecology's criteria. Based on this application, the TRC then makes a recommendation to Ecology as to which acceptance classification level to apply. Technologies receiving PLD or CULD designations are then required to implement a two-year monitoring study for installations in the Pacific Northwest. This monitoring study must be designed according to the important and often unique specifications and guidelines provided in the TAPE. One distinctive requirement of the monitoring study is the development of a Quality Assurance Project Plan (QAPP) that assures that the results of the study can be reconciled with the design and performance criteria outlined in the TAPE. Another important requirement is the selection of a qualified, 3rd party project evaluator (evaluator). The evaluator helps to assure the quality and integrity of the data, and on completion of the data acquisition phase of the project, the reviewer must certify the final report and present the recommendation to Ecology.

In 2002, Stormwater Management, Inc. (SMI) was granted a CUD by the TRC for The Stormwater Management StormFilter® (StormFilter). A QAPP was developed and implemented in early 2003 and data acquisition was initiated at five sites shortly thereafter. The SMI QAPP adds additional provisions beyond those detailed in the TAPE, such as provisions for the abandonment of the site during the course of the sampling phase should poor site hydraulics, low influent pollutant concentrations, or excessive organic loading produce operating conditions that either cannot be monitored, are outside of StormFilter design parameters, or are beyond the intent of the TAPE.

This paper discusses some of the issues and complexities that were found during the monitoring of a site used to support a GULD decision. In addition, the qualification data collected according to the QAPP for this project are presented to illustrate these findings.

Complexities of Designing and Executing an Extensive Field Evaluation

Data Quality Objectives

The TAPE provides both criteria and general guidelines that are to be used to define the data quality objectives (DQOs) to be used to qualify each storm capture during the data reconciliation process. These DQOs are intended to ultimately produce a data set that is representative and comparable relative to the Ecology approval process. These DQOs are summarized in Table 1.

The first issue encountered was the effect of the DQOs on the probability of capturing a qualifying event. During the course of the data acquisition period, the storm capture success rate was approximately 75%. This resulted in over 80 successful storm captures among 5 test sites. Of these events, approximately 20% met all of the TAPE DQOs except influent TSS-WA EMC range. Disregarding the DQOs based upon the TAPE guidelines increased this number to 60%. Clearly, if the TAPE criteria and guidelines are strictly applied as DQOs, a qualifying storm event capture success rate less than 20% greatly elevates the cost of such a project, the anticipated time to project completion, and requires perseverance and endurance of the project staff to relentlessly pursue the goal of satisfying the TAPE while maintaining the highest quality of data collection and analysis.

Table 1. DQOs based upon TAPE storm event criteria and guidelines (Ecology, 2002).

Criteria	Minimum Event Depth:	0.15-in
	Antecedent Dry Period:	6 hours with < 1 mm (0.04 inches) of rain
	Minimum Event Duration:	1 hour
Guidelines	Minimum Number of Aliquots:	10
	Average Event Coverage:	75%
	Average Intensity:	0.76 mm/hr (0.03 inches/hr)
	Influent TSS-WA EMC Range:	33 to 300 mg/L
	Influent Total Zinc EMC Range:	0.053 to 0.48 mg/L
	Influent Total Phosphorus EMC Range:	0.11 to 0.99 mg/L

To illustrate the impact of the literal use of the Ecology criteria and guidelines as DQOs for a project performed under the TAPE, data from an actual project site will be used. The site used is Heritage Market Place, which was ultimately one of the two sites that had enough data to meet the TAPE protocol. This paper presents and discusses issues from the time of site selection to the final data analysis and ultimate acceptance by Ecology.

Impacts of Site Specific Variables on Representative Data Collection

Site Representativeness and Maturity

Challenges with data collection start with the site. The site selected needs to be representative of the typical BMP application, such that the interpretation of the data is relevant. All too frequently, data from BMP's sampled in one application are translated to another with the presumption of equivalent performance. For example a site in the NE United States removing sanding material from road runoff in a high rainfall intensity is presumed to remove finer sediments in the Pacific Northwest. To prevent this, the TAPE required the monitoring to take place in the Pacific Northwest.

The site must also be mature. Newly developed sites have characteristics that are different from ones that have had at least one year to stabilize. Newer sites could have more mineral type soils exposed in the landscape areas or construction traffic from surrounding development. Over time the landscape materials stabilize soils but contribute more organic matter. Pores in the pavement begin to fill with particulates and seal with oil and grease drippings. The conveyance system builds a layer of armor while sumps, cracks and other areas accumulate sediment and reach some state of equilibrium. All of the sites select for this study were at least two years old.

Ability to Collect Samples

Facility access and hydraulics need to be issue-free to ensure actual storm capture and the accurate monitoring of storm attributes such as flow and rainfall. For example, on some sites, base flows have been encountered that prevented flow monitoring and the ability to collect flow-weighted samples. During the course of the study, a number of site-specific variables presented the project team with significant challenges in both data collection and data interpretation.

Defining Total Suspended Solids

Despite the best attempts by scientists and regulators to define TSS, the composition of the solid material originating from a site is entirely uncontrollable in the field environment. This is of particular importance when a substantial organic load exists, since organic materials are both lower in density than mineral solids and are also more prone to decomposition. Leaves, for example, will be transported into a system and not be recorded as part of the influent sample (the leaves are larger than the holes in the sample suction line and strainer). [Note: The issue of automatic samplers missing large debris, including leaves, is discussed in the Coarse Solids chapter.] Eventually, these leaves will decompose into constituent chemicals and fine particulate matter. Unlike the larger leafy debris in the influent, the effluent sampler can pick up these constituent materials. The transitory nature of organic material presents a problem in that it raises the question of how to account mass loads which are undetected on the influent. Clearly it is not possible to resolve the balance through analysis of the effluent samples alone, and therefore it may be necessary to somehow add in the mass of organic materials captured in the BMP in order to balance the mass load equation and accurately calculate removal on a load basis. Another confounding factor is that some BMPs may, for example, allow leaves to flow through during a storm. This mass is not accounted for in any way yet ultimately contributes nutrient loads, BOD, and TSS to the effluent.

Heritage Marketplace Stormwater Monitoring

As part of the performance assessment of the StormFilter for approval in the State of Washington, a StormFilter system using Perlite/Zeolite/GAC combination media installed at Heritage Marketplace was evaluated between May and December of 2003 (Figure 1). This system treats stormwater runoff draining from the rooftops and parking lots of the shopping center. It was evaluated to determine removal of a suite of pollutants, including TSS, total Zn, and total phosphorus. Out of over 25 storm event capture attempts during the 2003 monitoring period, 21 individual storm events were successfully monitored, resulting in only 2 events that qualified according to all of the TAPE Criteria and Guidelines.

Figure 1. View of part of the Heritage Marketplace site drainage area in proximity to the StormFilter system.

Site and System Description

The typical StormFilter unit is composed of three bays: the inlet bay, the filtration bay, and the outlet bay. Stormwater first enters the inlet bay of the StormFilter vault through an inlet pipe, which is plumbed directly to catch basins and rooftop downspouts throughout the drainage area. Stormwater in the inlet bay is then directed through a flow spreader, which traps some floatables, oils, and surface scum, and over the energy dissipater into the filtration bay where treatment will take place. Once in the filtration bay, the stormwater begins to pond and percolate horizontally through the media contained in the StormFilter cartridges. After passing through the media, the treated water in each cartridge collects in the cartridge's center tube from where it is directed into the outlet bay by an under-drain manifold. The treated water in the outlet bay is then discharged through a single outlet pipe to an open channel drainage way.

The StormFilter system installed at Heritage Marketplace consists of a standard, 2.4 m by 4.9 m (8.0-ft by 16-ft), pre-cast, concrete StormFilter vault with 23 StormFilter cartridges containing Perlite/Zeolite/GAC combination media and configured for a per-cartridge flow rate of 0.5 L/s (7.5 gpm). Using a per-cartridge flow rate of 1.0 L/s (15 gpm), the overall system is designed to treat a peak flow rate of 23 L/s (0.77 cfs) which equates to 64% of the 2-yr, 24-hr storm using the Santa Barbara urban hydrograph to determine peak flow. However, for the purpose of demonstrating the performance at 0.5 L/s, the design capacity of the system was effectively halved.

Water Quality Sampling Methods

The equipment and sampling techniques used for this study are in accordance with a Quality Assurance Project Plan developed by Stormwater Management Inc. (SMI) according to the Washington State Department of Ecology (WADOE) TAPE protocol. SMI personnel were responsible for the installation and maintenance of the sampling equipment, sample retrieval, system reset, and sample submittal activities. A general overview is provided.

Equipment Specifications

Samples were collected using two ISCO 6700 portable automated samplers, containing 24, 1000-ml, polypropylene, ISCO wedge shaped bottles. Each sampler was connected to individual 12VDC, deep cycle power supplies. Flow measurements were made using ISCO 750 area velocity modules with low profile area velocity sensors.

Sample intakes from each of the automated samplers' peristaltic pumps were connected to 3-m (10-ft) lengths of 10-mm (0.4-in) diameter Acutech Duality FEP/LDPE suction line. A stainless-steel, low-profile sample strainer (14-mm [0.56-in] diameter with multiple 6-mm [0.25-in] openings) was installed at the end of the suction line to protect the pump head. All fittings were polyethylene in composition.

Internal overflow was monitored using an Overflow Detection System (ODS) consisting of a float switch connected to a Onset Hobo State Logger. Rainfall was monitored using an Onset RG2 data logging rain gauge.

Equipment Installation

All sampling and flow monitoring equipment was installed inside the vault for security and protection reasons. The automated samplers and 12-VDC batteries within the vault were placed on top of a platform to achieve minimal suction line length and eliminate dips in the suction line. Maximum inline velocity ($\geq$ 2 ft/s) was maintained by avoiding extraneous suction line length, excessive bends, and kinks in suction lines. The rain gage was place atop an adjacent building within the shopping center with no overhead obstructions.

Individual automated samplers, suction lines, and flow sensors were used to monitor the influent entering and effluent exiting the StormFilter system; one sampler, flow sensor, and suction line was placed in the inlet pipe and another sampler, sensor, and suction line was placed in the outlet pipe. Thus, each sampler was independently controlled; the influent sampler by flow entering the system and the effluent sampler by flow leaving the system.

The flow sensors and suction lines were mounted on ISCO stainless steel spring rings sized to match the inner pipe diameter at the sampling locations shown in Appendix A. The spring rings were inserted as far up into the pipe as possible, keeping the suction lines and flow sensors in a fixed position at the invert of the pipe with no vertical offset. The suction lines and flow sensor cables were bound together and routed out of the pipe and up to the samplers. Samples were taken as water entered and exited the first StormFilter vault and did not measure pollutant removal associated with treatment by upstream structures or catch basins.

The Overflow Detection System (ODS) was attached to a cartridge located towards the rear of the filtration bay such that the switch would activate at a water surface elevation of 21-in from the floor of the filtration bay (design overflow elevation). The state logger was placed in a waterproof housing and secured to a cartridge inside the vault.

Equipment Operation

Flow meters were set to record measurements every 1 to 5 minutes, which was typically extended to allow for extended deployment and minimal power usage during colder weather. To further minimize power consumption and avoid false starts caused by dry weather flows, samplers were programmed to begin sample acquisition after a minimum flow rate condition was met. Once enabled, the equipment collected samples on a volume-paced basis.

Sample Collection Program

The sample program input into each of the automated samplers was a two-part program developed to maximize both the number of subsamples collected and the coverage of an individual storm event. The first part of the program sequentially filled the first set of bottles every X-gal. The second part of the program sequentially filled the remaining bottles every 2X-gal. This increased the probability of adequate coverage of both small and large precipitation events by allowing the use of subsamples collected according to X or 2X sample pacing. The sample pacing value was changed on an as-needed basis based upon anticipated storm size. A record of each of the program changes was made by the automated sampler and was reflected in the collected data.

Sample Retrieval and Analysis

The sampling equipment was inspected for samples following precipitation events. Sample bottles were capped, labeled, and transferred from the sampler base section directly to a cooler stocked with gel ice packs. The samples were then taken to the SMI lab and composited and split using an appropriately sized churn sample splitter (Bel-Art Products) to create flow-weighted, influent and effluent, event mean concentration (EMC) sub-samples for submittal to North Creek Analytical, Inc., Beaverton, OR (Oregon and Washington State accredited laboratory) for analysis using the methods listed in Table 2.

Table 2. Analytical methods for water quality sample analysis. Shading denotes extrapolation.

Analyte	Analytical Method
Total Suspended Solids (TSS)	ASTM D3977B
Total Volatile Suspended Solids (TVSS)	SM 2540E
Total Mineral Suspended Solids (TMSS)	[TSS] – [TVSS]
Washington Total Suspended Solids (TSS-WA)	500-um Filtration + ASTM D3977B
TVSS-WA	500-um Filtration + SM 2540E
TMSS-WA	[TSS-WA] – [TVSS-WA]
Coarse Solids (CS)	[TSS] – [TSS-WA]
Coarse Volatile Solids (CVS)	[TVSS] – [TVSS-WA]
Coarse Mineral Solids (CMS)	[TMSS] – [TMSS-WA]
Total Zn, Cu	EPA 200.8
Dissolved Zn, Cu	0.45-um Filtration + EPA 200.8
Hardness	SM 2340B
Total Phosphorus	EPA 365.1
Dissolved Phosphorus	Filtered + EPA 365.2
Ammonia-N	EPA 350.1
Nitrate/Nitrite-N	EPA 353.2
Total Kjeldahl-N	EPA 351.2
Total-N	[Total Kjeldahl-N] + [Nitrate/Nitrite-N]
pH	EPA 150.1/9040A

Field QC

To avoid contamination issues, disposable and certified clean materials were used whenever possible. Upon installation of the sampling equipment, new sampler tubing was used so as to avoid the need for decontamination and the associated equipment initial rinse blank. During the course of the project, wedge-shaped ISCO bottles were only used once and sent to North Creek Analytical for cleaning and acid-washing. Upon completion of the study, a sampling equipment field blank was collected from the influent sampling equipment.

The sampling equipment field blank was performed according to SMI (2004a), and involved pumping deionized water through the fully assembled samplers. Samples were then submitted to North Creek Analytical and analyzed for TSS, TVSS, total metals, total phosphorus, Nitrate/Nitrite-N, Ammonia-N, Total Kjeldahl-N, and Hardness. The results returned non-detects for all analytes except Total Zn (~3 times above PQL) and Nitrate/Nitrite-N (~10 times above PQL), suggesting potential contamination of influent samples by the sampling equipment.

Data Qualification

Results and data for each storm event were compiled into Individual Storm Reports (ISR's) as shown in Appendix 1 for the purpose of evaluation and qualification. The ISR was developed by Stormwater Management to allow for a uniform and concise method to present all of the important data resulting from a storm event. All too often, reports on BMP performance do not present all of the collected data or the formats are so different it is difficult to make comparisons. The ISR includes the critical site information, antecedent conditions, date of late maintenance, depth of rainfall, storm volume, influent and effluent hydrographs (with the Y axis set to the design flow of the BMP unless flows exceeded design) sampling points, and hyetograph. The ISR also presents all of the critical laboratory data and any notes regarding the storm event. Qualification of the storm events involved the selection of events representative of the design operating condition according to both the product design specifications and the TAPE Criteria and Guidelines. Qualification of each storm event with regard to the TAPE Criteria and Guidelines is shown in Table 1.

Table 3 presents event characteristics for 21 storm events captured from two different sites as part of a StormFilter monitoring effort used to pursue a GULD in the State of Washington. Of these 21 events, only two will qualify based upon strict interpretation of the criteria and guidelines as DQOs. Observations such as substantial runoff and pollutant loads associated with storm events that did not meet event depth objectives brings the significance and utility of these DQOs for evaluation purposes into question. Investigation of the events on an individual basis with regard to the TAPE goal of evaluating pollutant loads and water quality on an annualized basis led to the eventual qualification of 13 after clarification of the criteria and use of the guidelines as true guidelines as opposed to requirements in the determination of DQOs for the project.

This is an excellent example of how tight criteria severely reduces the sample population and can make data analysis nearly impossible due to the low probability of meeting all the criteria. Over the course of the data qualification process, the following observations were made:

Event Depth: The data clearly show substantial TSS contributions for depths less than the minimum criteria. To ignore these storms is to say they are not important to water quality.

Antecedent Dry Period: By itself, this appears to have little or no influence on concentration. For example a high energy storm immediately following a low energy storm can detach and transport higher TSS concentrations.

Event Duration: This may also not be important because a "cloud burst" can transport more pollutants than a long duration, low intensity, and low depth storm.

Number of Aliquots: This guideline is very important, however the number provided is not based upon empirical data and thus is not well supported. Guidance for minimum aliquot volume should also be provided. Maximum discrete sample volume was used for this study (1-L), and it could be argued that 5 high volume aliquots may prove equally—if not more—representative than 10 aliquots of a lesser and unspecified size.

Average Event Coverage: This guideline is also very important depending upon the purpose of the study. Sufficient coverage is critical to the generation of a true EMC, though this may still be affected by overall event dynamics and hydrograph shape. It is less critical for the determination of the performance of a flow-through treatment device with zero storage such as the StormFilter, where even an instantaneous performance assessment may be useful.

Average Intensity: Though the effects of intensity on pollutant mobilization and concentration are logical, average intensity does not seem to influence pollutant concentrations and transport due to masking of intensity peaks. Peak intensity may be more informative, however, its impact on pollutant mobilization and concentration cannot be reconciled with an EMC.

Influent Concentration: Data clearly show the majority of the TSS concentrations are very low and hence do contribute to the annual load. The data also suggest that larger events may have lower EMCs due to a dilution effect. To discount low concentration events regardless of volume is not compatible with an evaluation based upon load calculations.

Table 3. Summarized event characteristics and qualification decisions for monitoring events in the State of Washington of the StormFilter with ZPG media at a 7.5 gal/min/cartridge filtration rate.

	TAPE Storm Event Criteria and Guidelines							Qualification		Other Character	
Event ID	Event Depth (in) [minimum 0.15]	Antecedent Dry Period (hr) [minimum 6]	Event Duration (hr) [minimum 1]	Number of Aliquots [minimum of 10 (Inf:Eff)]	Average Event Coverage (%) [minimum of 75]	Average Intensity (in'hr) [minimum 0.03-in]	Influent TSS-WA* EMC (mg/L) [33 to 300]	Strict Application of TAPE Criteria and Guidelines	Ecology Final Decision	Influent Volume (gal)	Peak Operating Rate** (%)
HMP050303	0.26	103	8	15:16	89	0.03	66.4	✓	✓	21892	92
HMP050703	0.19	55	9	16:16	90	0.02	519		✓	26541	138^
HMP051603	0.10	16	5	3:9	63	0.02	987			11058	17
HMP090703	0.14	384	5	11:18	86	0.03	378		✓	7217	101
HMP090903	0.16	24	4	21:15	76	0.04	76.9	✓	✓	12965	85
HMP091603A	0.05	120	3	8:4	79	0.02	35.5			4878	15
HMP091603B	0.10	10	2	17:15	96	0.05	96.9		✓	8744	96
HMP100603	0.27	384	5	21:21	58	0.05	117		✓	17335	257^
HMP100803	0.07	36	3	7:8	93	0.02	43.4			3866	31
HMP100903A	0.15	18	2	14:13	52	0.08	83.6		✓	13581	142^
HMP100903B	0.25	3	2	21:21	39	0.13	58.2			28521	228^
HMP101103	0.15	36	4	21:21	71	0.04	7.53		✓	15570	71
HMP102203	0.17	62	2	18:18	73	0.09	22.1		✓	14681	125
HMP111003	0.14	264	4	14:17	83	0.04	30.6		✓	9193	97
HMP111503	0.23	96	6	18:18	74	0.04	6.85		✓	16901	96
HMP111903	0.96	26	7	18:18	12	0.14	29.4			104132	377^
HMP112103	0.08	17	4	11:9	86	0.02	85.2			8189	94
HMP120203	0.24	30	8	16:16	29	0.03	270			34988	412^
HMP120403	1.10	40	18	18:18	10	0.06	35.9			117340	104
HMP121003	0.26	42	7	13:16	79	0.04	28		✓	20814	78
HMP121603	0.22	22	5	10:8	54	0.04	45.9		✓	22981	79

* WA definition of total suspended solids (500-um pre-filtration, whole volume analysis)

** expressed as percentage of effluent design Q

^ internal bypass confirmed by ODS

shading = DQO met

Given these factors, it is likely that the criteria and guidelines do not represent reality and the data need to be evaluated from a makes-sense approach versus hard line criteria.

Data Interpretation and Performance Assessment

The purpose of this paper is not to analyze data but to provide a perspective on issues associated with collecting and interpreting data. Table 4, a summary of different analytical methods used, including descriptive statistics, regression analysis, and load analysis. Each of these methods is used by professionals throughout the US and arguments are made as to which one is better. To illustrate that one should use all the methods the 21-storm data set is used.

The expected removal rate is a function of the influent concentration. Ecology's criterion is 80% TSS removal above 100 mg/L and an EMC of 20mg/L for influent concentrations less than 100 mg/L. For example, one would expect a 50% removal given an influent EMC of 40 mg/L. From this perspective, given the descriptive statistics with a mean effluent EMC of 14 mg/L for all samples below 100 mg/L and mean TSS removal of 78% for samples greater than 100 mg/L. Hence these data indicate the performance meets Ecology's criteria.

Using regression analysis, the predicted removal rates are 76% for TSS. On an aggregate load reduction basis, however, the load reduction is 58%, which at first appears to be low, but with low influent concentrations associated with the influent, a load reduction of 58% is easily within expectations. As an extreme example if the EMC's of the influent and effluent were 20 mg/l there would be an expected removal of 0%. Previous work by De Ridder (Stormwater Management, 2001) discussed the numerous methods for calculating removal efficiencies.

This provides a clear illustration as to how important it is to consider both load and concentration. It is also possible for the reverse to happen. For example, a highly energetic storm conveys a high mass load of sand which greatly outweighs other less energetic storms which convey lower loads of finer particles that have poor reductions in concentrations. The analysis of loads may conclude that the technology meets the 80% criteria whereas on a concentration basis, the technology does not meet the 80% removal criteria. Hence, it is very important to use different methods of analysis to insure the system is operating within the expected operational performance envelope of the BMP.

Table 4. Summarized performance for the StormFilter performance evaluation of the Heritage Marketplace StormFilter.

		Descriptive Statistics			Regression of EMC				Aggregate Load Reduction	
	Analyte	*n*	Range of Influent EMCs (mg/L)	Median Influent EMC (mg/L)	Mean Removal Efficiency Estimate (%)	95% Confidence Interval for the Mean Removal Efficiency Estimate (%)	Median Effluent EMC Estimate (mg/L)	95% Confidence Interval for the Median Effluent EMC Estimate (mg/L)	Mean Removal Efficiency Estimate (%)	One-Tailed Sign Test* (H0=H1=0.5)
Solids	TSS	8	6.99 to 100	32	74*	56 to 91	15	9 to 21	58	R
	TVSS	7	7 to 34.3	18	76*	55 to 96	10	8 to 12	---	~
	TMSS	6	11.8 to 65.7	29	---	--- to ---	---	--- to ---	64	R
	TSS-WA	8	6.85 to 96.9	29	76*	59 to 93	14	8 to 20	58	R
	TVSS-WA	6	7 to 33	16	---	--- to ---	---	--- to ---	43	R
	TMSS-WA	5	12 to 64	14	---	--- to ---	---	--- to ---	66	R
	CS	2	4 to 14.9	9	---	--- to ---	---	--- to ---	---	---
	CVS	0	--- to ---	---	---	--- to ---	---	--- to ---	---	---
	CMS	0	--- to ---	---	---	--- to ---	---	--- to ---	---	---
Metals	Total Cu	8	0.00515 to 0.0203	0.0106	49**	22 to 76	0.00745	0.00605 to 0.0088	33	R
	Diss. Cu	8	0.00209 to 0.0118	0.00464	38*	-9 to 86	0.00423	0.00252 to 0.00594	---	~
	Total Zn	8	0.0523 to 0.192	0.0896	52***	35 to 69	0.0641	0.0562 to 0.0721	31	R
	Diss. Zn	8	0.0339 to 0.134	0.0599	26**	-9 to 62	0.0532	0.0427 to 0.0638	15	R
	Hardness	8	6.7 to 21.3	9.67	11***	-9 to 30	9.01	8.07 to 9.94214	9	R
Nutrients	Total P	8	0.0544 to 0.249	0.1013	38***	26 to 50	0.075876	0.0674 to 0.0844	28	R
	Diss. P	4	0.01 to 0.0175	0.0122	---	--- to ---	---	--- to ---	---	---
	NH4	8	0.0617 to 0.602	0.2385	34***	22 to 46	0.180	0.161 to 0.200	23	R
	NO2/NO3	8	0.0922 to 0.544	0.2705	-21**	-101 to 58	0.347	0.2192 to 0.475	---	~
	TKN	7	0.508 to 2.23	0.94	42**	10 to 74	0.80	0.63 to 0.97	---	~
	Total N	7	0.615 to 2.411	1.38	49*	2 to 95	1.23	0.98 to 1.48	---	~

*** = P < 0.001
** = 0.01 > P > 0.001
* = 0.05 > P > 0.01
--- = undeterminable due to insufficient data quantity
R = removal is significant at the 5% level or less
~ = no significant difference
A = addition is significant at the 5% level or less

(shaded) = Ecology Basic Treatment Parameter: 95% confident that performance goal met or exceeded

Issues Associated with Total Suspended Solids – Measurement and Definitions

The Definition of Suspended Solids

During the course of the study, issues associated with the definition of suspended solids were noted. This is important as it affects the grounds upon which expectations for solids removal performance are founded. For example, many BMP's are substantially more effective for highway sites with large, mineral-based solids than parking lot sites with finer, more diverse solids because these systems are not capable of removing small, organic particles. Also, de Ridder and Lenhart also showed that sampling and analytical methods have a significant impact on removal percentages. For this study the whole volume method for TSS analysis was used, realizing that other methods may provide different results.

As per the TAPE, TSS was defined for this study to be particles greater than 2-μm and less than 500-μm. This definition of TSS was selected by Ecology to eliminate the issue of performance bias imparted by coarse solids and debris which are often present to some extent in urban runoff. However, land use can dramatically affect performance by affecting the composition and particle size distribution of the suspended solids present in runoff. While the TAPE specifies a particle size distribution for silica materials (S.G 2.65) to be used for laboratory testing, this material is not universally representative. Roadways producing heavier solids from pavement decay will show high TSS removals as compared to parking lots with solids originating from atmospheric depositions, tire wear, a deciduous canopy, and possible runoff of soil fines from landscape areas and landscaping activities such as mulching. Thus, to appropriately review TSS removal performance, it is critical to tie performance expectations to particle size distribution.

Particle composition is equally important. Analytical data show that volatile—presumably organic—materials are frequently a large component of suspended solids. For the sites studied to pursue a GULD for the StormFilter and represented in Table 2, 20% to 25% of the suspended solids load is volatile and presumed to be organic. It follows that stormwater solids of organic origin (leaves, paper, mulch) have a lower specific gravity than stormwater solids of mineral origin (sand and grit) and thus are more difficult to remove. Thus laboratory testing conducted with a material of mineral origin may not be comparable to field data with suspended solids of both mineral and organic origin.

During the course of the study, a number of issues associated with TSS monitoring and removal evaluation have been noted. For example, it is frequently found that TSS loads coming from certain sites are near the limits of quantization and may be negligible. TSS concentrations ranging from 5 to 20 mg/L are common at stable, urban sites. Though this is desirable from a water quality perspective, these low concentrations provide little insight on facility performance, as 20 mg/L is commonly recognized as a irreducible concentration below which a BMP cannot be expected to remove additional solids (Schueler,1996). In this case, many storms must be rejected, even to the extent the site may have to be abandoned. De Ridder and Lenhart also showed that sampling and analytical methods have a significant impact on removal percentages. For this study the whole volume method for TSS analysis was used, realizing that other methods may provide different results.

For the purposes of this study, TSS was defined to be particles greater than 3 μm and less than 500 μm. This definition of TSS ignores the issue of coarse solids and debris which are often present to some extent in urban runoff. In many cases however a large fraction of the total solids load is composed of particles greater than 500 um. This typically requires a bedload sampler, which is not called for in the TAPE and was ultimately not required. However, it may be needed for a site in the North East US where road sanding is a common practice. From one perspective the BMP itself collects a large sample over time, and if one could characterize the effluent mass load relative to the total mass load captured, it may present a better method of evaluating performance. The behavior of coarse solids in treatment devices can greatly influence the final treatment performance data. It is the reason why other test protocols are requiring additional, complementary solids measurements such as the ASTM protocol on Suspended Sediment Concentration (SSC).

In addition, no specific particle size distribution was recognized by the TRC, hence land use can dramatically affect performance. Roadways producing heavier solids from pavement decay will show high TSS removals as compared to parking lots with solids originating from atmospheric depositions, tire wear, and possible runoff of soil fines from landscape areas. Thus, to appropriately review TSS removal performance, it is critical to tie performance expectations to particle size distribution. Many BMP's are substantially more effective for highway runoff than parking lot runoff because they are not capable of removing most particles less than 20 to 50 μm. Specific gravity is also important as analytical data show that TVSS is frequently a high fraction of the TSS. For this site 40-50% of the TSS is volatile and presumed to be organic. If this organic fraction consists primarily of leaves, then leaf decay creates problems – increased effluent nutrients, BOD and fine particulates. A lack of quantification of the coarse solids captured in treatment (because it is not required as part of the evaluation protocol) changes the performance ability of the treatment device.

Conclusions

BMP performance is probabilistic. Multivariate factors of particle size distribution, flow rate, antecedent conditions, etc., introduce the need for many samples to gain high levels of statistical confidence. Given the data collected, it is problematic as to which storms to accept as part of the data set. Once the data set is defined, it is problematic to establish the exact analytical method to use. Clearly, interpretation of results is highly dependent on the analytical methods used and there seems to be no definitive method from which to make objective decisions. While this paper has focused on the TAPE protocol of the Washington State Department of Ecology, other BMP evaluation agencies have significantly different protocols. Ultimately, it seems that evaluation of performance has a strong non-objective part that is subject to the interpretation of the individual. Hence it is extremely important to establish the measure of minimum acceptable performance at the onset of the QAPP rather than at the conclusion of the study.

Of the five sites initially tested only two had enough data to meet and satisfy the Ecology criteria. Heritage Market Place was one of the sites, and the other was a bridge deck located in Snohomish County, Washington. The three other sites either produced no substantial pollutant load, or were operating a higher filtration rate and did not produce enough data for a conclusion, or was using an alternative filter media. The StormFilter was accepted as a BMP for General Use Level Designation for Basic Treatment in January 2005.

References

State of Washington Department of Ecology (WADOE). (2002). Guidance for Evaluating Emerging Stormwater Treatment Technologies: Technology Assessment Protocol—Ecology (Publication No. 02-10-037). Lacey, WA. Author. Retrieved February 24, 2004 from: www.ecy.wa.gov/programs/wq/stormwater/newtech/02-10-037%20TAPE.pdf

State of Washington Department of Ecology (WADOE). (2001). Stormwater Management Manual for Western Washington. (WADOE Publication No. 99-11 through 99-15). Lacey, WA. Author. Retrieved on February 24, 2004 from: www.ecy.wa.gov/programs/wq/stormwater/manual.html

Schueler, Thomas. (1996). Technical Note 75: Irreducible Pollutant Concentrations Discharged from Urban BMP's. Watershed Protection Techniques, Vol 2, No 2. Author: Ellicot City, MD.

Stormwater Management, Inc. (2001). Silt loam TSS removal efficiency of a stormwater BMP: Coarse/fine perlite StormFilter cartridge at 7.5 gpm. (PD-01-001.1). Author: Portland, Oregon. Retrieved on February 24, 2004 from: www.stormwaterinc.com/literature/pdfs/PD-01-001.1.pdf

Stormwater Management, Inc. (2002a). Stormwater Management StormFilter® Interim Short-Term Use Application for Basic Treatment in Western Washington. Author: Portland, Oregon. Retrieved February 24, 2004 from: www.ecy.wa.gov/programs/wq/stormwater/newtech/media_filtration.html

Stormwater Management Inc (SMI). (2004). Performance of the Stormwater Management StormFilter Relative to Ecology Performance Goals for Basic Treatment (Report No. PE-04-007.0). Portland, Oregon: Author.

Appendix 1: Sample Individual Storm Report

General Information

Site:	Heritage Marketplace (2270), Clark County, WA
System Description:	8x16, ZPG, 23 cartridges, 7.5 gpm
Event Date:	9/9/03
Date of Last Maintenance:	3/3/03
Antecedent Conditions:	1 day since last rain event

Hydrology

Total Precipitation (in):	0.16
Peak Flow (gpm):	Influent 166, Effluent 147
Total Runoff Volume (gal):	Influent 12965, Effluent 10546

Event Hydrograph

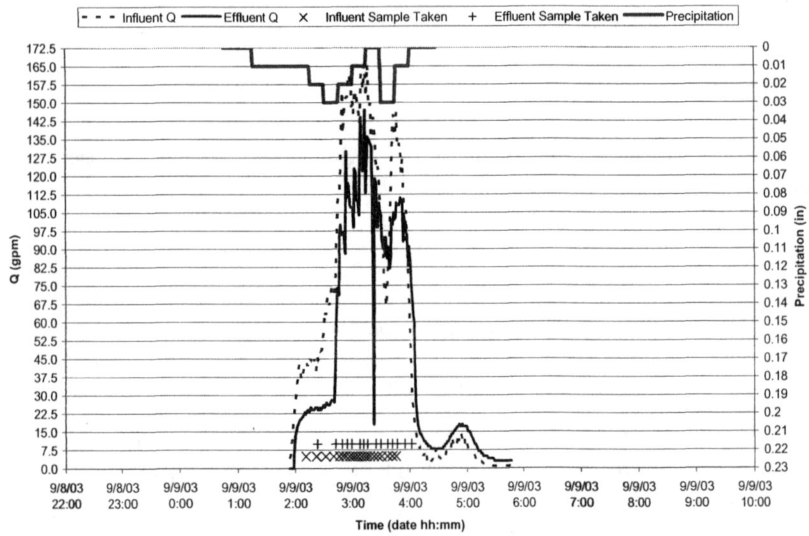

Analytical

Bottles Used:	Parameter	Concentrations (mg/L) Influent EMC	Effluent EMC	PQL	Dup. RPD	Discrete Removal Efficiency
IN 2,4,6,7-24	TSS	91.8	18.6	4.00	0.00%	80%
EFF 2,4,6,7-18	TVSS	31.0	11.0	10.0	20%	65%
	TSS (<500µm)	76.9	16.0	4.00	0.00%	79%
	TVSS (<500µm)	24.5	7.99	3.47	20%	67%
	Total Cu	0.0187	0.0120	0.00200	3.28%	36%
	Dissolved Cu	0.0118	0.0104	0.00200	3.39%	12%
	Total Zn	0.192	0.112	0.00500	5.50%	42%
	Dissolved Zn	0.134	0.107	0.00500	5.70%	20%
	Total P	0.249	0.167	0.0200	20%	33%
	Ortho-P	ND	0.0137	0.0100	5.32%	release
	NH4-N	0.602	0.419	0.0500	0.656%	30%
	NO2/NO3-N	0.410	0.610	0.0500	4.27%	release
	TKN	ND	ND	5.00	3.39%	undeterminable
	pH	6.51	6.78	---	0.765%	release
	Hardness	21.3	19.7	0.660	1.50%	8%

Notes

Shaded Dup. RPD values defaulted to 20% standard due to QC complications. Possible time sync issues between standalone rain gage and automated sampler.

Monitoring Data on Effectiveness of Sediment Control Techniques

Michael E. Barrett[1] and Joseph F. Malina, Jr., P.E., DEE[2]

[1]Center for Research in Water Resources, University of Texas, Austin, TX 78712; email: mbarrett@mail.utexas.edu

[2]Department of Civil Engineering, University of Texas, Austin, TX 78712; email: jmalina@mail.utexas.edu

Background

The mitigation of pollution from nonpoint sources is an important environmental issue. Increased public awareness and enactment of state and federal regulations have been the response to heightened concerns over the problem of nonpoint source pollution. Stormwater runoff from construction sites is an important contributor to nonpoint source pollution.

Silt fences are one type of temporary sediment control accepted by the construction industry and regulatory agencies. Silt fences are employed extensively on highway and other construction projects. These temporary controls are reinforced and supported geotextile fabrics that reduce solids loading through filtration and by impounding runoff, causing velocity reduction, and thus enhancing sedimentation.

Previous research on the sediment removal by silt fences showed high removal in laboratory settings (Kouwen, 1990; Crebbin, 1988). Total suspended solids (TSS) removal in these studies ranged from about 85% to 100%. However, the average diameter of the solids used in the sediment slurry was much larger than the particles generally encountered in construction runoff. Consequently, the sediment removal efficiency expected under field conditions may have been overstated.

There are several existing test methods for evaluating the performance characteristics of geotextile fabrics. Many of these methods measure index characteristics that provide information about the fabric but do not model field performance. Current industry accepted standard test methods are typically index in nature; consequently, most of the specifications for geotextile fabrics are based on the results of tests performed by manufacturers or state testing agencies.

The hydraulic properties of geotextiles often are characterized by apparent opening size (AOS) and permittivity (Ψ). Apparent opening size normally is reported as a sieve size and is the estimated largest pore size in the fabric. The diameter of the openings is commonly between 100 and 1000 μm. The permittivity (Ψ) of a geotextile is a measure of the ease with which water flows through the fabric. Permittivity is determined by placing a fabric specimen of known area in a horizontal orientation, applying a known head of water, and measuring flow rate through the fabric.

Field Monitoring of Temporary Controls

The construction of several new highways in the Austin, Texas area which cross the recharge zone of the Edwards aquifer led the Texas Department of Transportation (TxDOT) to support a four-year study of the effects of runoff from highway construction and operation on water quantity and quality(Barrett et al., 1995a). One element of this investigation was an evaluation of the performance of geotextile fabrics for temporary control of construction runoff. The performance of silt fences initially was evaluated in a comprehensive field monitoring; however, difficulties in obtaining representative water quality samples led to the development of a laboratory program.

Evaluation of the performance of temporary controls in an operational setting required an extensive field monitoring program. Details of the monitoring effort are described by Barrett et al. (1995b) and a summary of the results was presented by Barrett et al. (1998). The sediment removal efficiencies of silt fences were determined from grab samples taken during storm events on highway construction projects. The efficiency of the geotextile silt fences was based upon a comparison of the TSS concentrations in the upstream pond and the effluent downstream of the silt fence, and excludes removal attributed to sedimentation in the pool behind the fence. An example silt fence installation showing the ponded water behind the fence is presented in Figure 1.

Figure 1. Installation of Silt Fence on Highway Construction Project

The median TSS removal efficiency resulting from filtration by the fabric was 0%, with a standard deviation of 26%. The range in calculated efficiencies was -61% to 54%. A negative reduction signifies an observed increase in TSS downstream of the silt fence. Minor errors for in-situ sampling at construction sites are typical. Other sources of error that could result in negative removal efficiencies include: disturbance of bottom sediments during sample collection and untreated flows entering the sampling area below the silt fence.

The generally poor removal efficiency attributed to filtration can be explained by an analysis of the particle size of the samples. Silt and clay sized particles (particles with diameters less than 62 ∝m) comprised the majority of the solids collected from the pond and below the silt fence. The percentage of silt and clays ranged from 68 to 100%, with a median value of 96%. The predominance of small particles in the samples is attributable to the nature of the local soil and to settling of the larger particles in the ponds prior to sampling. The silt and clay sized particles remained in suspension and were able to pass through the silt fence because the diameters of the particles were smaller than the AOS of the fabric.

The turbidity reduction attributed to the silt fences was determined by comparing concentrations in the water impounded by the silt fence and concentrations in samples collected below the silt fence. The median removal for all samples was only 2%, with a standard deviation of 10%. Removals ranged in magnitude from -32% to 49%. Increases in turbidity below the temporary control fence probably are the result of the same sources of error that resulted in negative removals for TSS. Turbidity is a function of the number of small particles in a sample; therefore, these results are consistent with the findings that all of the particles remaining in suspension above the fence are smaller that the AOS of the fabric and consequently little reduction should be expected.

The geotextiles used in silt fence applications often are called "filter fabrics." However, comparison of water quality samples collected in the pond directly behind the fence with those collected immediately downstream of the fence indicate clearly that filtration by the fabric plays little or no role in the reduction of the suspended solids concentrations in construction runoff.

Laboratory Tests of Silt Fences

Flume Tests

The sediment removal performance of silt fences also was investigated under controlled conditions in an outdoor flume. Monitoring these devices in a flume allowed control of such variables as the influent flow rate and TSS concentrations. An outdoor flume at the Center for Research in Water Resources at the University of Texas was used for the sediment control experiments. A picture of the experimental setup is shown in Figure 2.

Top soil (Austin silty clay) was mixed with water to create simulated runoff with an influent TSS concentration of approximately 3,000 mg/L, which was in the upper range of TSS concentrations observed in runoff in the field and is the concentration used in ASTM D 5141 (VTM-51 test method, Wyant, 1993).

The runoff flowed through a baffle wall into the flume and down to the sediment control. The silt fences were placed in the flume at a distance of 7.6 m from the mixing tank. The resulting pool of runoff drained through the sediment control and the water level in the flume upstream of the sediment controls was monitored with an ISCO recording flow meter. Grab samples of the influent were taken from the mixing tank during the filling portion of each test. Effluent samples were collected downstream of the control during the filling and draining portions of each test on timed intervals.

The change in removal efficiency over time was observed by running a series of tests through the same control. Four types of silt fences were subjected to cycles of simulated runoff events. Properties of the fabrics as reported by the manufacturers are summarized in Table 1.

Table 1. Summary of Fabric Properties

Type of Fabric	AOS, # sieve size(µm)	Permittivity, sec-1 reported/observed	No. of Tests
Belton woven	#30 (600)	0.4/0.015	9
Exxon woven	#30 (600)	0.1/0.002	5
Mirafi non-woven	#100 (150)	1.5/0.0004	5
Amoco woven	#20 (850)	0.2/0.012	5

Figure 2. Flume for Determining Performance of Silt Fence Fabrics Table 1 Tested Silt Fence Fabric Properties (Barrett et al., 1998)

The water level upstream of the controls was monitored for the duration of each flume test. Changes in the level of the ponded water surface were converted into volumes of runoff passing through the sediment control in a given time interval. These volumes were used to calculate a detention time, effluent suspended solids load, and TSS removal efficiency for each test. In the filling portion of the test, the incremental volumes flowing out of the control were calculated as the difference between the volume of runoff going into the flume and the change in volume within the flume over that time interval.

One flume test was conducted with no control device in the flume. The influent flow rate matched that observed for the slowest filling test in order to determine the highest removal expected from sedimentation in the test apparatus itself. Some settling of the suspended particles occurred as the runoff passed from the highly agitated mixing tank to the relatively quiescent flume. This phenomenon is similar to sediment-laden runoff flowing down a slope in the field at a high velocity. Sediment is deposited at the toe of the slope when the slope gradient and runoff velocity decrease. TSS removal observed without any control in the flume was 34%.

Flow rates measured in the flume tests were approximately two orders of magnitude less than the values reported by the manufacturers. A comparison of reported and estimated permittivities is shown in Table 1. Much of the difference between measured and predicted flow rates appeared to be caused by sediment clogging the fabric openings.

Detention times were calculated for each test of the silt fences. An average detention time was determined for the controls. The relationship between detention time and TSS removal efficiency for each test is shown in Figure 3. A correlation between detention time and TSS removal efficiency is apparent. The highest removal efficiencies were observed for the non-woven Mirafi fabric. This fabric also had the lowest flow rates and the longest detention times of the fabrics tested. The Mirafi fabric produced longer detention times than the other fabrics in this series of tests even though this fabric had the highest reported permittivity (Table 1). This fabric also had the smallest apparent opening size, suggesting that clogging of the fabric with sediment was responsible for the unexpected hydraulic performance. This observation demonstrates that field performance may not be determined from current parameters used to characterize the hydraulic properties of these fabrics.

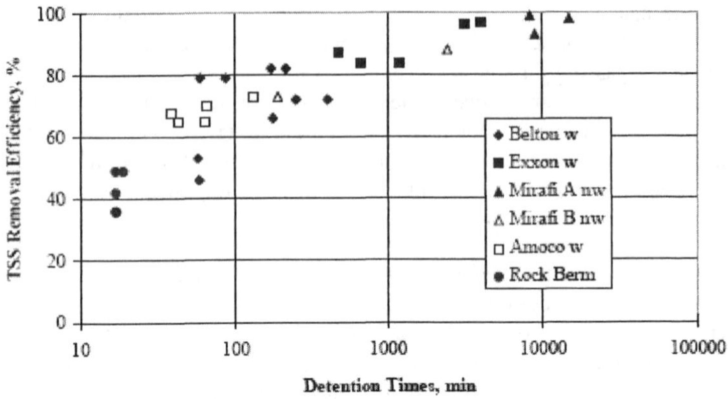

Figure 3. Relationship between Detention Time and TSS Removal (Barrett et al., 1998)

After major rainfall events, detention times decreased for the woven fabrics. This observation suggests that sediment which had accumulated in the woven silt fence fabrics was washed off resulting in increased flow rates. This washing of the fabric may also occur in the field when rainfall precedes the ponding of runoff. The nonwoven fabric appears to have retained more of the trapped sediment in the three-dimensional structure where it was not as easily removed by rainfall events.

Permeameter Tests

Samples of silt fence fabric also were subjected to constant head permeability tests because of the extremely low flow rates observed in the flume tests. Permeameter tests permitted a greater insight into the hydraulic behavior of silt fence fabrics than the flume tests. The relationship between head and flow rates was more apparent because constant heads were used and a wider range of heads were possible.

Three samples were chosen from the fabrics tested in the flume. A modified soil permeameter allowed the flow rates to be determined for heads of 0.05, 0.1, 0.2, 0.3, 0.4, 0.5, and 0.6 m. ASTM D 4491 (ASTM, 1992) is run at a head of 0.05 m, while silt fences in the field can experience heads of up to 0.6 m.

During a test the inlet valve was adjusted until a steady state flow rate was achieved at the specified head. A container of known volume was placed beneath the discharge from the column and the time required to fill the container was recorded. Effects of entrained air on flow through a geotextile were determined by using two sets of tests. In one test, the air was removed from the fabrics by applying a vacuum before each change in head. Water was drawn through the fabric prior to the test with a hose connected to a vacuum flask to remove any trapped air from the fabric. In the other test vacuum was not applied to the samples. Two tests were run for each of three fabric samples; i.e., with and without application of a vacuum to the fabric specimens. Constant heads varied from 0.05-0.6 m.

The permittivities (Ψ) observed in the tests at heads of 0.05 m were the same order of magnitude but slightly greater than the permittivities reported by the manufacturers. This observation may be the result of systematic differences between the permeameter test and ASTM D 4491. A typical plot of flow rate versus head is presented in Figure 4. These data demonstrate a nonlinear relationship between flow rate and head that diverges rapidly from the extrapolated value expected based on the ASTM test. The best-fit equations relating flow to head show that the flow is a power function of the head with exponents ranging from 0.46 to 0.71.

The ASTM method assumes that flow is laminar (i.e. viscous effects dominate); therefore, the flow rate should be a linear function of the head (exponent equal to 1). If inertial forces dominate (turbulent flow), the flow rate should be a function of the square root of the head. The data observed for the permeameter tests indicate that the actual drainage behavior lies somewhere between these two cases.

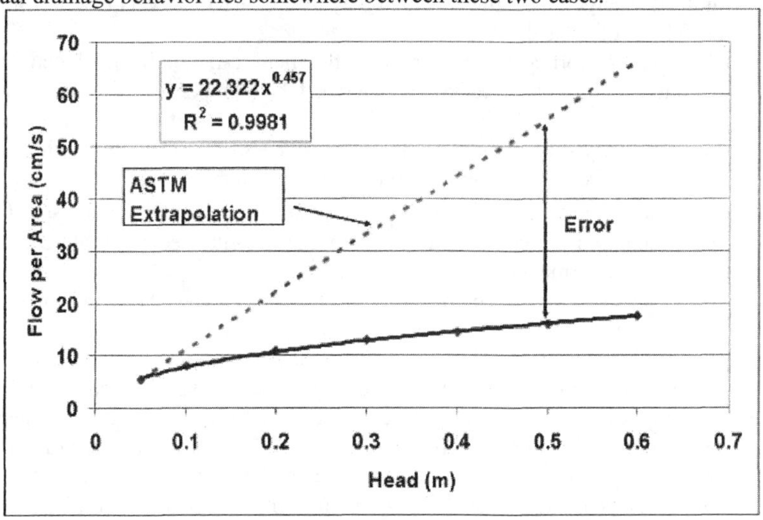

Figure 4. Relationship between Head and Flow Rate

Conclusions and Recommendations

1. Field evaluation of the efficiency of silt fences in removing sediment in runoff from highway construction sites showed that removal was not attributable to filtration by the fabric.
2. The median concentration of solids discharged from the silt fence controls was approximately 500 mg/L. This concentration is significantly lower than the 3000 mg/L of TSS often reported in construction runoff and suggests that approximately 75% of the particles in the runoff were removed by sedimentation in the ponds which formed behind the fences.

3. Geotextile silt fences were ineffective in reducing turbidity. The particles responsible for causing turbidity are mainly silt and clay size (diameter less than 62 μm) or an order of magnitude smaller than the size of the openings in the fabric (about 600 μm) .
4. The efficiency of silt fences observed in the field appears to be dependent mainly on the detention time of the runoff behind the control. The detention time is controlled by the geometry of the upstream pond, hydraulic properties of the fabric, and maintenance of the control.
5. The mean sediment removal efficiency observed in flume studies ranged from 68 to 90% and was correlated with the detention time of the runoff. The geometry of the flume created a large ponded area behind the controls resulting in long detention times and significant particle settling even with the fine-grained sediment used in the tests. This observation indicates that silt fences should be sited in the field to maximize the ponded volume behind the fence. The TSS concentration in construction runoff typically is 3000 mg/L or greater; therefore, at a removal efficiency of 75% the concentration of TSS in the discharge is 750 mg/L, which may lead casual observers to question the usefulness of these devices.
6. Hydraulic performance of fabrics in the field can not be predicted from parameters currently used to characterize fabric properties. Sediment-laden runoff flow rates through the controls were two orders of magnitude less than those typically specified by many regulatory agencies. The fabric, which had the longest detention times in this series of flume tests, had the highest reported permittivity (1.5 s^{-1}), but the smallest apparent opening size (150 ∝m), suggesting that clogging of the fabric with sediment affected the hydraulic performance.
7. The results of modified permeameter tests of silt fence fabrics indicated that flow rates through the fabric were not linearly related to head as is assumed in the ASTM definition. This discrepancy may be attributed to turbulent flow through the fabrics at high heads, resulting in much lower flow rates.
8. Development of a new test or series of tests to characterize the expected performance of geotextile fabrics as silt fences is needed. The use of current parameters results in an over-estimate of the area of a site from which the runoff can be treated without over-topping. Knowledge of the performance of silt fences under field conditions would allow the development of rational guidelines governing the placement of these controls. Testing of fabrics using sediment size distributions of construction site sediment yield should serve as the foundation for future studies.

Acknowledgments: This study was funded by the Texas Department of Transportation (Project #1943) through the Center for Transportation Research at The University of Texas at Austin. Project guidance during the study was provided by a project monitoring committee that included representatives of TxDOT and the Barton Springs/Edwards Aquifer Conservation District. Graduate students in the Environmental and Water Resources Engineering Program of the Department of Civil Engineering participating in the study included John Kearney and Terry McCoy.

References

ASTM, 1992, Annual book of ASTM standards. American Society for Testing and Materials, Philadelphia.

Barrett, Michael E., Malina, Joseph F., Jr., Charbeneau, Randall J., and Ward, George H., 1995a, Water Quality and Quantity Impacts of Highway Construction and Operation: Summary and Conclusions. Center for Research in Water Resources Technical Report 266, The University of Texas at Austin.

Barrett, Michael E., Kearney, John, E., McCoy, Terry G., Malina, Joseph F., Jr., Charbeneau, Randall J., and Ward, George H., 1995b, An Evaluation of the Performance of Temporary Sediment Controls, Center for Research in Water Resources Technical Report 261, The University of Texas at Austin.

Barrett, Michael E., Malina, Joseph F., Jr., and Charbeneau, Randall J., 1998, "An evaluation of the performance of geotextiles for temporary sediment control," Water Environment Research, Vol. 70, No. 3, pp. 283-290.

Crebbin, C., 1988, Laboratory evaluation of geotextile performance in silt fence applications using a subsoil of glacial origin. Master's Report, University of Washington.

Goldman, S. J., Jackson, Katherine, and Bursztynsky, Taras A., 1986, Erosion and Sediment Control Handbook. McGraw-Hill, Inc., NY.

Horner, R. R., Guedry, J., and Kortenhof, M. H., 1990, Improving the cost effectiveness of highway construction site erosion and pollution control. Washington State Transportation Center, Washington State Department of Transportation, Seattle, Washington.

Kouwen, N., 1990, Silt fences to control sediment movement on construction sites. The Research and Development Branch Ontario Ministry of Transportation, Report MAT-90-03.

Wyant, D. C., 1981, Evaluation of filter fabrics for use in silt fences. Transportation Research Record 832, pp 6-12.

Wyant, D. C., 1993, Developing VTM-51 into an ASTM test method. Virginia Transportation Research Council, Virginia Department of Transportation, and Federal Highway Administration, FHWA/VA-94-R2.

Performance and Design of Vegetated BMPs in the Highway Environment

Michael E. Barrett, Center for Research in Water Resources, University of Texas, Austin, TX 78712; email: mbarrett@mail.utexas.edu

Introduction

Many types of Best Management Practices (BMPs) are available for reducing the concentrations and loading of pollutants in stormwater runoff. These structural BMPs are often costly to construct, require substantial ongoing maintenance, and may be objectionable on an aesthetic basis. Consequently there is a need to develop and document the performance of sustainable treatment systems such as vegetated BMPs.

Vegetated BMPs include grassed swales and vegetated buffer strips that may mitigate the impact of stormwater runoff in two major ways, reduction of the concentrations of pollutants and reduction of the volume of stormwater discharged to surface waters as a result of infiltration into the soil.

Vegetated buffer strips, also known as filter strips, are relatively smooth vegetated areas with moderate slopes that accept stormwater runoff as overland sheet flow. Buffer strips are often present along highways, especially in rural areas. Although not designed specifically for water quality treatment, vegetated areas such as highway medians and shoulders may provide substantial pollutant load reduction compared to discharging untreated runoff directly to receiving waters.

Methods

Eight study locations were selected across California that represented the range of slopes, climate, vegetation coverage, soil characteristics, and other regional factors that might impact pollutant removal in roadside vegetated buffer strips. Concrete collection channels were constructed at various distances from the edge of pavement so that the effect of buffer width on pollutant removal could be quantified. A photograph of a typical system is presented in Figure 1. A critical requirement in this study was the collection of runoff samples directly from the edge of pavement. This was done to avoid problems associated with previous studies where the runoff had crossed at least some vegetated area before the first sample was collected (Welborn and Veenhuis, 1988; Kaighn and Yu, 1996; Dorman et al., 1996).

Figure 1. Typical Monitoring Configuration of California Sites

Slopes at the individual sites ranged from 5 to 52 percent, which is much higher at the upper end than normally recommended in design guidelines for vegetated buffer strips. Widths varied between 1.1 and 13 meters. Vegetation coverage at the sites was generally between 60 and 100 percent; however, at one site in an arid location in southern California, the coverage dipped to as low as one percent at times. Sites in southern California differed from those in the north in that they receive less rainfall and gophers were present at all but the most arid site. Barrett et al. (2004) provided a summary of the sampling methodology and results and a detailed description of all study aspects were provided by Caltrans (2003).

Water quality performance of roadside swales was documented by Barrett et al (1998) at two sites along freeways in Austin, Texas. The pollutant removal performance was determined by comparing concentrations of pollutants in samples of stormwater runoff that were collected directly off the road surface and in samples of stormwater discharged from the swale located in the highway median.

The median swales were designed to convey stormwater and not as a runoff treatment device. A picture of one of the sites is presented in Figure 2. The median cross-section is V-shaped with a rounded bottom. Runoff from the highway flows as sheet flow down the sides of the grassy slope and then along the center of the median. The sides of the swale varied in slope between 9 and 12 percent, while the longitudinal slopes were 0.7 to 1.7 percent.

Figure 2. Median Swale Monitored in Austin, TX

Results

Buffer Strip Performance

There are a number of possible measures of performance of storm water BMPs. The two most common report performance as a removal efficiency (i.e., a percentage change between the influent and effluent quality) or focus on the effluent quality achieved. A boxplot of observed total suspended solids (TSS) data at the Sacramento, California site is presented in Figure 3 and suggests that an irreducible minimum concentration occurs that is relatively insensitive to concentrations at the edge of pavement. Consequently, the primary measurements included in this report are the minimum concentration observed for each constituent and the distance at which it stabilizes. An analysis of variance indicated that the TSS concentrations at this site were not significantly different at distances of 4.6 meters and greater from the edge of pavement.

Monitoring of the eight sites indicated that concentration reductions occurred consistently for TSS and total metals and frequently for dissolved metals, and that nutrient concentrations were generally unchanged by the buffer strips. Concentrations of dissolved solids and total and dissolved organic carbon tended to increase as the runoff crossed the vegetated areas.

For the constituents whose concentrations were lowered by the buffer strips the median of the average values at each of the sites except Moreno Valley, which had no reduction in concentration because of the extremely low levels of vegetation coverage, is shown in Table 1.

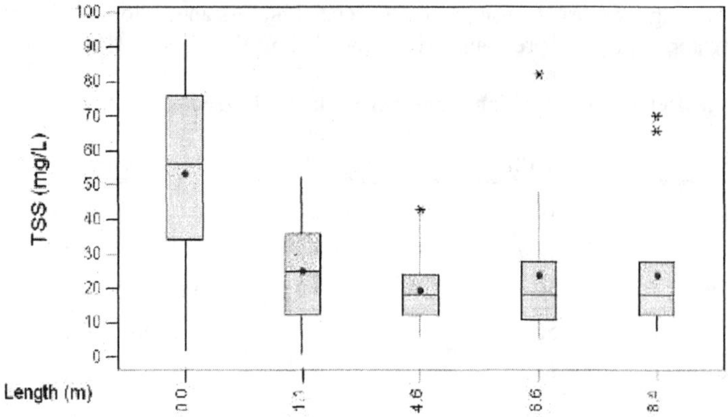

Figure 3 Relationship between Concentration and Width of Buffer Strip

Table 1 Comparison of Median Edge of Pavement and Discharge Concentrations

Constituent	**Median Edge of Pavement Concentration**	**Median Discharge Concentration**
TSS (mg/L)	88	25
Total Copper (∝g/L)	43	8.6
Total Lead (∝g/L)	23	3.0
Total Zinc (∝g/L)	127	25
Dissolved Copper (∝g/L)	16	5.2
Dissolved Lead (∝g/L)	1.4	1.3
Dissolved Zinc (∝g/L)	43	12

A summary of the shortest length observed to produce a constant (best) discharge quality for all constituents that decrease in concentration is presented in Table 2. For the sites with relatively few samples (Irvine and San Onofre) the distance presented is where the lowest concentrations are observed rather than where no statistical difference was demonstrated. The only site not to produce significant reductions was Moreno Valley, which is located in the arid region and had a very low level of vegetation coverage.

Figure 4 presents a comparison of TSS effluent concentrations at the four northern California sites, which are generally similar except for slope. These slopes ranged from 10 percent at the "299" site near Redding, California to approximately 50 percent at Cottonwood and San Rafael. Based on an ANOVA these concentrations are not statistically different at the 95 percent confidence level despite the differences in slope among the sites. Consequently, vegetated buffers appear to be effective at much greater slopes than previously recommended by Young et al. (1996) and others.

Table 2 Buffer Width at which Concentrations Stabilized

Site	Slope (%)	Vegetation Coverage (%)	Distance (m)
Redding	10	88	4.2
Sacramento	33	93	4.6
San Rafael	50	87	8.3 (shortest measured)
Cottonwood	52	78	9.2 (shortest measured)
San Onofre	11	78	5.3
Irvine	11	74	13
Yorba Linda	14	80	13
Moreno Valley	13	19	No concentration reduction

Although concentration reductions were observed on slopes of up to 50 percent, any concentrated flow down these slopes resulted in severe erosion. Consequently, runoff should be diverted away from slopes that exceed about 30 percent.

The results of this study were compared to those obtained in the recently completed California BMP Retrofit Pilot Study in southern California that included an assessment of the performance of vegetated buffer strips (Caltrans, 2004). The buffer strips in the pilot study were engineered and operated specifically for water quality improvement. These sites also had a very rigorous operation and maintenance protocol that kept grass height and coverage within strict requirements. However, the improvement in water quality was no better than that observed in this study of standard roadside shoulders.

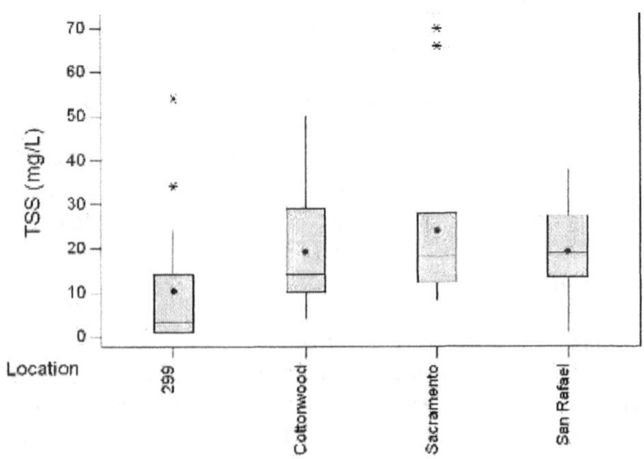

Figure 4 of TSS Effluent Concentration from Northern California Sites

Vegetated Swale Performance

The concentrations of selected constituents at the edge of pavement and at the discharge point of the swale are shown in Table 3. Note that the discharge concentration for TSS is the same in this study as in the California assessment of vegetated buffer strips. In contrast to the California study, reduction of nutrient concentrations was observed in the swale system.

Table 3. Concentrations Observed at Swale Sites (Barrett et al., 1998)

Constituent	Edge of Pavement Concentration	Discharge Concentration
TSS (mg/L)	174	25
Total Lead (∝g/L)	115	80
Total Zinc (∝g/L)	238	32
Nitrate (mg/L)	1.09	0.71
Total Phosphorus (mg/L)	0.39	0.24
TKN (mg/L)	2.39	1.46

Figure 5 presents the concentration measured along the length of the swale for several events. Given that the average edge of pavement TSS concentration is 174 mg/L, it is apparent that substantial reduction in concentration occurs on the sideslopes of the swale before the runoff reaches the centerline. For many events the change in concentration along the length of the swale is small, which suggests that an irreducible concentration is often achieved on the sideslope.

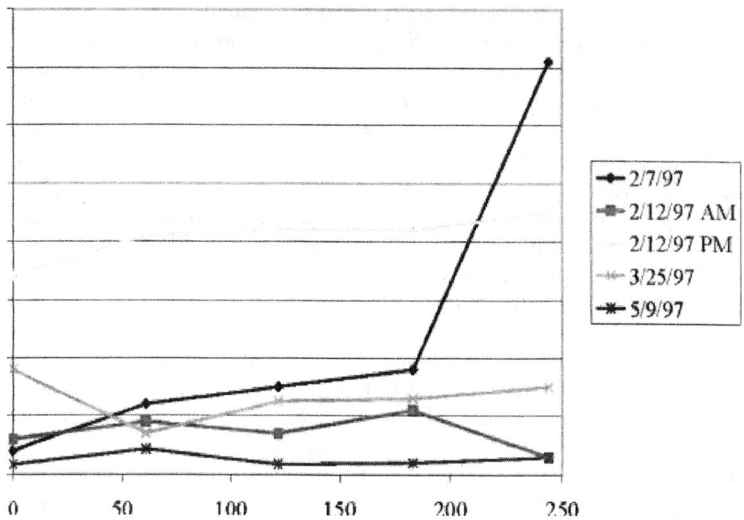

Figure 5. TSS Concentration along Center of Median (Barrett et al., 1998)

The amount of infiltration that occurred in both the California and Texas studies was substantial. No runoff from the longer buffer strips was observed during many events, especially in southern California where the antecedent moisture conditions were generally lower. In general, infiltration is responsible for an equal or greater portion of the load reduction than the change in concentration. The infiltration losses resulted in load reductions even for those constituents that did not exhibit a concentration reduction.

Vegetation Type

Vegetation type and relative quantity was similar at all the California sites except for Moreno Valley, which had less than 25% vegetation coverage for most of the study period. Non-native grasses (Italian rye and brome grasses primarily) dominate and comprise between 65% and 100% of the vegetative cover type. Average vegetation height varied between 11 and 22 cm, and was not correlated with performance. The vegetation at Redding, which produced runoff with the lowest constituent concentrations, consisted of 73% grasses with an average height of about 15 cm. This height is near the conventional recommendation for vegetated storm water controls.

Redding and Sacramento have average vegetation coverage exceeding 80% and with moderate slopes achieve irreducible minimum concentrations within 5 meters of the edge of pavement. Sites in southern California such as Irvine, Yorba Linda, and San Onofre have coverages of less than 80% and with similar slopes require about 10 meters to achieve minimum concentrations. This suggests that performance falls off rapidly as the vegetation coverage declines below about 75 or 80 percent.

The two swale sites monitored in the Austin had substantially different types of vegetation, with one consisting primarily of Prairie buffalo grass and the other a more diverse assemblage of Bermuda grass and various native wildflowers. Vegetation coverage was between 86 and 98 percent. Despite the differences, average TSS concentrations discharged from the two sites were very similar, 21 mg/L and 29 mg/L. These concentrations are similar to those observed in the California study, suggesting that vegetation type is not an important factor in performance. This conclusion is reinforced by the results of the earlier BMP pilot study conducted by the California Department of Transportation of buffer strips. The average minimum concentration of TSS observed at these sites, which had a monoculture of salt grass was about 27 mg/L (Caltrans, 2004).

In summary, these results indicate that substantial pollutant concentration and load reduction occurs in vegetated areas adjacent to highways, even when these areas were not designed as treatment systems. These reductions are comparable to those that would be expected in engineered systems for treating highway runoff. Consequently, maintaining vegetated areas adjacent to highways and eliminating curb and gutter conveyance systems where possible is a cost-effective, sustainable, stormwater treatment system for highways, especially those located in rural areas.

Design and Maintenance Recommendations

Vegetated Buffer Strips

- Vegetated shoulders exhibit substantial pollution removal at slopes up to 30 percent.
- A minimum vegetative cover of about 70 percent is required for concentration reduction to occur, although a rapid decline in performance occurs below about 80 percent.
- Vegetated shoulders with at least 80 percent coverage reduce pollutant concentrations to irreducible levels within 5 meters on slopes of less than 30 percent.
- At vegetation coverages of between 70 percent and 80 percent irreducible concentrations occur at about 10 meters.
- Runoff should be diverted away from slopes greater than 30 percent to prevent erosion.
- Vegetation type and height, highway width, and hydraulic residence time have little or no impact on discharge concentrations.

Vegetated Swales

- The greatest pollutant removal occurs when the geometry of the swale maximizes the length of the side slope, which is where most of the pollutant reduction occurs.
- The optimum cross-section geometry for highway medians is a broad V-shape or parabolic rather than the trapezoidal geometry normally illustrated in guidance manuals. The relatively constant side slopes facilitates mowing and is safer for the traveling public than channels with abrupt changes in side slope.
- The sides of swales along highways are often constructed at slopes of approximately 6:1 (H:V) for safety reasons and this is well within the range at which pollutant reduction occurs.
- Longitudinal slopes of the swales should not exceed 2 percent to prevent channelization or resuspension of previously accumulated material.
- A perfectly flat bottom is not required for effective operation of swales along highways, which are normally sized to convey large, infrequent events, which when combined with the low side slopes results in a broad bottom that reduces the concentration of runoff constituents.

Maintenance Requirements

The sites monitored in both the California and Texas studies were not subject to any special maintenance related to water quality. Primary activities included periodic removal of litter and mowing approximately twice per year. No particular effort was made to keep the vegetation height at some arbitrary level. Consequently, one can conclude that in areas with sufficient rainfall to maintain an adequate vegetation coverage no special maintenance is required to achieve water quality benefits. Occasional mowing for traffic safety reasons (i.e., to preserve visibility) or for fire suppression will produce roadside vegetation with characteristics that promote the removal of many stormwater contaminants.

References

Barrett, M. E., Walsh, P. M., Malina, J. F., Jr., Charbeneau, R. J. (1998). "Performance of vegetative controls for treating highway runoff." *ASCE J. of Env. Eng.*, 124 (11), 1121-1128.

Barrett, M.E., Lantin, A., and Austrheim-Smith, S. (2004). "Stormwater pollutant removal in roadside vegetated buffer strips." *Proceedings 83rd Annual Meeting of the Transportation Research Board*, Washington, DC, January 11-15, 2004.

Caltrans. (2003). *Roadside Vegetated Treatment Sites (RVTS) Study Final Report*, Report No. CTSW-RT-03-028, Division of Environmental Analysis, California Dept. of Transportation, Sacramento, CA.

Caltrans. (2004). *BMP Retrofit Pilot Program Final Report*, Report No. CTSW-RT-01-050, Division of Environmental Analysis, California Dept. of Transportation, Sacramento, CA.

Dorman, M.E., Hartigan, J., Steg, R.F., and Quasebarth, T.F. (1996). *Retention, Detention and Overland Flow for Pollutant Removal from Highway Stormwater Runoff.* Volume I: Research Report, Federal Highway Administration Report FHWA-RD-96-095.

Kaighn, R.J., Jr., and Yu, S.L. (1996). "Testing of roadside vegetation for highway runoff pollutant removal." *Trans. Res. Rec.*, 1523, 116-123.

Welborn, C.T., and Veenhuis, J.E. (1988). *Effects of Runoff Controls on the Quantity and Quality of Urban Runoff at Two Locations in Austin, Texas*, U.S. Geological Survey, Water-Res. Invest. Rpt. 87-4004.

Young, G.K. et al., (1996). *Evaluation and Management of Highway Runoff Water Quality.* Publication No. FHWA-PD-96-032, U.S. Department of Transportation, Federal Highway Administration, Office of Environment and Planning.

Evaluating Bioretention Areas from Two Field Sites in North Carolina

W.F. Hunt[1] and A.R. Jarrett[2]

[1] Assistant Professor and Extension Specialist, Department of Biological and Agricultural Engineering, North Carolina State University, Raleigh, NC 27695-7625, PH: 919-515-6751; email: bill_hunt@ncsu.edu

[2] Professor, Department of Agricultural and Biological Engineering, Pennsylvania State University, University Park, PA 16802, PH: 814-865-5661, email:arj@psu.edu .

Introduction

Stormwater runoff is a primary factor in the degradation of many streams and other water bodies (USEPA, 2000). The adverse impacts of this and other sources of pollution include closures of shellfish waters, fish kills, and reduction of aesthetics, which consequently reduces fishing and recreational values of downstream waters. Stormwater runoff may contain many pollutants, including nitrogen, phosphorus, and many metals (Barrett *et al.*, 1998, and Wu *et al.*, 1998). All impermeable surfaces, including rooftops and parking lots, contribute to the pollution loads found in stormwater runoff.

In coastal areas of the eastern United States, nutrients are a particular concern. Excess nitrogen and phosphorus loading in coastal North Carolina are blamed for fish kills in NC estuaries (Gray, 2000). Similar impacts were noted in the Chesapeake Bay, to which over one-half of PA drains (News & Observer, 2001). Government officials from Chesapeake Bay watershed states have declared a nitrogen reduction goal of 50% by 2010. State and federal regulations do force new developments to mitigate the effects of pollutants in stormwater by requiring new developments to install stormwater treatment devices called Best Management Practices (BMPs). Examples of stormwater BMPs include stormwater wetlands, sand filters, wet ponds, and, more recently, bioretention areas. Some states, such as North Carolina, require that all stormwater practices be proven effective to reduce the amount of nitrogen leaving developments in runoff (NCDENR, 1999).

Stormwater practices, particularly those installed in areas frequented by pedestrian traffic, not only need to be effective pollutant removers, but they must also maintain some aesthetic appeal. BMPs that can be integrated into the design of parking lots and neighborhoods are particularly important. The use of bioretention devices, essentially a vegetated sand filter planted with shrubs, trees or grass, is poised to expand because their use appears to improve water quality and they can be designed to be aesthetically pleasing. Bioretention can be used to treat commercial site runoff, particularly from parking lots and rooftops, because they can be sited within parking lot medians. Many communities require up to 15% of these large commercial developments to be set aside for medians and other open space (Avery, 2001).

However, current bioretention designs do not substantially reduce the levels (concentration and mass) of nitrate-nitrogen (Davis *et al.*, 2001 and 2003), especially compared to the performance standards some states will require BMPs to meet. For example, North Carolina requires all new stormwater practices to have nitrate removal rates approaching 30%. Davis's laboratory bioretention studies report that nitrate-nitrogen removal rates were only 16%. The study did show, however, that bioretention areas can substantially reduce concentrations of phosphorus and several metals in runoff. One main reason bioretention areas might not reduce nitrogen loads, particularly nitrate-nitrogen loads, lies in the nitrogen cycle. For nitrate to be reduced to nitrogen gas, it must be in an anaerobic environment. Without an anaerobic environment, nitrate, which is a very stable form of nitrogen, will remain in solution (Kadlec and Knight, 1996). Bioretention designs currently used do not incorporate an anaerobic zone (NCDENR, 1997). The research investigated ways to create an anaerobic zone in a bioretention cell and measured how the addition of an anaerobic zone increases denitrification rates.

Study Sites

Bioretention areas were installed at three locations in North Carolina: Wilson, Chapel Hill, and Greensboro (see Figure 1). A description of each of these three locations follows.

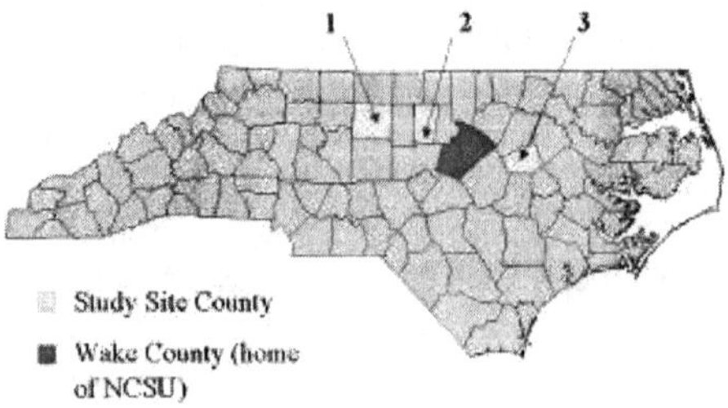

Figure 1. Map of North Carolina showing the three locations – 1)) Greensboro, 2) Chapel Hill, and 3) Wilson – where bioretention areas were studied.

Greensboro Battleground Crossing Shopping Center.

Battleground Crossing Shopping Center in Greensboro was constructed in 2000 and 2001. The county soil survey indicated that the soil on site was a Madison clay loam, relatively tight, with a very low permeability (0.0014 to 0.0042 mm/s) (0.2 to 0.6 in/hr) below the A soil horizon (Stephens, 1977). The soil typically had a perched water table near the surface (0.15-0.45 m or 6-18 in). This bioretention design included underdrains due to the soil's low conductivity. Underdrains were installed to ensure the bioretention areas drained freely and did not become a wetland. The watershed's use was commercial with nearly all site visits coming from leisure-commercial traffic that is cars, pick-up trucks, and SUVs. The shopping center was immediately upstream from Horsepen Creek, which eventually drains into the Cape Fear River, an important water supply for Fayetteville and Wilmington, NC. Two bioretention areas were constructed at the site. Both had 1.2-m (4-ft) deep soil media layers. As described earlier, bioretention areas are conventional treatment devices. Because of this, they were believed to produce NO_3^-. By introducing an anaerobic layer within the bioretention area, the ability of NO_3 to be denitrified into N_2 gas was expected to increase. This was done by a slight change in drainage configuration; in lieu of installing underdrains that simply flow by gravity to the outlet box, an elbow was installed so that an anaerobic zone was produced (see Figure 2 for a graphical view). An upturn in the drainage pipe forced the bottom 0.6 m (2 ft) of the bioretention cell to remain saturated. The second cell was conventionally drained. The watersheds of the bioretention areas were 0.20 ha (0.5 acres) each and the watershed size to bioretention surface area ratios were nearly the same (5%). An organic sandy soil was backfilled over the underdrains. The soil's hydraulic conductivity was tested two years after construction and was found to be around 0.11 mm/s (15 in/hr), which was higher than standard design guidelines.

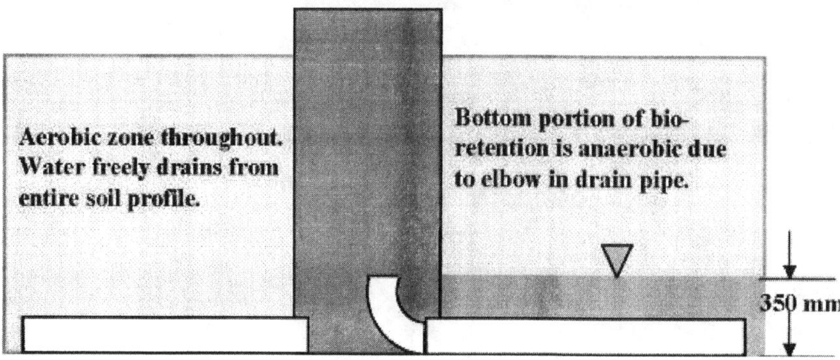

Figure 2. Creating an anaerobic zone in the bioretention device by using an upturned elbow.

Chapel Hill University Mall.

The University Mall in Chapel Hill has been open since the early 1970's. The county soil survey shows that the soil on this site was relatively tight: clay, clay loam, and silty clay (White Store-urban complex) (Dunn, 1977). The soil had a very low permeability with a hydraulic conductivity less than 0.42 ∝m/s (0.06 in/hr) below the A soil horizon. The soil typically had a perched water table near the surface (0.15-0.46 m or 6-18 in). Two separate bioretention areas were constructed at the mall during the summer and fall of 2001. As with the Greensboro site, the bioretention designed here included underdrains due to the soil's low conductivity.

The watershed's use was primarily commercial with nearly all site visits coming from leisure-commercial traffic. The shopping center drained to Little Creek, which eventually drained into Jordan Lake, an important water supply for the Raleigh-Durham metropolitan area. The watershed of the bioretention area C1 was 0.09 ha (0.28 ac); the watershed of C2 was 0.23 ha (0.75ac). The bioretention surface area to watershed size ratios indicate that these two bioretention areas occupied a substantial portion of the watershed (6.6% to 14.9%) and were not the same. The design parameter tested at the Chapel Hill site was soil depth, as nominal depths of 0.6 and 1.2 m (2 and 4 ft) were tested. No anaerobic conditions were designed into the system. A sandy media was used to backfill the bioretention cells. The hydraulic conductivity of the sandy medium cores was calculated 18 months after the sites were constructed using the constant head permeability test and were found to range from 0.009 to 0.021 mm/s (1.3 to 3.1 in/hr). This permeability was within the target design range. Figure 3 shows cell #1 eight months after it was constructed.

Figure 3. Bioretention cell #1 in Chapel Hill eight months after construction.

Field Data Collection and Analysis

At each site the following sampling scheme was followed. Two tipping-bucket rain gauges were installed in an area clear of interference within 10 m (30 ft) of the Greensboro site; at Chapel Hill only one rain gauge was installed. Automated samplers were installed at the invert of each underdrain as it emptied into the drainage. Some of the outflow samplers were flow activated. Samples for outflow water quality were designed to be collected on a set time interval from non-flow-activated samplers, because flow through the soil was expected to be relatively constant. The volume of flow (or flow rate) from the underdrains was either automatically or manually recorded, depending upon the type of automated sampler used. Equipment types are reviewed in the next section. The inflow runoff was collected for water quality purposes at two cells in Greensboro, one in Chapel Hill, and two in Wilson. In Greensboro, runoff data were collected at the entrance of both cells. In Chapel Hill, inflow runoff was collected at one of the two cells studied, and in Wilson, 2 of the four cells were outfitted with equipment to sample inflow runoff. Not every cell had inflow captured because of financial and spatial constraints. The study did not have enough money to purchase samplers for every cell's inflow, and several sites did not have enough elevation to install a v-notch weir to effectively measure overflow. This was not viewed as a particular liability because the treatments were located very close to each other, 90 m (300 ft) in Chapel. It was assumed that the water entering the median at the sampling location would be characteristic of that across the entire bioretention watershed. For water quantity purposes, two flow gauges were installed in Greensboro, with only one being installed in Chapel Hill. Because the bioretention cells served a uniform watershed (nearly 100% impervious) and the watershed areas were clearly defined, it was the author's intent to calculate runoff volumes utilizing data from the rain gauges, which measured both depth and intensity. These calculations were hoped to be calibrated by comparing them to the actual measured inflow at a limited number of locations at each site. The samples were composited on a flow basis, which captured first flush and excess runoff. The total volume of water captured for each runoff event was dependent upon the event size. The sampler bottles stored up to 11.4 liters (3 gallons) of runoff.

Results

Estimating Annual Pollutant Mass Inflow and Outflow

Sufficient data were gathered at two bioretention cells to estimate annual pollutant mass inflow to the bioretention area in runoff and outflow from the device to the storm drain network. Data collected for Greensboro Cell #2 and Chapel Hill Cell #1 were used for this estimation.

Greensboro Cell G2

In Greensboro, flow was measured continuously from June 1, 2002 through May 15, 2003 with the exception of two periods running from July 21, 2002 through August 17, 2002 and January 9, 2003 through February 6, 2003. Precipitation amounts were collected either on site or at the nearby U.S. Geological Survey (USGS) gauging station (Horsepen Creek USGS Gauging Station # 020939399200, 0.40 km (0.25 miles) from the site) or airport (Piedmont Triad International Airport, GSO, approximately 6 km (4 miles) from the site). An annual precipitation record was combined. Rain gauges at the site malfunctioned during much of the winter months due to several icy periods. Data from 11 storm events were collected during the one-year period from June 2002 through May 2003, with three storms per season except for the summer (which only had two).

The inflow and outflow masses were summed allowing an annual reduction (or increase) in pollutant load to be calculated. As seen in Table 1, cell G2 had high removal rates for most metals, TN, and NO_3-N. Lead's removal rate is "only" 81%, because concentrations entering the bioretention cell were nearly negligible. As found in Davis *et al.* (2001, 2003), this bioretention cell had very high metal removal rates. Additionally, the reduction of Total Nitrogen was very high, even though this design did not include an anaerobic configuration. During a field investigation, when soil samples from the site were being taken to determine the media's permeability and lateral conductivity tests were performed, pocket saturated zones were found on site. As the cell was augered, much of the bottom layer of soil was dry, i.e., there was no presence of a water table. However, there were portions in the middle of the soil column that were clearly filled with water, and probably saturated.

Table 1. Annual inflow and outflow mass loading for various pollutants at G2 cell in Greensboro.

Pollutant	**Annual Mass in (Kg)**	**Annual Mass out (Kg)**	**Percent Reduction (or increase)**
TN	2.53	1.51	40%
NO_3-N	1.194	0.297	75%
TKN	1.22	1.28	-4.9%
NH_3-N	0.595	0.601	-0.99%
TP	0.129	0.442	-240%
Ortho-P	0.048	0.052	-9.3%
Zn	0.459	0.009	98%
Cu	3.68	0.021	99%
Pb	0.016	0.003	81%
Fe	0.128	16.845	-13,000%
TSS	17.9	48.4	-170%

The bioretention cell emitted substantially amount more total phosphorus, total suspended solids, and iron to the storm sewer network than that which entered the cell. The high TP outflow was directly attributable to (1) a high P-Index (86-100) of the fill soil and (2) the unexpectedly high TSS outflow. Because the watershed was no longer under construction, there was very little TSS inflow into the bioretention device in stormwater runoff. The cell, however, did emit TSS and the phosphorus that was bound to it. Iron effluent masses were evident from the first samples, which were always rusty in color. It was possible that iron from the fill media was leaching from the bioretention cell as it became saturated. This study indicates that high iron levels in effluent may be expected when the bioretention media is comprised partly of clay fines, particularly if the media becomes saturated.

Chapel Hill Cell C1

The process was repeated for Cell #1 in Chapel Hill (C1). Like Greensboro's cell G2, C1 was nominally 1.2 m (4 ft) deep with a conventional drainage configuration. C1 had a smaller contributing watershed than was initially designed due to site development changes during construction. Instead of the prescribed 0.2 ha (0.5 acre) watershed, C1 served a drainage area of 0.06 ha (0.15 acres). Consequently, instead of a bioretention surface area to watershed area ratio of 4-5% (similar to Greensboro's), the actual ratio for this bioretention cell to its watershed was close to 15%, much higher than is typically used in design.

The data collected in Chapel Hill was limited compared to that of Greensboro. Nutrient concentrations were collected (TKN, NH_3-N, NO_3-N, TP, and OP) and flow volumes were measured from enough storms to estimate the reduction of pollutants over a 10-month period (June 2002- April 2003). A summary of these data is shown in Table 2.

Removal rates associated with the Chapel Hill bioretention device were generally better than the site in Greensboro. It was observed in a few select events that, unlike Greensboro, Chapel Hill did not have high TSS rates leaving the bioretention device. Moreover, Chapel Hill's TP removal rate was much better, probably due to the low P-Index (4-12) of the fill soil. It was not surprisin g that the TP removal rates were not only better in Chapel Hill, but that C1 actually removed TP, while G2 adds total phosphorus.

Table 2. Calculated loads and removal efficiencies over a 10-month period (June2002 through April 2003) for bioretention cell C1 in Chapel Hill, NC.

Pollutant	10-Month Mass In (Kg)	10-Month Mass out (Kg)	Percent Reduction (or increase)
TN	0.608	0.365	40%
NO_3-N	0.093	0.081	13%
TKN	0.515	0.284	45%
NH_3-N	0.328	0.045	86%
TP	0.139	0.048	65%
Ortho-P	0.059	0.016	69%

While the data did not show a difference between location of TP and Ortho-P loadings, there was a large difference in the mass of nitrogen constituents between the two devices. While TN removal rates were nearly the same between C1 and G2 (40% each), the variables that did show a difference were TKN and NO_3-N. TKN removal rates in C1 were substantial (45%), while G2 slightly added TKN (5%) to the storm drainage network. Conversely, NO_3-N removal levels were very high in G2 (75%);, whereas, they were minimal in C1 (13%). This may be explained partially by the fact that the Chapel Hill bioretention soils were more of a uniform sand and were absent of much organic matter, unlike Greensboro's bioretention soils in which organic matter was readily evident. Perhaps there was enough organic matter to provide food for microbial action to convert NO_3-N to N_2 in Greensboro, but enough organic matter was not present to do likewise at Chapel Hill.

Examining Variation in Flow

Statistical analyses were conducted to compare outflow to runoff seasonally. The first two events listed in June 2002 were grouped with the dates from late March, April and May of 2003 to comprise the spring season. A standard ANOVA test was used to compare the seasons individually (i.e. Summer to Fall, Summer to Spring, Summer to Winter, etc.). A significance of $p<0.10$ was chosen for the comparison. Comparisons including the mean, standard deviation and the p-value from the ANOVA tests are listed in Table 3. The three warm seasons in North Carolina (Spring, Summer, and Fall) when individually compared to the cold season (Winter) showed a significant difference in runoff reduction. During the warm seasons less water flowed through bioretention areas, therefore, less water entered the storm sewer network than did in the winter. There were no significant differences in runoff reduction among the three warm seasons.

Table 3. Comparison of flow reduction (as measured by % outflow) between seasons. A significance of $p<0.10$ was selected. Any p-value less than 0.10 is italicized in the table below.

Season	**Mean** *(%)*	**St.Dev** *(%)*	**Comparison of Percent Outflow to that of other Seasons. P-value from ANOVA is shown.**			
			Spring	**Summer**	**Fall**	**Winter**
Spring	0.14	0.18	n/a	0.25	0.85	*0.002*
Summer	0.07	0.07	0.25	n/a	0.14	*<0.001*
Fall	0.13	0.10	0.85	0.14	n/a	*0.002*
Winter	0.54	0.21	*0.002*	*<0.001*	*0.002*	n/a

Conclusions

There are five major conclusions which may be drawn from the research presented herein:

(1) Annual TN pollutant load reduction estimated from two conventionally-drained bioretention cells with 1.2 m (4 ft) of fill media from central North Carolina each showed a mass removal rate of 40%. NO_3-N loads were reduced by varying degrees by the two conventionally-designed cells mentioned above. The Greensboro cell removed 75%, while the Chapel Hill cell removed 13%. This may have been due to the possible formation of an anaerobic zone in G2 and not in C1.

(2) Bioretention removal rates for TP at the field sites were not as uniform as they were for TN. This may have been due to the type of media used in the

bioretention cell. The P-index of the media in cell G2 was high (86-100) indicating that the media was saturated with phosphorus. In Chapel Hill the P-index was low (4 to 12) indicative of a media that could accept more phosphorus. Consequently, the site in Greensboro (G2) released to the storm drainage system a TP load 242% higher than the inflow load, while the cell in Chapel Hill (C1) removed 65% of the TP mass that entered the bioretention cell. Another important difference was a relatively high TSS loading in the effluent from the site in Greensboro. The soil media leaked into the drawdown pipes and is believed to have carried with it a higher TP load. Drainage configuration had a significant effect ($p<0.10$) on outflow concentrations at the sites in Greensboro. Outflow from the anaerobic design had a reduced TP concentration when compared with the outflow from the conventional design.

(3) Bioretention areas reduced the mass of many metals from inflow to outflow to a very high degree. Mass metal removals from a bioretention cell studied in Greensboro, NC, had Zn and Cu removal rates over 98%. Pb removal rates over 80% were found during the year-long period beginning in June 2002 and ending in May 2003. The lower percentage of mass removal for Pb, 80%, was due to the fact that lead inputs into the bioretention cells were extremely minimal, since outflow concentrations were almost exclusively below the limit of detection. Iron mass amounts increased dramatically, 13,000%, perhaps due to the bioretention cell's location adjacent to iron-rich in-situ soil or more likely to the fill soil becoming saturated and having its Fe content leach from it and entering the storm sewer system.

(4) Drainage configuration had a significant impact ($p < 0.10$) on Fe outflow concentrations. Outflow from the anaerobic design had lower Fe concentrations than those of the conventional configuration, perhaps since some of the Fe may have settled in the bottom layer of the anaerobic device, although, this is a supposition.

(5) Outflow reduction, was addressed by research at the Greensboro field site. Seasons, and their relative temperatures, had a significant impact ($p < 0.10$) on outflow volume from a bioretention cell studied in Greensboro. The warm seasons (Spring, Summer, and Fall) each had a significantly lower ratio of outflow to inflow than that of the winter. There was not a significant difference in the outflow to inflow ratio among the warm seasons. Because mass removal rates are based upon inflow and outflow, the mass removal rates were much higher in the warm seasons, particularly in the summer, than in the winter.

References

Avery, Donna. 2001. Personal communication. Ms. Avery is the Chief Planner in the Stormwater Services Department in Wilson, NC.

Barrett, M. E., L. B. Irish, J. F. Malina, and R. J. Charbeneau. 1998. Characterization of highway runoff in Austin, Texas, area. *Journal of Environmental Engineering*124(2): 131-137.

Davis, A. P., M. Shokouhian, H. Sharma, and C. Minami. 2001. Laboratory study of biological retention for urban stormwater management. *Water Environment Research* 73(1): 5-14.

Davis, A.P., M. Shokouhian, H. Sharma, C. Minami, and D. Winogradoff. 2003. Water Quality Improvement through Bioretention: Lead, Copper, and Zinc Removal. *Water Environment Research* 75(1): 73-82.

Dunn, J. 1977. *Soil Survey of Orange County, North Carolina*. U.S. Department of Agriculture, Soil Conservation Service. Washington, D.C.

Gray, J. 2000. What are we trying to protect? *Stormwater* 1 (November/December): 46-49.

Kadlec, R. H. and R. L. Knight. 1996. *Treatment Wetlands*. Boca Raton, FL: CRC Press

NC DENR. 1997. *Stormwater Best Management Practice Design Manual*. Raleigh, NC: North Carolina Department of Environment and Natural Resources – Division of Water Quality.

NC DENR. 1999. *Tar-Pamlico River Basin: Proposed Text of Rules*. Raleigh, NC:North Carolina Department of Environment and Natural Resources – Division of Water Quality.

News and Observer (Raleigh, NC). 2001. Chesapeake Bay cleanup in along. Sunday, July 22, 2001 edition, page 5A.

Stephens, R.B. 1977. *Soil Survey of Guilford County, North Carolina*. U.S. Department of Agriculture, Soil Conservation Service, Washington, D.C.

USEPA. 2000. Storm Water Phase II Final Rule: Small MS4 Storm Water Program Overview. Fact Sheet 2.0. Washington, DC: EPA Document Number 833-F-00-002.

Wu, J. S., C. J. Allan, W. L. Saunders, and J. B. Evett. 1998. Characterization and pollutant loading estimation for highway runoff. *Journal of Environmental Engineering* 124(7): 584-592.

Chapter 9

BMP Modeling Techniques

Introduction

By Scott Struck and Robert Traver

There are many models available that incorporate steady-state dynamic simulation of mass transport and water quality processes in all types of surface water environments, including overland flow, small creeks, rivers, lakes, estuaries, coastal embayments, and offshore areas.

As the practice of stormwater management moves toward addressing runoff volume and quality, new design practices and innovative technologies are being explored. Similarly, the requirement for the development of total maximum daily loads (TMDLs) has caused a significant increase in the need for sophisticated modeling approaches to predict the outcome of water quantity and quality controls. Also, the U.S. Environmental Protection Agency determined that it is necessary to establish a level of consistency and defensibility for TMDL modeling tools.

In urban, suburbanizing, and agricultural areas, water quality is often impacted by changes in land use and/or runoff of nonpoint source pollution. Watershed managers often employ the use of best management practices (BMPs) to ameliorate the impacts in these areas. In fact, BMPs are quickly becoming the norm for treating runoff volume and quality, as shown by the success of the ASCE-sponsored International Stormwater BMP Database (http://www.bmpdatabase.org).

Bioretention, a technique incorporated in many BMPs, is becoming a major component of many stormwater quantity and quality control efforts and can include practices such as rain gardens, bio-infiltration, bioswales, or other low impact development (LID) techniques. It works simply by capturing and removing a portion of the storm runoff, through evapotranspiration and/or infiltration / exfiltration. The water quality function is performed as the device captures the first flush of the initial portion of frequent storms, removing it completely from the surface waters. The constituents captured are then treated through soil chemistry and biological action.

While bioretention devices (along with other LID and structural BMPs) are relatively simple, many questions remain on their function, performance, design, maintenance, longevity, and how to best utilize them as part of a watershed plan. Of great interest is how to accurately predict effluents from BMPs from both a quality and quantity perspective. The papers in this chapter are of interest because they address ongoing modeling efforts by incorporating laboratory and field research to improve wet weather flow modeling techniques whether to predict TMDLs or for meeting current water quality standards.

Modeling Concepts for BMP/LID Simulation

Wayne C. Huber[1], Fu-hsiung Lai[2], LaMarr Clannon[3] and Matt Stouder[4]

[1]Department of Civil, Construction, and Environmental Engineering, Oregon State University, 202 Apperson Hall, Corvallis, OR 97331-2302; wayne.huber@oregonstate.edu; (541) 737-6150
[2]U.S. Environmental Protection Agency, Office of Research and Development, Urban Watershed Management Branch, 2890 Woodbridge Ave., Edison, NJ 08837; lai.dennis@epa.gov; (732) 321-6632
[3]Nonpoint Source Education for Municipal Officials, South Portland, ME 04106; lcannon@maine.rr.com; (207) 771-9020
[4]Dept. of Public Works, City of Springfield, OR; mstouder@ci.springfield.or.us; (541) 736-1035

Introduction

A number of control strategies and best management practices (BMPs) are used to mitigate runoff volumes and associated nonpoint (diffuse) source pollution due to wet-weather flows (WWFs), including ponds, bioretention facilities, infiltration trenches, grass swales, filter strips, dry wells, and cisterns. A BMP concept called "low impact development" (LID) – or source control – strives to retain a site's pre-development hydrologic regime, reducing WWF and the associated nonpoint (diffuse) source pollution and downstream treatment needs. Common LID practices include rain gardens, bioretention, rooftop gardens, sidewalk storage, bioswales, porous pavement, buffers and filter strips, roof leader disconnection, rain barrels, and cisterns. Many BMP and LID strategies also apply to dry-weather urban runoff, such as from lawn irrigation and washing of cars. Within this paper, BMP is used generically to represent urban runoff controls, whether for wet or dry weather (but not including municipal or industrial waste treatment). LID will ordinarily be mentioned separately, even though it is included in the overall suite of urban runoff controls that might be used to manage urban discharges.

Methodologies are needed to evaluate these BMPs, their effectiveness in attenuating flow and pollutants, and for optimizing their cost/performance since most simulation models only partially simulate BMP processes. Enhanced simulation capabilities will help planners derive the least-cost combination for effectively treating urban discharges. There is currently a confusing array of options for analyzing hydrologic regimes and planning for LID. Integrating available BMP/LID processes into one model is highly desirable. This paper describes several modeling options for evaluation of BMP/LID performance, with the intention of facilitating the integration of improved BMP/LID modeling methods into the EPA's Storm Water Management Model, version 5 (SWMM5, http://www.epa.gov/ednnrmrl/swmm/index.htm). Suggested improvements are based both on capabilities of other contemporary models and on simulation of the environmental processes themselves as found in sources such as Kadlec and Knight (1996), Minton (2002), and Metcalf and Eddy (2003).

Fundamental Process Categories

A fundamental process category (FPC) is essentially the same as a unit process in environmental engineering (Metcalf and Eddy 2003). There is no single universal list of FPCs; Minton (2002) provides a useful taxonomy oriented toward stormwater treatment. Common structural BMPs are grouped by six FPCs in Table 1, but alternatives are possible. One BMP may fall under different FPCs, especially BMPs such as wetlands and bioswales that use multiple processes.

Modeling of BMPs by simulation of FPCs requires extensive data on stormwater treatability (settling velocity distribution and/or distribution of particle sizes and specific gravities) as well as site and design descriptions. Data availability becomes an issue. Most BMPs are modeled by a heuristic combination of FPC simulation and empirical performance measures, such as percent removal.

LID emphasizes distributed source controls, leading to a reduction in stormwater volume through storage, infiltration and evapotranspiration (ET). Such controls ideally are implemented at the individual lot or parcel level, assuming that institutional issues about the efficacy of homeowner maintenance can be overcome (Prince George's County 2000). Several authors have suggested that stormwater quality can be managed by treating the flows associated with small storms (Heaney et al. 1977, EPA 1983, WEF and ASCE 1998, Wright et al. 2000, Pitt and Voorhees 2000). These are storms that occur many times a year, with depths less than or equal to about 0.5 in. (13 mm), and are responsible for the majority of the pollutant discharges. A combination of hydrologic source control for reduction of volumes and BMP implementation for additional volume reduction and/or pollutant removal will likely be the control method of choice in the future. Such controls may be implemented in series ("treatment train") and in parallel (bypass and parallel treatment options). Simulation models such as SWMM need the flexibility to represent this wide variety of urban runoff controls.

BMP Modeling with SWMM

Research (Huber et. al. 2003), synopsized in this paper, identifies several BMP simulation needs with regard to improvements to SWMM, versions 4.x (1988 – 2004) and the improved SWMM5 about to debut in Spring 2004. These are grouped by BMP category in sections below. In addition, an effort has been devoted to identification of both hydrologic and water quality processes (i.e., FPCs) to be included in the simulation, since SWMM is a conceptual process model.

Table 1. Structural BMPs categorized by fundamental process categories.

Unit Process	BMPs
Gravity settling/ sedimentation/flotation	Extended detention basins
	Retention/detention ponds
	Wetlands
	Deep sump hooded catchbasins
	Tanks/vaults
	Oil/water separators
Filtration and sorption	Media/sand/compost filters
	Screens
	Bar/trash racks
	Biofilters
	Wetlands
	Bioswales
	Manufactured bioretention systems
Infiltration	Infiltration basins
	Infiltration trenches
	Porous pavement
	Dry well
	Bioretention cells
	Dry swales, filter strips
Hydrodynamics	Vortex separator
	Oil/water separator
Biological treatment	Wetlands/wetland channels
	Bioswales, biofilters
Chemical treatment	Precipitation/flocculation
	Disinfection

Wetlands and Ponds. Extended detention ponds are drawn-down between storms, typically within 24 – 72 hours after the end of the storm event (Urbonas and Stahre 1993). These areas are often used for multiple activities, such as recreation. Extended detention or "wet" ponds have a permanent pool and provide treatment by sedimentation and through additional biological and filtration mechanisms, as do wetlands.

Currently in SWMM (i.e., in versions preceding SWMM5), removal can be simulated through use of a generic removal function, in which removal can be a function of detention time, influent concentration, and other state variables. In addition, within the old Storage/Treatment (S/T) Block of SWMM, it has been possible to route and settle particulates described by a settling velocity (or particle size and specific gravity) distribution. Within SWMM5, routing of such a distribution should be possible through all flow "objects" (overland flow, channels, pipes, storages).

However, it is also desirable to incorporate FPCs involving classic design of sedimentation facilities on the basis of overflow velocity, for example. In order to facilitate this activity, a compromise has been sought between the extremes of a continuous-flow, stirred-tank reactor (CFSTR) and a plug-flow reactor (PFR).

One common heuristic solution is to use Fair and Geyer's (1954) "tanks in series" (TIS) method for analysis of imperfect sedimentation basins for water and wastewater treatment, in which the fraction of particles captured that has settling velocity v_s. is:

$$R = 1 - \frac{C_{out}}{C_{in}} = 1 - \left(1 + \frac{v_s}{NQ/A}\right)^{-N} \tag{1}$$

where

R = fraction captured or retained in pond
Cout = effluent event mean concentration (EMC)
Cin = influent EMC
Q = pond outflow rate
A = pond surface area
N = number of CFSTRs in series (empirical measure of hydraulic efficiency)

The hydraulic efficiency factor, N, can reflect short-circuiting, for example, and ranges from 1 for "very poor performance," to 3 for "good performance," to 5 or higher for "very good performance" (Fair and Geyer 1954, Pitt and Voorhees 2000, Minton 2002). Another interpretation of N is the number of CFSTRs or "tanks" in series, as an approximation leading up to plug flow (Kadlec and Knight 1996, Chapra 1997). That is, Equation 1 is the removal efficiency for N CFSTRs in series, all with the same flow rate, and each with 1/N of the total pond area A. The ratio, vsA/Q, can be expressed in several forms, all related to detention time:

$$\frac{v_s}{Q/A} = \frac{v_s t_d}{h} = k\,t_d = \frac{kV}{Q} = \frac{k'}{q} \tag{2}$$

where

k = first-order decay coefficient (1/time)
k' = k h = rate constant (depth/time)
q = Q/A = overflow velocity (depth/time)
V = volume
t_d = detention time
h = pond depth

When N = 1, the equation gives the steady-state performance of one CFSTR. The case $N \rightarrow \infty$ corresponds to perfect horizontal plug flow representation for a tank or pond (Chapra 1997):

$$\lim_{N \to \infty}(\text{Eqn. }1) = 1 - e^{-\frac{v_s}{Q/A}} \qquad (3)$$

The removal efficiency, R (fraction), of particles having settling velocity vs in quiescent settling (e.g., in the permanent pool of a wet pond), is given by:

$$R = \frac{v_s t_d}{h} \leq 1 \quad , \qquad v_s \leq h/t_d$$

$$R = 1 \quad , \qquad v_s \geq h/t_d.$$

All particles with settling velocity vs will be removed as long as the storm inter-event time is greater than the detention time (neglecting resuspension and several other side effects). Note that Equation 4 is a special case of Equation 1 with N = -1. The effect of N is shown in Figure 1. Plug flow has a higher removal for the same dimensionless treatment rate than one or any number of CFSTRs in series. Another way of stating this result is that for the same treatment efficiency or removal, a PFR requires less detention time (Chapra 1997).

Equation 1 provides a useful transition between one extreme of plug flow to the other extreme of complete mixing in a storage device (or chemical "reactor"). This could be implemented as an option for treatment in a SWMM5 "object" that includes storage, such as a channel, pipe, or storage unit. However, it will still require user judgment as to whether mixing is "good" or "very good," thus retaining the empiricism that seems difficult to avoid. If tracer data are available, quantitative inferences about the degree of mixing can be made (Kadlec and Knight 1996, Persson et al. 1999). Wetlands can be modeled similarly. SWMM is well-equipped to deal with hydrologic design of detention and extended detention. The relationship of storage availability and draw-down to local meteorology and catchment conditions (Urbonas and Stahre 1993) may be analyzed quantitatively via continuous simulation. Principal enhancements for SWMM would include more useful statistical methods to analyze time series of volume (or depth), in addition to inflows and outflows from storage. This will require SWMM to have the ability to store additional time series of model state variables.

Swales and Infiltration Channels. One of the biggest needs for SWMM with regard to infiltration trenches and swales is the ability to simulate infiltration from channels as well as from overland flow planes. All conveyance and storage options within the model need to include a full vertical water balance that includes precipitation, ET, and infiltration. Within this framework, SWMM conveyance and storage elements should be able to simulate quality processes characteristic of infiltration devices and bioswales. To the extent that flow through swales behave like a sedimentation device for which performance is a function of hydraulic overflow velocity (Minton 2002), swales can be simulated in the current SWMM S/T Block. However, a ready means of simulating infiltration from S/T storage is still needed. Notwithstanding the need to be able to simulate the infiltration process from channels, infiltration trenches can be simulated within the current SWMM as small, deep overland flow planes, with the ability to receive runoff from an upstream overland flow plane.

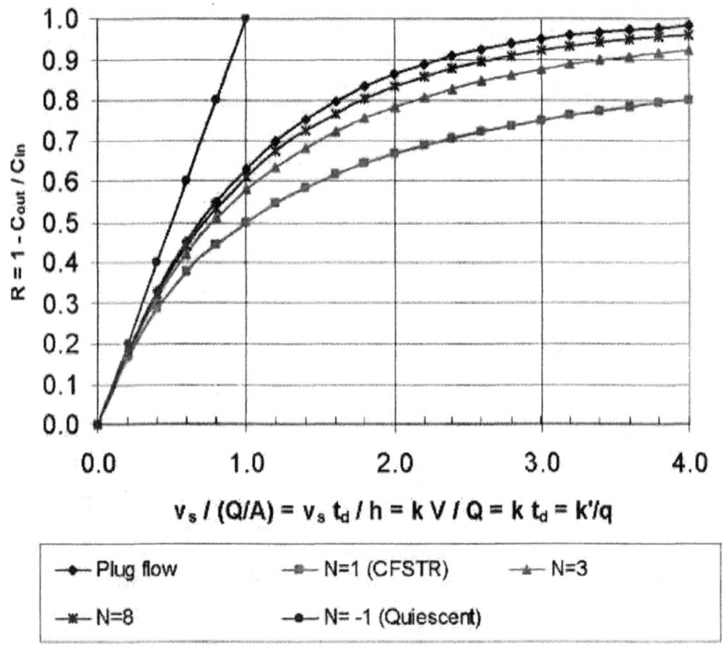

Figure 1. Removal efficiency, R, as a function of dimensionless forms of detention time or loading (Huber et al. 2003).

Filter Strips and Vegetated Buffers. The distinction between a filter strip or buffer and a swale or bioswale is that the swale is distinctly designed for drainage and flow routing, whereas the filter strip or buffer is not. Flow in the latter may occur off a roadway, for instance, into a swale. Routing from one overland flow plane to another already exists as an option in SWMM (Huber 2001). The question is whether a more detailed representation of the quality processes is needed, as in the Riparian Ecosystem Management Model, REMM (USDA 1999). The linkage with multiple subsurface layers and extensive vadose zone and groundwater chemistry is probably more than is warranted for SWMM. If such processes must be simulated, REMM or a similar model might be used.

Dry Wells and Cisterns

Modeling of dry wells can involve complex analytical techniques for both quantity and quality for simulation of the groundwater flow regime. Under suitable soil circumstances, though, it may be possible to infiltrate the entire design inflow over the approximate storm duration. SWMM is currently capable of accepting flows into a device that represents a dry well, but not capable of simulation of the complex flow net that results as water enters the soil. This latter effort is probably not necessary to simulate dry wells as BMPs as long as their capacity can be provided as a function of time or other simple relationship (e.g., a rating curve of head versus outflow rate). For cistern modeling, SWMM and similar models will need an overall water use simulation capability that includes all sources (e.g., cisterns, city water supply) and an irrigation schedule, as in SLAMM (Pitt and Voorhees 2000) in order to dispose of runoff collected in cisterns. The irrigation schedule might in turn be linked to a soil moisture accounting model, leading to greater complexity. SWMM is capable now of simulating the storage associated with dry wells and cisterns.

Porous Pavement

SWMM currently can simulate porous pavement by using its Horton or Green-Ampt infiltration equations, and the resulting infiltrated water can be tracked through use of the groundwater routines, except for water quality. James et al. (2001) demonstrated how the current version of SWMM can be used to simulate roadways and parking lots with porous pavement or permeable pavers.

Hydrodynamic Devices

Because of the variability between devices in this category (e.g., proprietary oil and sediment vortex separators), the effects may be simulated with removal equations based on the manufacturer's specifications, or with data from the EPA/ASCE BMP Database (http://www.bmpdatabase.org/). On the other hand, removal as a function of overflow velocity may work as well. Mechanisms need to be incorporated to determine if devices are overloaded for the design treatment rates. SWMM dynamic flow routing (i.e., Extran prior to SWMM5) is capable of simulating the hydraulics of some common hydrodynamic devices through proper use of orifice settings in particular, but water quality has not been tracked in the past. Water quality simulation will be possible with all flow routing options in SWMM5.

Stormwater Treatability Data Needs

Almost all BMP design is linked to a description of the influent stormwater. A characterization of solids is especially important, i.e., treatability data on settling velocities and/or particle size and specific gravity distributions. Such data are essential in order to simulate a "treatment train," in which downstream performance depends on upstream performance (e.g., upstream devices remove the larger, heavier particles leaving a more difficult influent for downstream devices). This information must be input and/or generated upstream in the model – a very complicated process since it relates to erosion, scour, deposition, and sediment transport, all of which are poorly represented in SWMM and in any alternative model. Typical particle size distributions are available from several sources (e.g., Minton 2002) that can be provided as a default, in lieu of much preferable site-specific data. SWMM can be improved to track particle sizes by treating a size range as a separate constituent subject to a constant settling velocity. Even without improved scour and deposition routines, simply linking the constituents grouped by particle size (or settling velocity) to downstream channels and/or storage devices would be a huge enhancement to the overall SWMM simulation capability. If SWMM5 is improved to perform this linkage, then it will immediately provide an improved BMP simulation capability.

SWMM Improvement Needs for BMP/LID Simulation

When modeling BMPs, the ability of SWMM to redirect flow from impervious areas to pervious (e.g., simulating the effects of rooftops draining to lawns) greatly enhances the model's ability to simulate LID (Huber 2001, Lee 2003). However, for the model to be able to characterize treatment based on solids settling, improvements in the overall ability of SWMM to simulate sediment erosion, transport, deposition and scouring (i.e., to incorporate scarce treatability data) need to be provided, as well as the ability to simulate infiltration in channels and ponds. The vertical water balance must be computed for all conveyances and storages, not just for overland flow planes. Since much of the effectiveness of hydrologic source control depends on infiltration and ET, a good accounting of both is required within the model. Simulation of transpiration from the vadose zone can be improved through seasonal coefficients for various vegetation. Above all, site-specific data for infiltration parameters are required, since such data for disturbed and compacted soils are highly variable.

Other Improvement Needs

BMP performance is often characterized by the nature (e.g., frequency distribution) of the effluent EMCs, in lieu of abortive attempts to define "percent removal." Hence, a useful addition to SWMM would be the ability to prescribe a lognormal (or other) frequency distribution to represent EMCs in surface runoff and/or effluent quality from BMPs. The modeling challenge would be in part to ensure conservation of constituent mass if EMCs and hence loads are inserted into the effluent (or nonpoint [diffuse] source runoff) from a prescribed frequency distribution.

Another issue is the malfunctioning or maintenance issues associated with BMPs. Reduced removal efficiencies are likely without regular maintenance and effects of clogging or other reduction in proper performance should be included in any modeling procedure.

Neither SWMM nor any single model will perform all the functions that a user might need. Alternative models, such as WASP (Wool et al. 2001), HSPF (Bicknell et al. 1997), REMM (USDA 1999) or WETLAND (Lee et al. 2002) should be considered, particularly for complex nutrient dynamics and subsurface water quality. In order to interface one model with another, standards for time series transfer are needed. Such standards would facilitate many multi-media modeling needs within the environmental community in general and U.S. Environmental Protection Agency in particular.

Summary and Conclusions

Most models, including SWMM, are able to simulate the hydraulics of storage very well. Some models, not necessarily including SWMM, are able to simulate treatment and/or fate and transport processes within certain kinds of storage, especially processes at work in receiving waters and channels (e.g., WASP and HSPF). In an urban setting, a model user will look to be sure that BMP-specific and LID-specific options are available. Past versions of SWMM are capable of simulating many such options, through a generic removal function. But SWMM5, the "next generation" version of SWMM, should enhance the model's overall status for use by practitioners in stormwater and urban runoff flow management.

Acknowledgments

The authors acknowledge the assistance of EPA Contract No. 68-C-01-020 in support of this research. The authors also thank colleagues Dr. James Heaney and his team at the University of Colorado at Boulder for leadership and support during the course of the project. This paper has also benefited from BMP evaluation efforts on projects sponsored by the National Cooperative Highway Research Program and by the Water Environment Research Federation. The authors acknowledge the cooperation of their colleagues on these two projects as well.

Disclaimer

Any opinions expressed in this paper are those of the author(s) and do not, necessarily, reflect the official positions and policies of the USEPA. Any mention of products or trade names does not constitute recommendation for use by the USEPA.

References

Bicknell, B.R., Imhoff, J.C., Kittle, J.L. Jr., Donigian, A.S., Jr. and Johanson, R.C. (1997). *Hydrologic Simulation Program – Fortran (HSPF): User's Manual for Release 11*, USEPA, Office of Research and Development, Athens, GA, (Available from: http://www.epa.gov/ceampubl/softwdos.htm.)

Chapra, S.C. (1997). *Surface Water-Quality Modeling*, McGraw-Hill, New York.

Fair, M.F. and Geyer, J.C. (1954). *Water Supply and Waste-Water Disposal*, John Wiley and Sons, New York.

Prince George's County (2000). *Low Impact Development Design Strategies: An Integrated Design Approach*, Department of Environmental Resources, Prince George's County, Largo, MD.

Heaney, J.P., Huber, W.C., Medina, M.A., Jr., Murphy, M.P., Nix, S.J. and Hasan, S.M. (1977). Nationwide Evaluation of Combined Sewer Overflows and Urban Stormwater Discharges, Volume II: Cost Assessment and Impacts. EPA-600/2-77-064b. NTIS PB-266005, Environmental Protection Agency, Cincinnati, OH, March.

Huber, W.C. (2001). "New Options for Overland Flow Routing in SWMM," *Urban Drainage Modeling*, R.W. Brashear and C. Maksimovic, eds., *Proc. of the*

Specialty Symposium of the World Water and Environmental Resources Conference, ASCE, Environmental and Water Resources Institute, Orlando, FL, May, pp. 22-29.

Huber, W.C., Clannon, L. and Stouder, M. (2003). *BMP Modeling Concepts and Simulation.* Draft Report to the U.S. Environmental Protection Agency, EPA Contract No. 68-C-01-020, Urban Watershed Management Branch, National Risk Management Research Laboratory, Edison, NJ.

James, W.R.C., James, W. and van Langsdorff, H. (2001). "Stormwater Management Model for Environmental Design of Permeable Pavement," Chapter 26 in *Models and Applications to Urban Water Systems*, Proc. Conference on Stormwater and Urban Water Systems Modeling, W. James, ed., Computational Hydraulics International, Guelph, Ontario, pp. 423-445.

Kadlec, R.H. and Knight, R.L. (1996). *Treatment Wetlands*, Lewis Publishers, Boca Raton, FL.

Lee, E.R., Mostaghimi, S. and Wynn, T.M. (2002). "A Model to Enhance Wetland Design and Optimize Nonpoint Source Pollution Control." *Journal of American Water Resources Association*. 38(1):17-32, February.

Lee, J.G. (2003). *Process Analysis and Optimization of Distributed Urban Stormwater Management Strategies*, Ph.D. Thesis, Department of Civil, Environmental and Architectural Engineering, University of Colorado, Boulder.

Metcalf and Eddy, Inc. (2003). *Wastewater Engineering*. Fourth Edition, McGraw-Hill, New York.

Minton G.R. (2002). Stormwater Treatment; Biological, Chemical and Engineering Principles, Resource Planning Associates, Seattle, WA.

Persson, J., Sommes, N.L.G. and Wong, T.H.F. (1999). "Hydraulics Efficiency of Constructed Wetlands and Ponds" *Water Science and Technology*, 40(3):291-300.

Pitt, R.E. and Voorhees, J. (2000). *The Source Loading and Management Model (SLAMM), A Water Quality Management Planning Model for Urban Stormwater Runoff.* University of Alabama, Department of Civil and Environmental Engineering, Tuscaloosa, AL.

Urbonas, B.R and Stahre, P. (1993). Stormwater: Best Management Practices and Detention for Water Quality, Drainage, and CSO Management, Prentice Hall, Englewood Cliffs, NJ.

USDA-ARS-Southeast Watershed Research Laboratory (1999). *Riparian Ecosystem Management Model*, U.S. Department of Agriculture, Agricultural Research Service, Tifton, GA. http://www.cpes.peachnet.edu/remmwww/remm/remmoldwww/default.htm

U.S. Environmental Protection Agency (EPA) (1983). Final Report for the Nationwide Urban Runoff Program. Water Planning Division, Washington, DC, December.

Water Environment Federation and American Society of Civil Engineers (1998). *Urban Runoff Quality Management*, WEF Manual of Practice No. 23, ASCE Manual and Report on Engineering Practice No. 87, Water Environmental Federation, Alexandria, VA.

Wool, T.A., Ambrose, R.B., Martin, J.L. and Comer, E.A. (2001). *Water Quality Analysis Simulation Program (WASP), Version 6.0 Draft User's Manual.* Environmental Protection Agency, Environmental Research Laboratory, Athens, GA.

Wright, L.T., Heaney, J.P. and Weinstein, N. (2000). "Micro-scale Modeling and Monitoring of Low Impact Development Management Strategies for Wet weather Flow Control." *Proc. ASCE 2000 Joint Conference on Water Resources Engineering and Water Resources Planning and Management-Building Partnerships*, Minneapolis, MN, American Society of Civil Engineers, Reston, VA, CD-ROM, 11 pp.

Hydrologic Characteristics Simulation for BMP Performance Evaluation

Angela Brown[1] and Wayne C. Huber[2]

[1]Black and Veatch Engineering, 1111 3rd Ave. Seattle, WA 98101, brownam@bv.com

[2]Department of Civil, Construction, and Environmental Engineering, Oregon State University, 202 Apperson Hall, Corvallis, OR 97331-2302, wayne.huber@oregonstate.edu, 541-737-6150 (Contact Author)

Introduction

Runoff may occur from urban and highway locations both in wet weather, due to the obvious consequences of rainfall, but also during dry weather due to baseflow resulting from irrigation and other urban water sources as well as release of bank storage accumulated during a storm event. Within this paper, urban runoff controls, for both quantity and quality impacts, are generically termed "best management practices," in accordance with common terminology, even though such controls may not be "best" in some objective sense. Rather, they represent a suite of practices suitable for management of urban runoff. Such controls may include hydrologic source controls, otherwise known as "low-impact development" (LID), with which an attempt is made to reduce runoff on a spatially distributed and localized basis through fundamental hydrologic processes such as infiltration and evapotranspiration (ET). Reduction of runoff volume ordinarily results in a reduction of pollutant load, since load = concentration x volume.

BMP performance may be characterized, based on the following four questions (Strecker et al., 2001):

1. *How much catchment runoff is averted by a control measure, e.g., by infiltration or evapotranspiration?*

2. *Of the runoff that enters the BMP, how much is treated and how much is bypassed?*
3. *What are the quality characteristics of the treated BMP effluent?*
4. *How much downstream flow management is provided by the BMP?*

Event mean concentration (EMC) data may be used to address the third question, e.g., through the use of the effluent probability method advocated by the authors of the ASCE/EPA BMP Database (GeoSyntec and UDFCD, 2002). But hydrologic modeling is necessary to characterize the stormwater runoff volumes that might be diverted and/or treated by various BMPs, such as ponds, swales, etc.

The performance and effectiveness of a BMP for use in treating and/or controlling stormwater runoff are dependent upon numerous variables, related not only to the design and mechanics of the system, but also the conditions of the site, techniques related to sampling, and constituents found in the water (Strecker et al. 2001). Site-specific, regulatory, O&M (operation and maintenance), and regional factors etc. may be considered through a "practicability analysis" that encompasses a broad list of tangible and intangible considerations. Assessment of hydrologic and quality performance may be performed through evaluation of existing sites on a regional and BMP-specific basis (e.g., storage and sedimentation devices such as ponds, flow-through and infiltration devices, such as swales, filtration and sorption devices, such as sand filters, etc.). Hydrologic simulation can provide good estimates of BMP effectiveness regarding prevention and/or bypass of runoff as well as other conclusions, such as design guidelines (whether flow or volume based) and estimates of downstream impacts caused by flows of a given magnitude and frequency. By combining the results of hydrologic simulations of urban catchments and BMPs with the quality-based evaluations of BMP performance, a distinct methodology may be developed for engineers throughout the country to provide guidelines for BMP selection within distinct regions.

As part of a project directed specifically at control of highway runoff but also applicable to the broader urban environment, an evaluation methodology for BMP characterization has been developed and tested on four site-BMP combinations. The methodology includes:

- characterization of site-specific, institutional, regulatory and similar criteria through a practicability analysis;
- characterization of quality performance through a variety of statistical analyses of influent and effluent EMC data; and
- characterization of regional hydrologic conditions through continuous simulation of rainfall and runoff processes for the catchment and specific BMP.

The third component is the focus of this paper. The methodology described herein is primarily for use for evaluation of structural BMPs during the course of the project. The results will then be synthesized and presented in the form of BMP selection guidelines for use by the urban runoff professional and highway drainage engineer.

The methodology also strongly emphasizes incorporation of fundamental unit processes of environmental engineering during the evaluation of individual BMP types as well as integration of hydrologic source control techniques.

Site Descriptions

The Search for Intra-event Data

In order both to calibrate simulation models and to characterize fundamental unit processes at work during BMP use, hydrographs and pollutographs (concentration vs. time) within storm events are required. Such data are exceedingly sparse. A thorough search of the EPA/ASCE BMP Database (http://www.bmpdatabase.org/) revealed only 11 sites in six U.S. cities with the desired data (and including documentation through reports, papers, and optional contact with the original investigators). Other sources investigated included recent CalTrans monitoring activities, USGS studies, and numerous studies conducted by a variety of transportation and other agencies around the country. Although there are many promising studies from which EMC and modeling data may become available in the future, only the EPA/ASCE BMP Database provided enough information for this project at this time. From these 11 sites, two were chosen for development of the hydrologic evaluation methodology, described below. In addition, two other sites (Seton Pond, a dry-detention area in Austin TX and a second filter strip or swale along Highway 183 in Austin) were included in the EMC evaluation portion of the research. The focus of this paper is the hydrologic evaluation; hence, the latter two areas are not considered further in this paper. Full details of the hydrologic and water quality evaluations are provided by Brown (2003).

Moyewood Pond Study Area in Greenville, North Carolina

Moyewood Pond, a 0.75-ac (0.07-ha.) extended dry detention pond was installed in November 1991 to treat runoff from a 200-ac (81-ha.) watershed area along the Tar River, west of downtown Greenville, North Carolina. Land use of the watershed area consists mainly of medium density single family and low-income multi-family residential with a small percentage of commercial area (Stanley 1994). Approximately 31% of the total watershed area is impervious, and about 65% of the total impervious is directly connected impervious area, from which runoff directly discharges to other impervious areas or the storm sewer itself (Stanley 1994).

Moyewood Pond was designed to capture runoff from the first 0.5 in. (13 mm) of rainfall over the site and discharge at a peak rate of 2.6 cfs (74 L/s). Runoff volumes exceeding 0.5 in. bypass the pond by flowing over a 45-ft (14-m) long trapezoidal concrete spillway. The total storage capacity of the pond is 338,026 ft^3 (9,579 m^3) with a maximum drawdown time of 74.8 hours. Eight storm events from February to August, 1992 were used for the hydrologic calibration and effluent quality assessment, and 15-min rainfall data for the available period of record (May 13, 1971 to January 1, 2003) were used for the continuous simulation. A map of the entire catchment area is shown in Figure 1.

Walnut Creek Filter Strip in Austin, Texas

The Walnut Creek filter strip is located in the highway median of the MoPac Expressway, south of Walnut Creek itself, in northwest Austin, Texas (Walsh et al. 1997). The 3460-ft (1054-m) long strip collects runoff from the north and southbound lanes of the expressway, and was originally intended for hydraulic conveyance, not water quality control (Walsh et al. 1997). Runoff enters the filter strip laterally, primarily as sheet flow from the north and southbound lanes, entering the filter strip along the sides of the vegetated slope, where it then flows northward (except for the northernmost 490 feet [150 meters] that drains southward) along the median to four drop inlets configured along the centerline. The runoff is then directly conveyed to Walnut Creek via a 48-in. (1220-mm) diameter concrete storm drain. Overall, the total drainage area of the storm drain system is 25.85 ac (10.5 ha) , approximately 38% impervious, consisting of the roadway and median, in addition to the grassy shoulder area that does not drain to the filter strip itself (Walsh et al. 1997). Five storm events between November 1996 to April 1997 were used for the hydrologic calibration; 26 storm events between December 1995 and May 1997 were used for the effluent quality assessment, and 15-min rainfall data from the entire available period of record (December 1, 1975 to December 1, 2000) were used for the continuous simulation. A schematic of the catchment is shown in Figure 2.

Figure 1. Area of Moyewood Pond catchment (M. Algae, City of Greenville, obtained 2003). Moyewood Pond is the oblong shape in top center, to left of bridge. No map (outline) of the catchment (just below the pond) or definition of the color-coding was available.

SWMM Calibration Methodology

Using initial site and flow parameters, including soil, ET, and infiltration characteristics and rate constants, specific storm events used within the scope of the EMC analysis were modeled within the SWMM program (Huber and Dickinson, 1988) to compare modeled storm volume and flow patterns with those monitored. Specific 15-min (for Moyewood Pond) and 5-min (Walnut Creek) storm rainfall information was acquired by the respective parties responsible for the initial monitoring of the BMP (Dr. Donald Stanley for Moyewood Pond and Dr. Michael Barrett for Walnut Creek). Thus, SWMM calibration runs could be compared to monitoring data. GeoSyntec Consultants supplied the measured storm hydrographs from data supplied to the EPA/ASCE BMP Database for the Moyewood Pond site, and measured hydrographs were calculated for the Walnut Creek site, based upon flow information provided by Dr. Michael Barrett. Five-minute time steps were used for the Moyewood Pond catchment simulation, while 1-min time steps were used for the highway site.

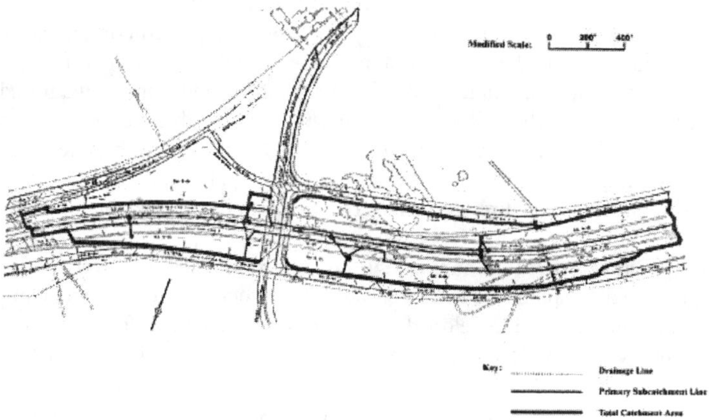

Figure 2. Walnut Creek catchment along MoPac Expressway in Austin.

Following each storm simulation, total modeled runoff volumes were plotted against the actual runoff volumes. Fitting a linear trend line through the data points, including the origin, the basis for volumetric calibration was determined by the overall slope of the trend line. When the slope of the trend line fell between 0.9 and 1.1, the calibration fell within the goal of ±10% agreement between monitored and simulated volumes. Hydrographs for each modeled storm event were compared to evaluate flow timing and peaks. Based upon results of the hydrograph comparison, characteristics such as baseflow and depression storage could be revised to better simulate the observed flow pattern. When the hydrograph simulation was improved, volumetric calibration was again conducted, based upon the improvements made to the model.

SWMM Continuous Simulation Results for Hydrologic Characterization

With the use of 15-min rainfall data, downloaded from the National Climatic Data Center (NCDC, data obtained August 2003), a continuous simulation of storm events at Moyewood Pond and Walnut Creek was performed with SWMM for the past 32 and 25 years respectively. From the continuous simulation, frequency curves for a variety of sizing parameters including total flow (inches), average flow (in./hr), peak flow (in./hr), and event and interevent durations were developed for each site. From these frequency relationships, conceptual BMP sizing parameters (both volumetric and flow-based) for urban areas may be developed and compared for the two sites in North Carolina and Texas. In addition, runoff averted by control measures (infiltration, ET) is normalized and reported for each site following the continuous simulation, in order to characterize hydrologic controls per typical catchment on a regional basis. Finally, by using the frequency relationships in conjunction with current system sizing, a hydrologic basis for BMP evaluation and design may be described in terms of flow treated and bypassed. Thus, taking typical urban sites or sites with representative imperviousness percentages, sizing criteria on the basis of frequency analysis can be used as a measure of how much runoff bypasses a particular system on a continuous basis. As the flow and volume magnitudes are reported in terms of inches per hour and total inches over the site, sizing criteria are normalized per unit area of a given catchment. One difference between this effort and similar valuable frequency-based analyses (e.g., WEF and ASCE, 1998) is the emphasis upon storm event runoff analysis, rather than just a precipitation analysis.

Frequency Curve Fluctuations Based upon Imperviousness.

Using generated frequency curves for the Moyewood Pond site, idealistic BMP design guidelines can be developed from the simulated flow frequency curves at variable imperviousness for the catchment area. As expected, with a consistent 6hour minimum interevent time (MIT) for separation of runoff events, total, average, and peak flow and event and interevent duration increase with increased imperviousness. For example, the total flow frequency curves are shown in Figure 3. By selecting an appropriate imperviousness value (three values are shown in the figure) design

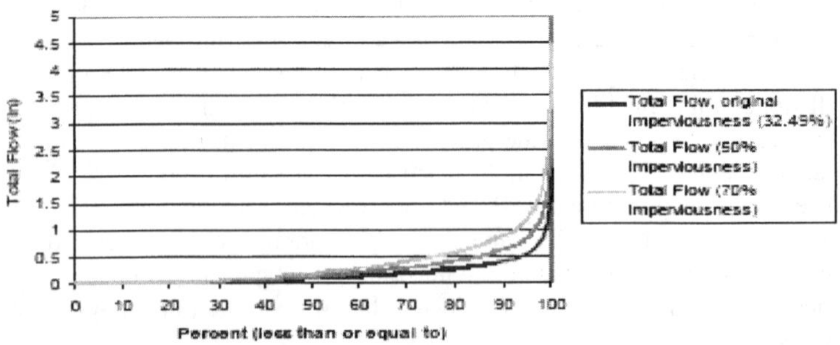

Figure 3. Total flow frequency curves for the Moyewood Pond catchment in Greenville, North Carolina. (1 in. = 25.4 mm).

criteria required to meet a suggested level of control (e.g., 90% volumetric control) in the given region under the defined conditions (imperviousness, MIT) for a specific system type (volume or flow based) may be determined. Moyewood Pond was designed to capture the first 0.5 in. (13 mm) of runoff from the catchment. This would result in control of approximately 94% of the annual runoff volume (at 32.5% imperviousness).

Frequency Curve Fluctuations Based upon MIT and Location. The selection of an MIT for use in the SWMM simulation greatly influences the frequency curves generated for particular design criteria. For instance, dry detention ponds are often designed to draw down within 48 to 72 hours of cessation of the storm event. Storms arriving before the pond has emptied may be "seen" as the same event by the pond if it still has water from the last storm. Thus, an MIT of 72 hours was compared with an MIT of 6 hours; the latter might be more appropriate for flow-through devices such as swales or filter strips, for example. Setting the imperviousness for both the Greenville and Austin sites to a consistent 38% (the original imperviousness for the Walnut Creek highway catchment), and varying the MIT from 6 hrs to 72 hrs (the designed draw-down time for Moyewood Pond itself), one may observe the variations in frequency curves due to location and MIT. Total and peak flow frequencies are compared in Figures 4 and 5.

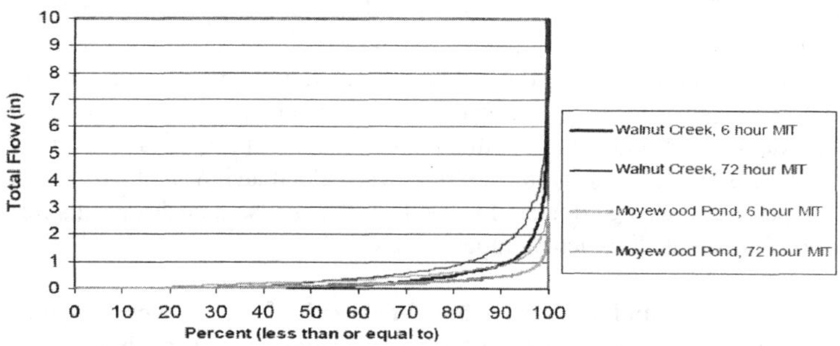

Figure 4. Total flow comparison between regions and MIT values based upon Walnut Crk and Moyewood Pnd site simulations at constant 38% imperviousness.

Increasing the MIT increases the magnitude of the total and peak flows for a given frequency for any location, as seen in the two figures, although average flow remains fairly consistent (not shown). As shown in Figure 4, the Walnut Creek site experiences a greater storm event runoff volume compared to the Moyewood Pond site. Increasing the MIT approximately doubles the event volumes for both sites. Hence, volumetric sizing criteria are hypothesized to be higher for the Austin region as opposed to the Greenville, North Carolina region at a consistent MIT, and may influence the types of BMP systems preferred for a given region.

Peak flow frequencies (Figure 5) are more distinct between the two locations. Differing from the results of the total volume analysis, the Greenville site shows a

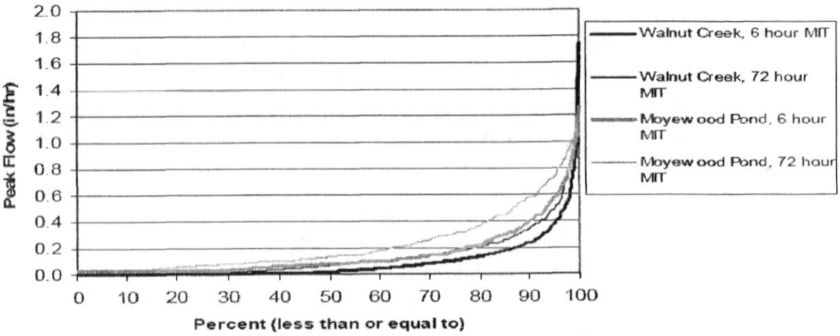

Figure 5. Peak flow comparison between regions and MIT based upon Walnut Creek and Moyewood Pond site simulations at constant 38% imperviousness.

higher storm event peak flow rate than the Austin site. Increasing the MIT increases the peak flow for each site by about 60%. Thus, flow-based sizing magnitudes are hypothesized to be higher for the Greenville site compared to the Austin site and may be an influencing factor controlling typical BMP selection processes in these regions.

Inferences from Continuous Simulation

Projection of Runoff Reduction Due to Control Measures. Continuous simulation of long-term rainfall data within a calibrated hydrologic model can provide a statistical interpretation of catchment runoff averted by a number of control measures including evapotranspiration and infiltration on a normalized basis. Continuous simulation results (water budget) for the Moyewood Pond and Walnut Creek sites at an equal imperviousness are compared in Table 1, which also includes results for the Moyewood Pond site with its actual imperviousness.

Table 1. Continuous simulation water budget statistics for Moyewood Pond and Walnut Creek sites (consistent 6-hr minimum inter-event time).

Continuous Simulation Characteristics	Moyewood Pond 32.5% Impervious	Moyewood Pond 38% Impervious	Walnut Creek 38% Impervious
Precipitation- total (in.)	911.25	911.25	751.70
Average yearly (in./yr)*	28.48	28.48	30.09
Infiltration- total (in.)	610.30	560.49	346.21
Average yearly (in./yr)	19.07 (66%)	17.52 (62%)	13.85 (45%)
Evaporation- total (in.)	43.87	50.98	70.09
Average yearly (in./yr)	1.37 (5%)	1.59 (6%)	2.80 (9%)
Runoff- total (in.)	267.63	310.88	342.28
Average yearly (in./yr)	8.36 (29%)	9.72 (32%)	13.69 (46%)

*Note, total precipitation is less than actual for the simulation period because of missing months of data from the 15-min NCDC records. (1 in. = 25.4 mm)

As expected, increased infiltration and ET occur at the Moyewood Pond site when the imperviousness is decreased. Results from the regional comparison, however, indicate how control measures may influence the frequency curve generation and thus affect not only sizing criteria but also the selection of a particular type BMP for a region. As seen in Table 1, average annual precipitation in each region is fairly similar (allowing for missing rainfall data); however, with equal imperviousness, differences in infiltration and drainage parameters cause less total runoff at the Greenville site than at the Austin site. Thus, BMP design specifications must reflect not only the rainfall characteristics but overall catchment characteristics for a site.

Approximation of Runoff Treated and Bypassed for Existing BMP Systems. The frequency curves can be used for an analysis and approximation of runoff treated and bypassed for a particular BMP, based upon the system's design standards. As previously mentioned, the frequency curves developed for the Moyewood Pond site at the actual 32.5% imperviousness can be compared to the original sizing of the pond system. For this site, the original design called for the capture and treatment of the first 0.5 in. (13 mm) of runoff over the site. For a 6-hr MIT, Moyewood Pond is predicted to treat 94% of the cumulative runoff from the site and bypass 6% of the runoff. Simulated with a higher MIT, the percentage of runoff treated will decrease, as seen in Figure 4 (for a different imperviousness).

Continuous simulation and frequency curve generation are helpful when defined sizing standards for existing BMPs are reported, particularly for volume-based systems with a defined drawdown time. Defined sizing standards for the Walnut Creek filter strip were not found within the literature search, but as a filter strip is generally utilized for both treatment and conveyance, all runoff routed to the system is assumed to be captured and treated by the BMP. However, the magnitude with which particular storms are treated is a function of storm size, and additional information would be needed about the precise design of the system in order to further analyze this site.

Regional BMP Sizing Correlations based upon Frequency Curve Generation. Based upon the continuous simulation results for total flow (Figure 4), while maintaining a consistent imperviousness and MIT, the volumetric design requirements for capture of 90% of the runoff volume for the site are smaller for the Greenville, NC region compared to the Austin region. This is due to the increased infiltration capacity of the sandy soils at the Greenville site versus the predominately clay soils at the Austin site. Thus, for these two sites, volume-based sizing at the Texas highway site may not be as cost effective as a BMP based on flow rate. Flow rate-based systems are typically designed on the basis of percentage of peak flow captured. On this basis, the 90th percentile peak flow rate for the Greenville region is about 67% greater than for the Austin region (Figure 5). The greater peak flow rate is likely the result of differing rainfall characteristics, representing more intense and shorter duration storms in the Greenville, NC region. Thus, sizing flow-through facilities in the Greenville region may not be as cost effective as in the Austin area.

Conclusions

Whether sized based upon a target volume, peak flow, or other hydrograph characteristic, the frequency curves resulting from continuous simulation can be used on a regional basis for sizing of urban BMP facilities because the simulation uses the typical rainfall and infiltration parameters applicable to a particular area. This paper demonstrates sizing criteria based only on catchment characteristics such as imperviousness. Were the BMP itself (e.g., a pond or swale) and/or a pollutant (e.g., to simulate a possible first flush) also simulated, additional sizing criteria could be developed based on flows through the device, frequency distribution of loads, EMCs, etc. for typical BMP applications. For existing BMP systems, previously designed according to specific standards, continuous simulation also allows for prediction and estimation of runoff captured and bypassed by a BMP.

Acknowledgments

The authors gratefully acknowledge the support of National Cooperative Highway Research Program Project 25-20(01) (Project Officer Chris Hedges) during the conduct of this research as well as project team colleagues James Heaney (University of Florida), Eric Strecker and Marcus Quigley (GeoSyntec Consultants) and Neil Weinstein (Low Impact Development Center). In particular, the GeoSyntec team conducted the search of the EPA/ASCE BMP Database that resulted in the case study data used in this paper. Michael Barrett of the University of Texas and Donald Stanley of East Carolina University were extremely helpful in augmenting the information contained in project reports.

References

Brown, A. (2003). *Development of a BMP Evaluation Methodology for Highway Applications*. MS Project Report, Dept. Civil, Construction, and Environmental Engineering, Oregon State University, Corvallis.

GeoSyntec Consultants and Urban Drainage and Flood Control District (2002). *Urban Stormwater BMP Performance Monitoring, A Guidance Manual for Meeting the National Stormwater BMP Database Requirements*. Environmental Protection Agency, Washington, DC. EPA-821-C-02-005 (CD-ROM) or at: http://www.bmpdatabase.org/docs.html.

Huber, W.C. and R.E. Dickinson (1988). *Storm Water Management Model Version 4, User's Manual*, EPA/600/3-88/001a (NTIS PB88-236641/AS), Environmental Protection Agency, Athens, GA.

Stanley, D.W. (1994). *An Evaluation of Pollutant Removal by a Demonstration Urban Stormwater Detention Pond*, Institute for Coastal and Marin Resources, Dept. of Biology, East Carolina University, Greenville, NC.

Strecker, E.W., M.M. Quigley, B.R. Urbonas, J.E. Jones, and J.K. Clary (2001). "Determining Urban Storm Water BMP Effectiveness," *J. Water Resources Planning and Management*, Vol. 127, No. 3, pp. 144-149.

Walsh, P.M., M. Barrett, J.F. Malina, and R.J. Charbeneau (1997). *Use of Vegetative Controls for Treatment of Highway Runoff*, Center for Research in Water Resources, Bureau of Engineering Research, University of Texas at Austin.

Water Environment Federation and American Society of Civil Engineers (1998). *Urban Runoff Quality Management*, WEF Manual of Practice No. 23, ASCE Manual and Report on Engineering Practice No. 87, Water Environmental Federation, Alexandria, VA.

Modeling Effectiveness of Bioretention Cells for Control of Stormwater Quantity and Quality

Christianson[1], R. D., B. J. Barfield[2], J. C. Hayes[3], and G. O. Brown[4]

[1]Research Engineer, Oklahoma State University, Stillwater, OK 74075
[2]Professor Emeritus, Oklahoma State University, Stillwater, OK 74075
[3]Professor, Clemson University, Clemson SC 29634-0357
[4]Professor, Oklahoma State University, Stillwater, OK 74075

Introduction

In 1972 the United States Environmental Protection Agency (USEPA) added the National Pollutant Discharge Elimination System (NPDES) to Section 402 of the Clean Water Act. This gave the USEPA authority to set effluent standards on certain chemicals for any point source pollution entering the nation's water bodies (USEPA, 2003a). In 1977 more chemicals were added to the list to be controlled and in 1987 the EPA started to monitor water quality in lakes to ensure standards were being met (USEPA, 2003a). Recently, focus has moved to Non-Point Source (NPS) pollution with the addition of Phase II of the NPDES. Phase II holds individual states, tribes, and territories accountable for controlling their NPS pollution (USEPA, 2003b).

With this recent move to control NPS pollution, there is a need to develop technologies that can be retrofit into existing developed areas, as well as, installed in developing areas as part of construction and landscaping. There are many different types of structures that may be investigated; however, bioretention cells may have the greatest economic potential. Suggested in a publication by the USEPA (1999) and Prince George's County Department of Environmental Resources (PGDER) (1993), costs of installing a bioretention cell as part of a new development are only slightly higher than normal landscaping. Coffman et al., (1999) estimated \$3 to \$4 per square foot for residential use and \$10 to \$40 for commercial. These upfront costs may actually be less than traditional alternatives when an economic investigation is completed.

Along with being relatively inexpensive, these techniques do not require a great deal of space or disturbance to the surrounding area. It is generally thought that a bioretention cell would be five to seven percent of the size of the land area draining into it (PGDER, 1993) and not be on greater than a 20% slope (USEPA, 1999). Sediment is normally trapped to a large degree (up to 99%) due to the filtering nature of a bioretention cell; however, removals for nitrogen have ranged from around 80%

to actually adding nitrogen to the treated water. Because of this variability in experimental data there is a need for a process-based model to approximate what is actually happening in the body of the bioretention cell and make bioretention cell sizing site specific.

Background

Bioretention cells are basically the integration of infiltration basins to remove water, sand filters to remove and degrade pollutants, and phytoremediation to provide plant uptake and increased degradation in the rhizosphere of plant roots. The initial removal of pollutants will be a quick chemical reaction or a physical removal due to sorption. Once the pollutant has been sorbed microbial degradation starts to occur. Aerobic bacteria degrade most pollutants, however, there are a few petroleum products that are only degraded with anoxic bacteria. Denitrification also occurs in an anaerobic state turning nitrate (NO_3) and a carbon source into nitrogen gas (N_2) and carbon dioxide (CO_2). Having multiple layers in the bioretention cell allows for multiple processes to occur simultaneously.

The proposed setup for this project is a three-layer system with a sand/soil mixed layer on top, followed by a layer with high organic matter, and concluded by an anoxic zone. This setup can be seen in Figure 1. The sand layer will provide infiltration and filtration of large particles. This layer may also have mulch directly preceding it. The organic layer will be used for sorption of metals and other pollutants, as well as, for bioremediation of pollutants. The anoxic zone will be constantly saturated allowing for the degradation of pollutants that are resistant to aerobic biodegradation and denitrification.

A major concern people have with bioretention cells is the fear of having them being a breeding ground for mosquitoes. To remove this threat the USEPA (1999) suggests all standing water should infiltrate within 24 to 48 hours. A proper design will insure rapid infiltration.

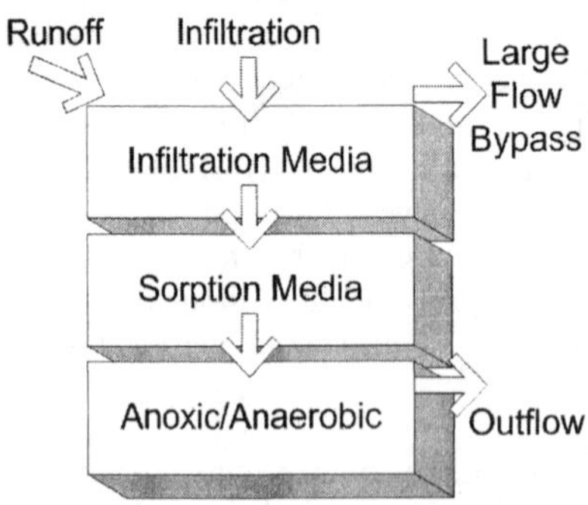

Figure 1. Schematic of the three-layer bioretention cell.

Current Technology

There are many projects around the country working on the concept of bioretention. Early work was done by Prince George's County in Maryland and by Dr. Allen Davis at the University of Maryland. Davis and his students are working on a number of experiments including laboratory and field instillations of bioretention cells. Removal rates for bioretention cells have been shown to be as high as 90% for copper, lead, and zinc. These removal rates were shown in the lab and reproduced in the field (Davis et al., 2003). Davis et al., (2001) believes that the mulch layer on top of the cell has the biggest impact to the removal of metals in the cell.

Prince George's County Department of Environmental Resources (PGDER) in Maryland is also doing bioretention cell instillations around the state. Costs for installing this technology is apparently only slightly higher than those of landscaping that would be done regardless and reduces the costs of treating stormwater runoff at a central location downstream of the cause (UAEPA, 1999). Though there are slight increases in the landscaping cost, these practices have the potential to reduce the cost of constructing traditional stormwater conveyance as well by reducing the amount of drainpipe required. Prince George's County is on the forefront of low impact development (LID) and in 1998 received an award from the USEPA as having the most Outstanding Municipal Stormwater Program (Coffman, 2003).

North Carolina State University, headed by Dr. Bill Hunt, is also doing work with bioretention cells. They have done field installations in around the state as well as done some initial work with an anaerobic zone for removing nitrates. Results on the anaerobic zone have shown to be inconclusive at this point (Hunt et al., 2003); however, the potential associated with an anaerobic or anoxic zone are motivating and warrant further research.

Other states that are taking steps involving bioretention cells, and LID in general, are California, Idaho, Michigan, Minnesota, Mississippi, Massachusetts, New Jersey, Pennsylvania, Tennessee, Texas, Virginia, and Washington (Low Impact Development Center, 2003). These states have shown initiative by advocating LID. Some have even produced manuals on suggested design specifications.

Modeling

The model currently being worked on at Oklahoma State University consists of an infiltration and sorption component. The objective of this model is to determine the efficiency of a bioretention cell setup given rainfall distribution, watershed size, and the size of the proposed bioretention cell. Currently the model takes into account the amount of water received during a number of storms and the standard levels of pollutants that are entering the cell. This model will be integrated into two existing models, the Integrated Design and Evaluation Assessment of Loadings (IDEAL) model and the Simple Urban Detention System (SUDS) model. Some of the basic inputs such as rainfall and flow are handled by each of these models.

When complete the infiltration and sorption module will be combined with evapotranspiration (ET), rainfall probability, water distribution, phytoremediation (plant uptake of pollutants and rhizosphere activity), and biodegradation. Ultimately, a single storm process based model for sizing bioretention cells using mathematics and theory along with empirical evidence will be created. This model will work for any location and be site specific rather than sizing based off of rules of thumb. The rainfall probability, say eight days, will be added into the overall single storm model to serve as a safety factor to ensure the cell will not be undersized. The Penman-Monteith method is used for ET with temperature, rainfall, solar radiation, and plant type as input. Water distribution is necessary to determine where there is available water for evaporation and if water is available for microbial needs and plant uptake. Phytoremediation will be based off of plant type and water use. Biodegradation will be assessed using methods similar to current models available.

The infiltration and sorption module is a Visual Basic model using the Green-Ampt method for infiltration and Freundlich isotherms for sorption. Water infiltrates into the soil profile through the different layers of media. As water containing pollutants moves through the media, sorption is evaluated and a mass balance constructed. Once sorbed, these pollutants are occupying exchange sites in the soil making the possibility of additional sorption more difficult. This is important when looking at the potential life of a bioretention cell and the performance to be expected over time. Plant uptake and Degradation of these pollutants will free up exchange sites. Values for isotherms are being determined as part of laboratory studies and will be implemented into the model as experiments are completed. Currently work with phosphorus isotherms is being done because of the importance of the nutrient in the environment.

With large rain events the infiltration rate of a given system may not be large enough to infiltrate all of the water entering the cell so a bypass component allows the appropriate depth of ponded water in the cell to be evaluated. This depth is a key factor in determining the infiltration rate as it adds to the pressure head for wetting at the wetting front. Ponded water along with the outlet structure specifics determines the rate at which water will move from the cell in a means other than infiltration.

The current infiltration module allows users to select the type of media in each of 4 layers. This selects the hydraulic properties of the selected media to be used in routing water through the system. Soil hydraulic properties were developed by Rawls, Brakensiek, and Miller (1983), while properties for the other media types were estimated based off of an article by McCuen, Rawls, and Brakensiek (1981). The depth of each of the layers is the user input that defines the hydraulic properties at a given depth. Initial water content is also taken into account at the start of the storm.

For sorption, the system is evaluated once per defined layer using Freundlich isotherms. This bathtub method seems to work well when modeling. Future work will be done to better evaluate the soil profile with a greater number of layers and pull out more detailed information.

The key to this portion is having reliable Freundlich constants and being able to evaluate the system over a number of storm events. Considering the starting conditions in the bioretention cell will have a large impact on the efficiency of it. If, for instance, the system is 4 years old there will be fewer exchange sites available and the cell may have a lower removal rate. Having a balance between the amount coming in and degradation will increase the life of the system.

Experimental Setup and Validation

Small columns (48x16cm) shown in Figure 2 are being used to test various organic media with surrogate chemicals to find associated sorption isotherms and to determine flow characteristics of the media. Flow profiling will allow for an estimate of the actual volume of the bed that will be effective in the removal of pollutants. This is necessary because the heterogeneous nature of some materials allows for preferential or macro pore flow. Heterogeneity of flow may pose a problem when designing, and modeling these structures because enough of the media must be in contact with the inflow constituents for adequate sorption to take place and for assumptions of even water distribution to be adequate. A way around this is to use media that is largely homogeneous with no major conduits for water to travel.

To profile flow, each column is back saturated with tap water then connected to a constant head tank. The height of the outflow of the column is varied to change the head and allow for differing flowrates through the column media. Because of the heterogeneity of the flow with some types of multi porosity media, four monometer ports were needed on each end of the column to measure the pressure difference between top and bottom. A pressure flowrate relationship is then developed to predict the homogeneity of the flow and how much media will be contacted by incoming pollution.

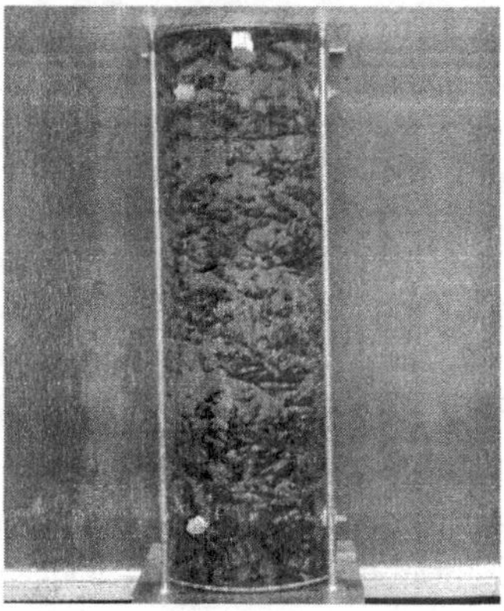

Figure 2. Column to determine flow characteristics and sorption isotherms.

A greenhouse experiment is setup three replicates with a Bermuda grass cover. Bermuda grass is being used because of the rapid growth and even coverage. The results from the experiment will help quantify plant uptake and water use as well as be used to validate the existing model. The grass will be started on a sand/soil mixture inside the boxes to insure a homogeneous cover. Once established, testing will proceed with a measured volume of water and a known concentration of pollutant. Leachate will be collected and analyzed giving a final concentration to be compared with model results. Phosphorus will be the pollutant of choice for the greenhouse portion to stay consistent with lab scale testing.

Once the phytoremediation and bioremediation components of the model have been complete, the greenhouse tests will serve as a validation tool and may also serve as a process-building tool. The setup and size of the test bioretention cell will be similar to Davis et al. (2001) with a wooden box apparatus and appropriate bioretention cell layers. Figure 3 shows the setup of the greenhouse boxes before the addition of soil. Three replicates of each of the layers will be evaluated then the boxes will be stacked to evaluate all three layers together.

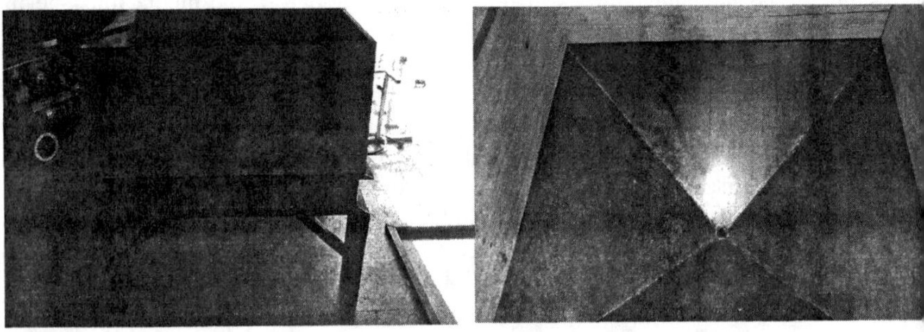

Figure 3. Stackable greenhouse boxes to validate model.

Conclusion

A process-based model for bioretention cells is needed to make installations predictable as well as site specific. Having correctly sized stormwater control devices will save money for instillation and reduce risk of an unexpected failure. There will also be an estimate on the life of the device and what steps would need to be taken in order to maintain or replace media within the cell. Oklahoma State University, with local cooperators, has the opportunity to install multiple bioretention cells in the state of Oklahoma. Data from these instillations will strengthen the model. Validation with data from field instillations is a key step in making a design model successful.

The proposed configuration in figure 1, along with the completed model will be a powerful tool in urban stormwater management. Applications of the model and the bioretention cell setup will be seen in retrofitting and new development to optimize performance and cost giving instillation sites another option in stormwater control. Bioretention cells also have the opportunity to be an integrated part of the landscape.

References

Barr Engineering Company (2003) "*Minnesota Urban Small Sites BMP Manual: Stormwater Best Management Practices for Cold Climates*", www.metrocouncil.org/environment/watershed/BMP/manual.htm.

Coffman, L. S. (2003) "Wastewater Infrastructure Needs", *FDCH Congressional Testimony*, March 19, 2003. Item: 32Y1283880085.

Coffman, L. S., Goo, R., and Frederick, R. (1999) "Low impact development: an innovative alternative approach to stormwater management", *Proceedings of the 26th Annual Water Resources Planning and Management Conference ASCE*, June 6-9, Tempe, Arizona.

Davis, A. P., Shokouhian, M., Sharma, H., Minami, C., and Winogradoff, D. (2003) "Water Quality Improvement through Bioretention: Lead, Copper, and Zinc Removal", *Water and Environmental Research*, 75(1): 73-82.

Davis, A. P., Shokouhian, M., Sharma, and H., Minami, C. (2001) "Laboratory Study of Biological Retention (Bioretention) for Urban Stormwater Management", *Water and Environmental Research*, 73(1):5-14.

Fetter, C.W. (1999) *Contaminant Hydrogeology*, Prentice Hall, Inc., Upper Saddle River, New Jersey.

Haggerty, R., and Gorelick, S. M. (1995) "Multiple-Rate Mass-Transfer for Modeling Diffusion and Surface-Reactions in Media with Pore-Scale Heterogeneity", *Water Resources Research*, 31(10):2383-2400.

Hunt, W. F. III, Jarrett, A. R., and Smith, J. T. (2003) "Field Study of Bioretention Areas in North Carolina", *Proceedings of the 2003 ASAE Annual International Meeting*, Las Vegas, Nevada, Paper Number: 032302.

Low Impact Development Center (2003) http://www.lowimpactdevelopment.org/links.htm. Accessed 8-1-2004.

McCuen, R. H., Rawls, W. L., Brakensiek, D. L. (1981) "Statistical Analysis of the Brooks-Corey and the Green-Ampt Parameters Across Soil Textures", *Water Resources Research*, 17(4):1005-1013.

Prince George's County Department of Environmental Resources (PGDER) (1993) "*Design Manual fur Use of Bioretention in Storm water Management*", Division of Environmental Management, Watershed Protection Branch. Landover, MD.

Rawls, W. J., Brakensiek, D. L, and Miller, N. (1983) "Green-Ampt infiltration parameters from soils data", *Journal of Hydraulic Engineering, Proceedings of the American Society of Civil Engineers*, 109(1):62-70.

Reedy, O. C., Jardine, P. M., Willson, G.V., Selim, H. M. (1996) "Quantifying the diffusive mass transfer of nonreactive solutes in columns of fractured saprolite using flow interruption", *Soil Science Society of America Journal*, 60(5):1376-1384.

USEPA (2003a). http://www.epa.gov/region1/npdes/history.html. Accessed 8 December 2003.

USEPA (2003b) www.epa.gov/OWOW/NPS/facts/point7.htm. Accessed 2 January 2003.

USEPA (1999) *Storm Water Technology Fact Sheet: Bioretention* EPA 832-F-99-012. Office of Water, Washington, D.C.

A Comprehensive Approach to Engineering LID BMPs- Bioretention Facilities

William C. Lucas

TRC Omni Environmental Corporation, Research Park, 321 Wall Street, Princeton, NJ 08540-1545; PH (609) 924-8821; email: wlucas@TRCSolutions.com

Introduction

Bioretention is acknowledged as one of the most effective BMPs for treating stormwater. Captured runoff flows through a soil profile formulated to provide a high cation exchange capacity, largely due to organic content. The resulting TPHC, TSS and metals removal efficiencies in bioretention facilities have been documented at well over 95%. Studies by Hsieh and Davis (2002) show 100 percent removal of TPHC by various types of bioretention media. The TP removal rate is estimated at 60 percent, per the findings of Davis et al (2000).

In addition to these excellent filtering capabilities, bioretention facilities have been reported as reducing the temperature of the discharged runoff by as much as 12 degrees compared to extended detention facilities (Davis et al, 2000). This Section presents the elements involved in the design of a large in-line Bioretention Facility.

Modeling Approach

In all hydrologic analysis, to properly model the response of the subareas under the storm event of interest for quality storm design, the contributory subareas should be divided into pervious and impervious subareas, with separate flow paths for each. By not aggregating the pervious and impervious areas into a "composite" CN, a more accurate runoff response is generated for small storm events. This is particularly important for events where a composite CN would give no runoff, even when it is obvious that the impervious areas will generate runoff.

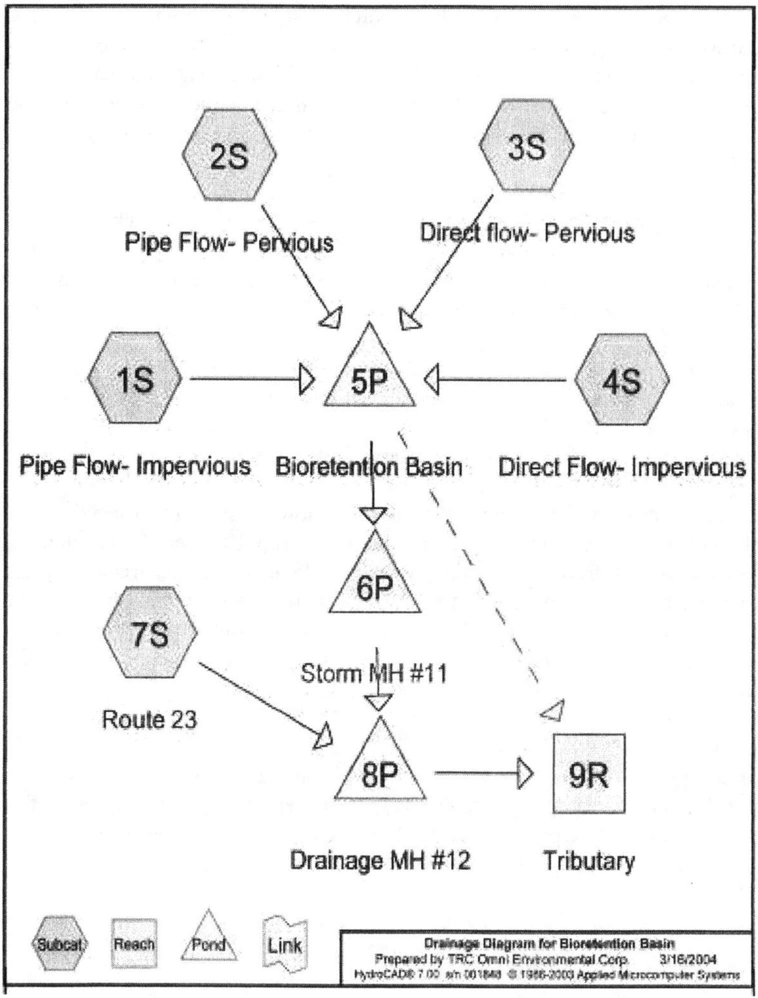

Figure 1

This routing approach requires a greater depth to treat the events of greatest interest from a quality standpoint. On the other hand, since the Tc's are different, desynchronizing the subarea peak flows often reduces the combined peak flow in larger events. Therefore, even while it may increase the storage required for the quality event, this discrete approach often does not increase the storage required for attenuating the peak event.

The storage-indication method is used for detention routing, with storage computed by the conic method. For outlet structure hydraulic routing, HydroCAD calculates the entrance, exit, and friction losses in the outflow pipe as if it were a culvert. This procedure thus determines the hydraulic gradient inside the structure. Standard orifice and weir flow equations are then applied to determine flows into the structure according to the difference in head across the orifices. HydroCAD then matches the flows and hydraulic elevations at each time step to arrive at the computed routing through the structure. HydroCAD also allows for exfiltration computations as described below.

Tailwater is accounted for by using the dynamic routing method, which carries tailwater from downstream structures up through the system for each time step, set at the default minimum of 0.01 hours. The downstream conveyance inlets and the associated pipes are also modeled in this manner.

As a result, the capacity of the conveyance system can be modeled more accurately than computations that do not take account the influence of downstream structures. The final discharge is the Tributary, for which a rating table was created to generate its tailwater elevation as a function of event peak flow.

Figure 1 shows how the contributory areas and flow paths are arranged. The Bioretention Facility is designed to intercept the runoff from Subareas 1S through 4S, with all of the runoff from the quality event of 1.25 inches to be treated by the bioretention media. In addition, much of the runoff from larger events will also be treated by the bioretention media; in particular the most polluted first flush. Flows from downstream subareas (7S) that may affect the capacity of the downstream conveyance structures (6P and 8P) are also addressed in this analysis.

Bioretention System Layout

As shown in Figure 2, the bioretention Facility is located next to a parking lot, from which runoff enters the Facility through a level spreader. Being constructed on fill, the Facility is lined with impermeable liner to prevent potential destabilization of the underlying slopes.

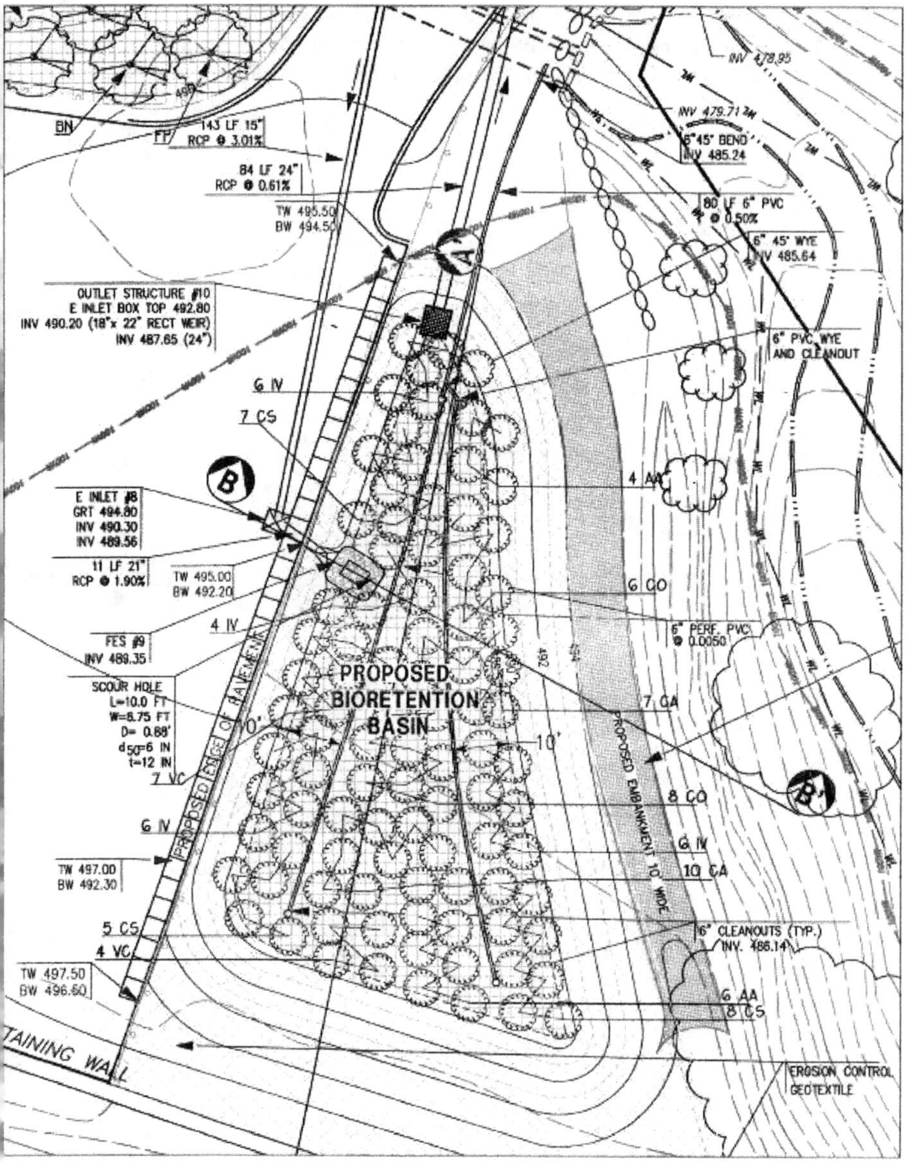

Figure 2. Design of the bioretention Facility

Figure 3 shows a vertically exaggerated profile across the facility with an underdrain system provided to convey the filtered runoff from the bioretention facility. This underdrain system terminates at the top of a 60-inch concrete culvert.

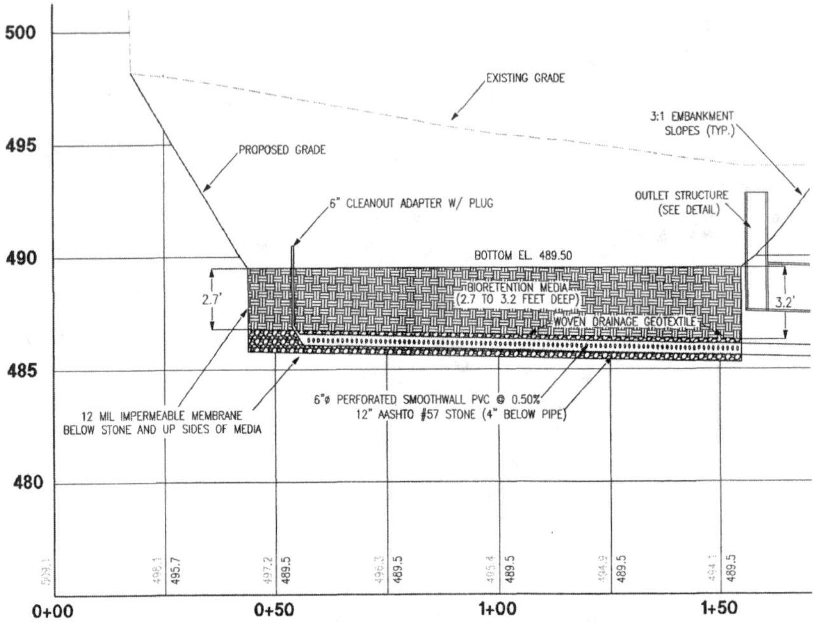

Figure 3

Flow from the underdrain is controlled by a clean out cap, into which orifices are cut to ensure that the drainage rates are restricted if they are too rapid, as shown in Figure 4. Peak flows from larger events are conveyed through a separate conventional outlet structure and pipe system. Figures 5 and 6 show additional cross-section details of this system. Peak flooding elevations from the routing are included in these figures.

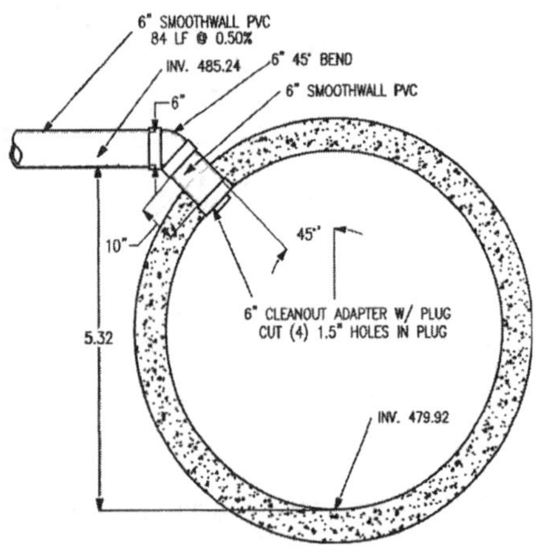

Figure 4

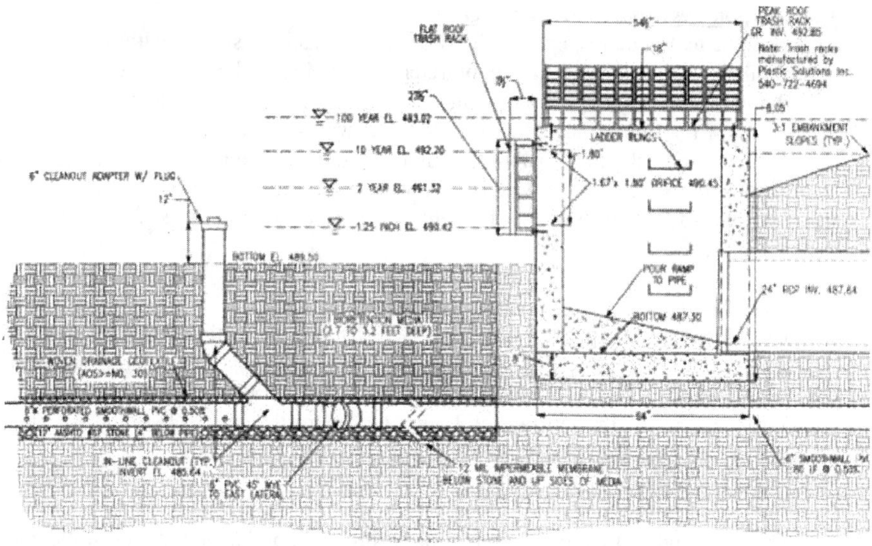

Figure 5 Outlet Structure Section

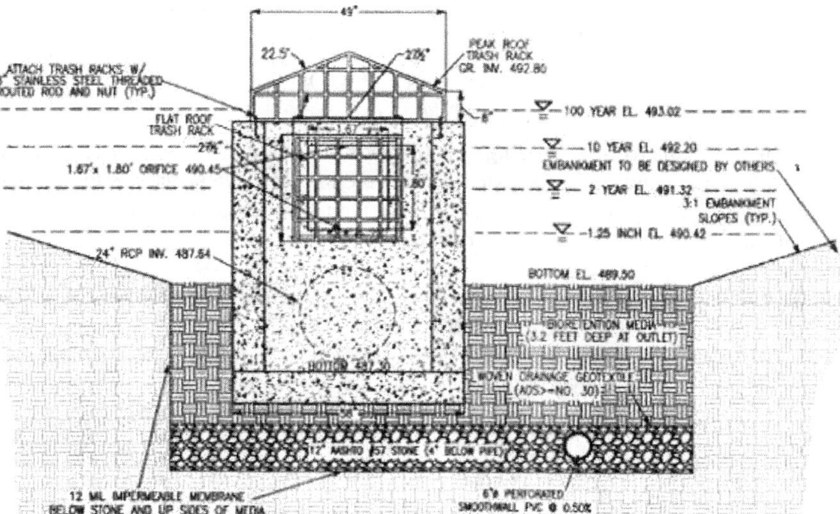

Figure 6 Outlet Structure Elevation

Bioretention System Design Parameters

To model the hydraulic response of the Bioretention Facility, storage within the voids is computed at 30 percent in the bioretention media, and at 40 percent in the underlying stone. To route flow through the media, HydroCAD treats the infiltrated component of total runoff as exfiltration over the wetted surface area. This increases the infiltration rate slightly as water ponds up in the basin, although at a lesser rate than would be expected from Darcy's Law. To take a conservative approach for modeling the surface outflow requirements, exfiltration was modeled at a rate of 1 inch per hour.

Exfiltration is then routed through the lateral perforations, thence flowing into two laterals, which are modeled as culverts. The laterals then discharge into a pipe modeled as culvert, regulated by four orifices cut into an end cap located where it discharges. Depending upon hydrostatic head, the capacity of these orifices is roughly twice the exfiltration rate of 1 inch per hour allocated to the media. This ensures that the Facility will not discharge at a faster rate than 2 inches per hour, even if the media infiltration rates should exceed this rate.

Since it is intended that the system not infiltrate into the underlying soils, the underdrain system is set at a 0.5 percent slope to ensure free drainage. (Note that the most underdrains are designed to operate under pressure flow conditions, where pipe slope would not be a factor in capacity computations.) The underdrain system is designed according to the ellipse equation from the Engineering Field Handbook, Section 14 (NRCS, 1992):

$$S = \sqrt{\frac{4P(b^2 - a^2)}{Q_d}} , \qquad (1)$$

in which S is the required spacing, P is the permeability of the stone, b is the distance from the high point of the saturated zone to the impermeable layer, a is the distance from the perforations (the low point of the saturated zone) to the impermeable layer, and Qd is the minimum acceptable drawdown rate.

For the purposes of this design, P of stone is estimated at 30 inches/hour, b is 1.33 feet (representing the thickness of one foot of stone, plus the 4 inches retained by capillary action in the media), a is 0.33 feet, and Qd is 0.5 inch per hour. Under these circumstances, (1) reduces to 20 feet. Note that the required spacing S increases as the saturated zone depth b increases.

Rearranging the terms in (1), note how the effective drawdown rate at a given spacing also increases as the saturated depth b increases:

$$Q_d = \frac{4P(b^2 - a^2)}{S^2} , \qquad (2)$$

Given a 20-foot spacing and a depth b of 1.85 feet (10 inches into the media), Qd increases to the design rate of 1.00 inches per hour. This represents an average depth of less than 6 inches in the media. Since this is very close to the depth of 4 inches where drainage stops due to capillary retention, this is effectively the minimum rate.

When half full, b is 2.5 feet, and Qd is 1.84 inches per hour. At full depth, b is 4.0 feet, and Qd is 4.78 inches per hour. Accordingly, by placing the underdrains 10 feet from the sides of the Facility and no more than 20 feet apart for any appreciable distance, the average drawdown rate is well over 1.0 inch per hour.

The underdrain discharge is modeled as secondary flow. The exfiltration rate of 1 inch per hour controls the system dynamics, while the capacity of the rest of the conveyance system is at least several multiples of this desired rate. This ensures that the system will function as designed, even if over well half the conveyance capacity of the orifices or laterals was to become impeded.

The remaining flows are allocated as primary flows through the main outlet structure, which is designed as an inlet box with a single rectangular orifice. For flows larger than the quality event, a rectangular orifice is set at an elevation slightly above the quality event peak elevation, as shown in Figures 5 and 6.

The summary presented in Figure 7 displays the resulting input parameters of this design, as well as the resulting allocation of flows through the various orifices and weirs during the 1.25-inch water quality event and the 7.5-inch 100-year event. Exfiltration through the media is routed through the lateral perforations (conservatively allocated as 300 ½-inch orifices), thence flowing into the two 6" laterals, which are modeled as culverts at a flat slope (so that the underdrain will drawdown completely between events). The laterals then discharge into a 6" PVC pipe to the Tributary, modeled as reach 9R. The flow capacity of the PVC pipe, modeled as culvert, is regulated by four 1½-inch orifices cut into an end cap located at the Tributary.

Depending upon hydrostatic head, the capacity of these orifices varies from 0.57 to 0.66 cfs, or roughly twice the exfiltration rate of one inch per hour estimated for the media. This ensures that the Facility will not discharge at a faster rate than two inches per hour, even if the media infiltration rates should exceed this rate. This issue is an important consideration, since the potential benefits of adsorption and biological uptake/immobilization are less likely at the short contact time under higher infiltration rates. By being located in an accessible location, the clean out cap can be replaced with a different orifice configuration if circumstances warrant.

This underdrain discharge is modeled as secondary flow. As shown in figure 7, the exfiltration rate of one inch per hour controls the system dynamics, while the capacity of the rest of the conveyance system is at least several multiples of this desired rate (except for the outlet orifices, as discussed above). This ensures that the system will function as designed, even if over well half the conveyance capacity of the orifices or laterals was to become impeded.

Pond 5P: Bioretention Basin

```
Inflow Area =     6.160 ac, Inflow Depth =  0.44"  for 1.25 inch event
Inflow      =     3.91 cfs @   1.27 hrs, Volume=      0.224 af
Outflow     =     0.27 cfs @   2.05 hrs, Volume=      0.223 af, Atten= 93%, Lag= 46.8 min
Primary     =     0.00 cfs @   0.00 hrs, Volume=      0.000 af
Secondary   =     0.27 cfs @   2.05 hrs, Volume=      0.223 af
```

Routing by Dyn-Stor-Ind method, Time Span= 0.00-36.00 hrs, dt= 0.01 hrs
Starting Elev= 485.80' Surf.Area= 1,327 sf Storage= 99 cf
Peak Elev= 490.42' @ 2.05 hrs Surf.Area= 4,980 sf Storage= 8,601 cf (8,501 cf above start)
Plug-Flow detention time= 313.8 min calculated for 0.221 af (99% of inflow)
Center-of-Mass det. time= 309.4 min (386.4 - 77.0)

#	Invert	Avail.Storage	Storage Description
1	485.64'	32,374 cf	**Custom Stage Data (Conic)**Listed below

Elevation (feet)	Surf.Area (sq-ft)	Voids (%)	Inc.Store (cubic-feet)	Cum.Store (cubic-feet)	Wet.Area (sq-ft)
485.64	100	40.0	0	0	100
486.14	3,934	40.0	311	311	3,935
486.33	3,934	40.0	299	610	3,977
486.34	3,934	30.0	12	622	3,979
489.50	3,934	30.0	3,729	4,351	4,682
489.51	3,934	100.0	39	4,390	4,684
490.00	4,568	100.0	2,081	6,471	5,328
491.00	5,544	100.0	5,048	11,519	6,336
492.00	6,586	100.0	6,058	17,577	7,414
493.00	7,409	100.0	6,993	24,570	8,288
494.00	8,205	100.0	7,804	32,374	9,144

#	Routing	Invert	Outlet Devices
1	Secondary	485.24'	**1.5" Vert. Orifice/Grate X 4.00** C= 0.600
2	Device 1	485.64'	**6.0" x 80.0' long Culvert** RCP, rounded edge headwall, Ke= 0.100 Outlet Invert= 485.24' S= 0.0050 '/' n= 0.010 Cc= 0.900
3	Device 2	485.64'	**6.0" x 100.0' long Culvert X 2.00** RCP, rounded edge headwall, Ke= 0.100 Outlet Invert= 485.64' S= 0.0000 '/' n= 0.010 Cc= 0.900
4	Device 3	485.83'	**0.5" Vert. Orifice/Grate X 300.00** C= 0.600
5	Device 4	0.00'	**0.002778 fpm Exfiltration over entire Wetted area**
6	Primary	487.64'	**24.0" x 80.0' long Culvert** RCP, square edge headwall, Ke= 0.500 Outlet Invert= 487.14' S= 0.0063 '/' n= 0.013 Cc= 0.900
7	Device 6	490.45'	**1.67' x 1.80' Vert. Orifice/Grate** C= 0.600
8	Device 6	492.85'	**4.00' x 3.50' Horiz. Orifice/Grate** Limited to weir flow C= 0.600

```
Primary OutFlow Max=0.00 cfs @ 0.00 hrs HW=485.80' TW=486.03' (Dynamic Tailwater)
  6=Culvert ( Controls 0.00 cfs)
    7=Orifice/Grate ( Controls 0.00 cfs)
    8=Orifice/Grate ( Controls 0.00 cfs)

Secondary OutFlow Max=0.27 cfs @ 2.05 hrs HW=490.42' TW=478.98' (Dynamic Tailwater)
  1=Orifice/Grate (Passes 0.27 cfs of 0.53 cfs potential flow)
    2=Culvert (Passes 0.27 cfs of 1.55 cfs potential flow)
      3=Culvert (Passes 0.27 cfs of 2.71 cfs potential flow)
        4=Orifice/Grate (Passes 0.27 cfs of 4.21 cfs potential flow)
          5=Exfiltration (Exfiltration Controls 0.27 cfs)
```

Type III 24-hr 100 year Rainfall=7.

```
Primary OutFlow Max=22.08 cfs @ 12.21 hrs HW=493.02' TW=489.19' (Dynamic Tailwater)
  6=Culvert (Passes 22.08 cfs of 29.60 cfs potential flow)
    7=Orifice/Grate (Orifice Controls 18.49 cfs @ 6.2 fps)
    8=Orifice/Grate (Weir Controls 3.59 cfs @ 1.4 fps)

Secondary OutFlow Max=0.38 cfs @ 12.21 hrs HW=493.02' TW=483.92' (Dynamic Tailwater)
  1=Orifice/Grate (Passes 0.38 cfs of 0.66 cfs potential flow)
    2=Culvert (Passes 0.38 cfs of 1.93 cfs potential flow)
      3=Culvert (Passes 0.38 cfs of 3.44 cfs potential flow)
        4=Orifice/Grate (Passes 0.38 cfs of 5.28 cfs potential flow)
          5=Exfiltration (Exfiltration Controls 0.38 cfs)
```

Figure 7. Input parameters used for Figures 5 and 6.

Note also the difference in tailwater elevations between the events, thus confirming that potential tailwater effects are addressed. In this instance, the underdrain and main outlet culvert discharge into two separate systems, so the tailwater elevations are different. However, note that the culverts are passing flows that are less than their capacity in all events, so the tailwater did not have any effect in this particular design. If the culvert had been under designed, such tailwater could have resulted in the inlet controlling flows.

Bioretention System Hydraulic Responses

The Hydrographs shown in Figures 8 and 9 show the system operates during the 2-year (3.3-inch) and 10-year (5.2-inch) events, as well as the quality and 100-year events. The elevation chart shows how the entire depth of the facility draws downs between events. The surface is located at elevation 489.50, so ponding occurs only at elevations above this value.

Examination of the hydrograph of the water quality event shows that virtually all of the runoff is drained through bioretention media, with a maximum ponding elevation of less than a foot. This filtered runoff is not only much cleaner, it is much cooler than the incoming runoff, and it is released at a rate that minimizes the contribution of this runoff toward downstream erosion and bank scour.

In the two-year event, over half the runoff volume is treated by the media, and it is only the less polluted peak flow of the hydrograph that bypasses the media. There is a temporary surge of ponding of less than two feet during the peak, but this persists for only several hours, after which ponding of less than a foot extends for the 12-hour duration of the recession limb of the hydrograph. In the 10-year and 100-year hydrographs, the temporary peak is slightly higher, but it persists for a similar time, with ponding drained several hours after the recession limb.

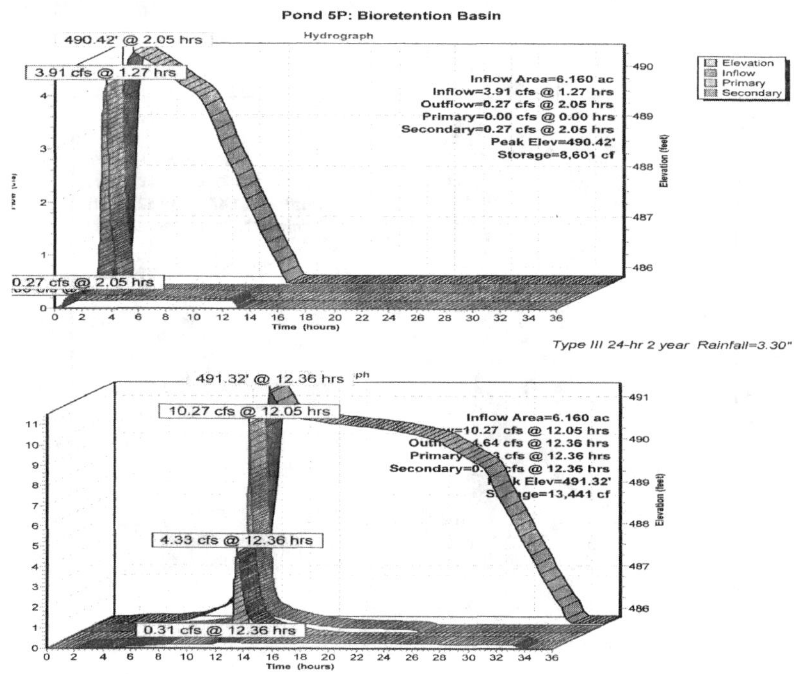

Figure 8: 1.25 – Inch and 2 – Year Hydrographs

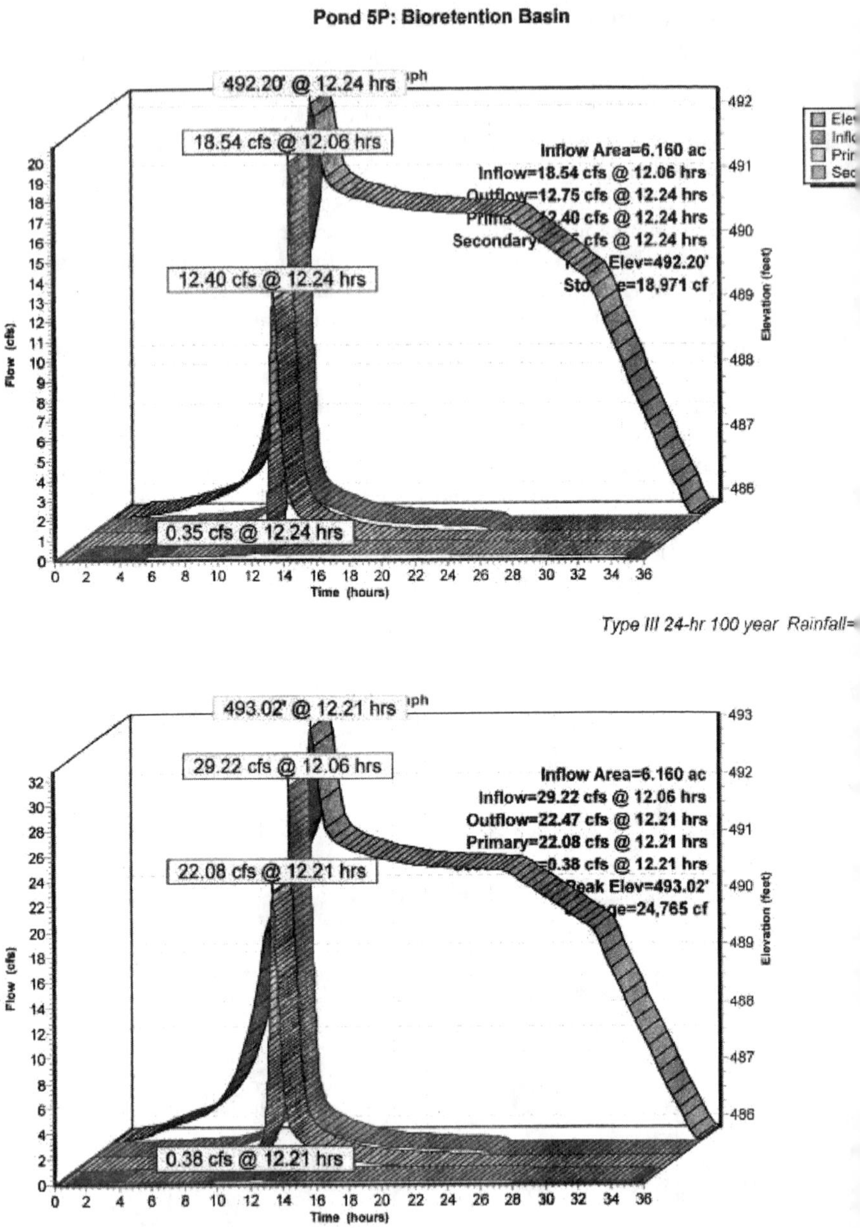

Figure 9: 10 – Year and 100 – Year Hydrographs

Bioretention System Planting Details and Media Specifications

Note that the hydrograph responses are similar to that found in floodplains throughout the region, so this type of in-line design does not pose unacceptable stress on the landscaping. In order to provide shade for the facility, a densely planted shrub layer is proposed in the bottom. Trees were specifically excluded from consideration, since they are more likely to extend their roots through the impermeable membrane under the underdrain. Furthermore, they are also much more likely to be subject to potential windthrow, given the loose sandy media. On the other hand, there are many shrubs that are well adapted to these conditions that are also capable of providing dense shade. These shrubs are to be complemented with native warm season grasses on the slopes.

Recent efforts in bioretention design have focused upon the composition of the media. It has been found that media with excessive fines (silts and clays over 25 percent) seem to fail with regularity, unless optimal vegetative growth is present to maintain tilth (Hsieh and Davis, 2003; DNREC pers. comm.; and Calvert County, MD, pers. comm.). On the other hand, recent installations by the Ocean County Conservation District show that media with a very high percentage of sand can be quite effective, and still maintain plant growth (C. Smith, pers. comm.).

Media with a high sand percentage have also been shown to have high pollutant removal rates (Hsieh and Davis, 2002), as is confirmed by the extensive literature on sand filters. However, it is important that there be enough organic matter to initiate plant growth. Once established, plants will maintain the organic matter balance (C. Smith, pers. comm.), largely due to the copious inputs of nutrients from urban runoff.

For this reason, the media proposed for the Facility is comprised of 90 percent bank run sand, which already has some organic content as well as some silt and clay. To this sand, aged compost is to be added at the shipping facility, and then delivered to the site. This eliminates the potential for contractor error in mixing the media. It is to be placed by excavator so that no compaction of the media will occur, which would otherwise greatly reduce infiltration rates (Pitt, 2000; Lucas, 2003; C. Smith, pers. comm.). It is expected that the stabilized infiltration rates would be at least two inches per hour.

In addition to ensuring that the media is blended and placed properly, it is necessary to ensure that proper maintenance is carried out, particularly during the establishment period. Figure 11 displays the notes that are necessary, at a minimum, to be placed upon a plan to ensure the proper media composition, installation and long-term success of the Facility.

A Comprehensive Approach to Engineering LID BMPs - Biofiltration Swales

Introduction

Biofiltration swales (also referred to as bioswales) are wide flat swales that are designed to convey quality event flows at depths well below the height of the vegetation. In this manner, flows are filtered in a manner similar to a filter strip. Provided there is enough retention time, bioswale sediment trapping efficiency can be quite high. In essence, while they are narrower than a filter strip with higher flow depths, they are much longer. Based upon flow depth and retention time, it is possible to project the reduction efficiencies in bioswales using the DURMM model (Lucas, 2004a). The interested reader is encouraged to review the Technical Manual supporting this model (Lucas, 2004b) for specific details of the bioswale hydraulics and trapping efficiency computations.

I addition to filtration, there are many cases where detention is necessary to meet post-development release rate criteria. Typically, such storage would be provided in a detention pond. However, bioswales can extend over a considerable area. Given a ponding depth of several feet, this area can then be utilized for detention storage. Placed at regular intervals, check dams in bioswales can be used to create a series of cascading pools that will provide substantial detention. This Section presents the application of the DURMM model to a specific site, and how the bioswales are modeled to compute peak rate reductions.

Modeling Approach

Subarea allocation follows the methods outlined in Section 1. The individual pools created by each check dam in Bioswale 1 are routed as separate ponds 4P, 5P and 6P. Likewise, Bioswale 2 is modeled as ponds 12P and 13P. The routing diagram in Figure 10 displays how the contributory subareas and their associated bioswale detention facilities are routed to allocate flows.

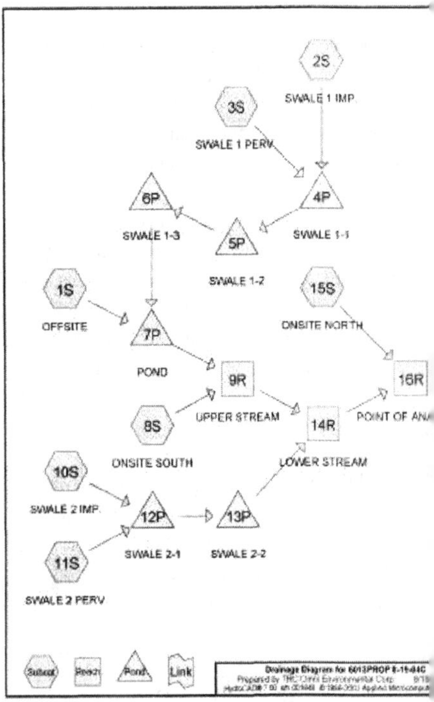

Figure 10: Routing Diagram

This diagram also addresses offsite subareas and an existing pond, as well as the stream reaches that flow through the site.

Figure 11 displays how the swales are located on the site, and how they are graded. Note how the swales represent a footprint that is no larger than would be required for a typical pond facility. However, by being relatively long, there is a retention time of 30 minutes in the upper swale and 15 minutes in the lower swale. These retention times result in sediment trapping efficiencies in the range of 85%.

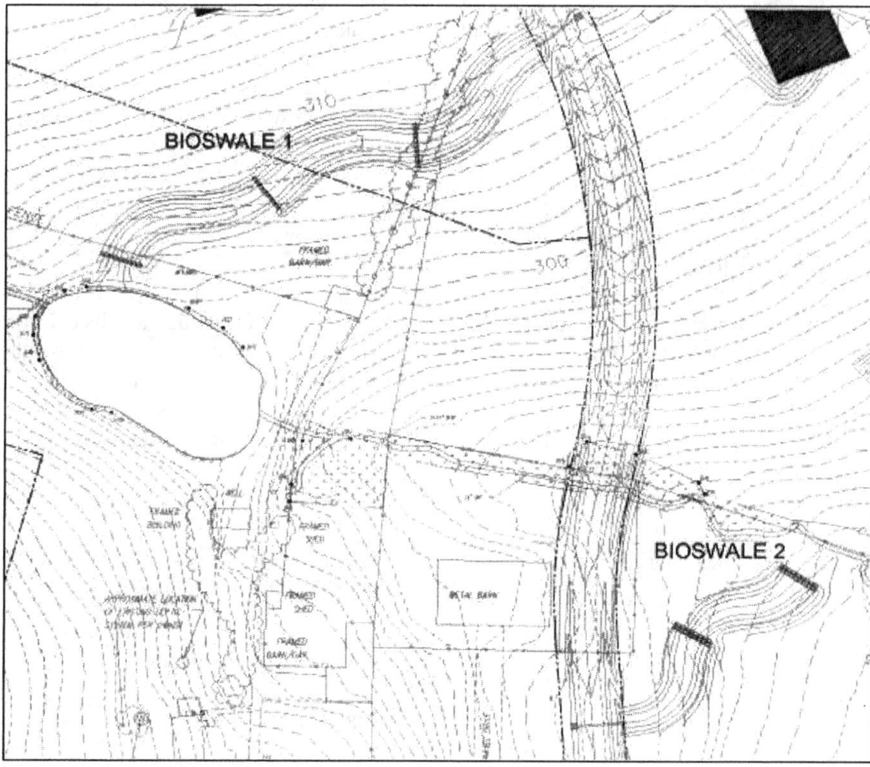

Figure 11: Biofiltration Swale Site Plan

To utilize the bioswales as detention facilities, a series of gabion check dams is proposed in each bioswale, providing shallow storage at a maximum depth of 2.8 feet. Flow through the stone check dams is routed according to the method set forth in the NAHB/NRC H.R.C. Research Series, Report No. 10 (McIntyre and others, 1992), in which unit flow rates are a function of head h, length L and stone size D:

$$q = \frac{h^{1.5}}{(L/D + 2.5 + L^2)^{0.5}} \qquad (3)$$

where D is the average stone diameter in feet. The stone gradation is intended to be relatively uniform in size. Designed for flow paths of up to 6 feet, (3) addresses the range of flow lengths involved in check dam flows. Unless constricted to fixed width by plugging portions of the check dam with concrete, the effective flow width is calculated as being the bottom width w of the swale, to which is added ponding depth h times swale side slopes ss:

$$W = (ss \times h) + w$$

If a specific width were to be specified, (4) would not apply. In the proposed gabion dams, the check dams involve two stages, a lower opening height set at the difference in elevation between the dams, and an upper opening that is the height of a standard gabion size. The lower opening is intended to route flows approaching the 2-year event, while runoff peaks from larger events flow through the upper opening, which is usually wider and has coarser stone to permit greater flows.
Typically, the lower outlet is 1.5-inch filter stone outlet with height that matches the fall between dams, and a flow length of 2.3 feet (equal to a typical 3-foot gabion, less 8 inches of 4-inch stone placed on the outside to contain the filter stone). Its width is specified in the Plans. The upper outlet is typically comprised of 4-inch stone, but 1.5-inch stone could be used in situations where the flows rates need to be substantially reduced.

Due to its relatively large subarea, Bioswale 1 is wider (12 feet wide), and its check dams have less flow restrictions, with larger (4-inch) stone used across the full width of the upper opening, as shown in Figure 12. On the other hand, due to its smaller contributory area, Bioswale 2 is 10 feet wide, and its openings are restricted to widths of 3 feet for the lower opening, and 12 feet of 4-inch stone used for the upper opening. Figures 12 and 13 display details of the check dams.

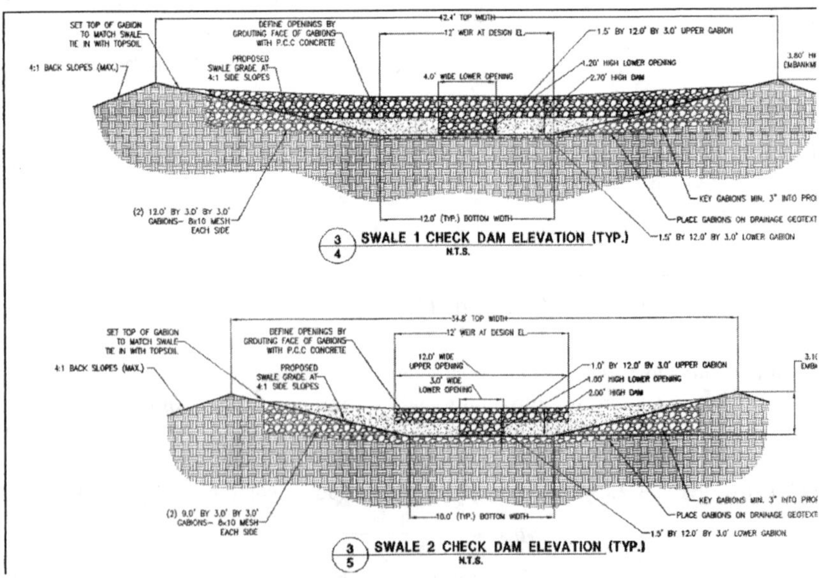

Figure 12: Bioswale and check dam elevations

Adapted from the latest version of the DURMM model (Lucas, 2004a), the NJURMM (New Jersey Urban Runoff Management Model) spreadsheet was used to develop the stage/outflow responses based upon the check dam design parameters. To account for the tailwater effects of downstream check dams, NJURMM has an option to set tailwater equal to the drop between dams. By entering "NO" at the "DROP?" prompt, flows through the filter stone in the lower opening remain constant once the elevation in the upper pool exceeds the drop between dams. This is based upon the assumption that there will be a constant drop across the dams once the pools fill up. As will be seen, there is actually a "pulse" of peak flows that translates down the system.

Since its existing slope is steeper, grade drop check dams are used to flatten the slope of Bioswale 2, increasing its pollutant filtration potential and storage volume.

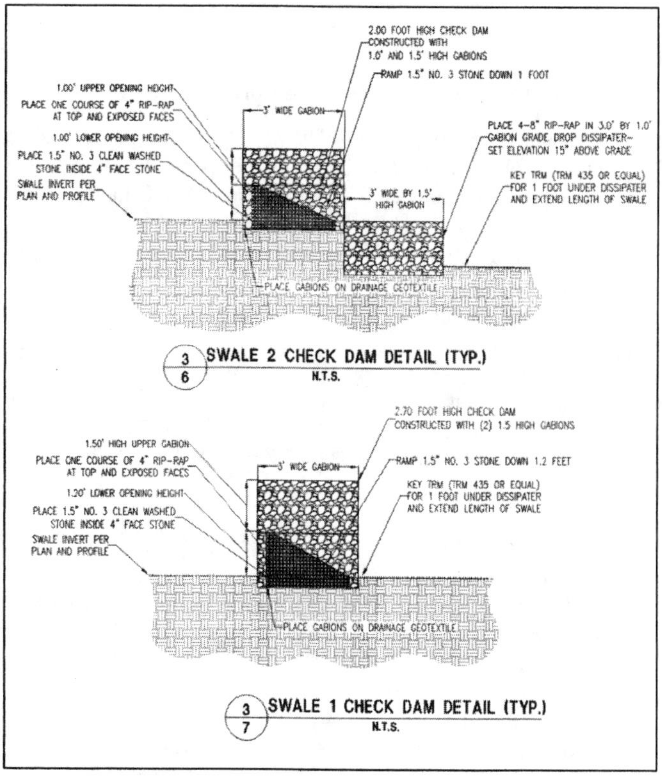

Figure 13: Check Dam Sections

The preceding stage/outflow relationships are then entered as a special outlet in HydroCAD®. To provide relief flow during extreme events, the crest of the check dam is a 12-foot wide level section set at an elevation 2.7 or 2.0 feet above the invert of the check dam for Bioswales 1 and 2 respectively, with side slopes extending up at least 1.1 feet up to the embankment of the swale. This outlet is modeled as sharp-crested trapezoidal weir. Given this design, outflow through the check dams was partitioned into primary flows through the check dam stone, and secondary flows that pass over the check dams.

The NJURMM spreadsheet also displays the stage/ surface area relationships based upon swale geometric parameters of length, width, side slopes, and slope. Due to the shallow depths involved, this method is more accurate than attempting to develop stage/area relationships from interpolation of coarser contour intervals. Figure 14

BIOSWALE QUALITY DESIGN	CN	84	LENGTH	425	SIDES:1	4.0	BOTTOM	12.0	SWALE OK?	OK
	SLOPE	0.83%	COVER	4	DENSE GRASS		VELOCITY	0.23	DEPTH	0.33
	INPUT LOAD		23,403	133.2	104.7	397.1	99.3	95.6	2.24	13.40
	OUTPUT CONC.		16.7	0.21	0.33	0.76	0.27	0.33	0.005	0.017
% FLOW	OUTPUT LOAD		4,297	54.6	85.8	196.5	68.5	84.2	1.30	4.44
100.0%	PERCENT REMOVAL		82%	59%	18%	51%	31%	12%	42%	67%
5950	RESIDENCE TIME (min.)		30.8	TO BMP	9091	FROM BMP	9104	RUNOFF REDUCTION		0%
BIOSWALE CAPACITY, STABILITY & VOLUMES		CHECK DAM DESIGN		POND EL.	AREA	VOLUME	STAGE	OUTFLOW	STAGE	OUTFLOW
		FILTER (in.)	1.50	0.59	3,050	598	0.20	0.07	1.37	1.38
MAX. VEL.	2.61	LENGTH (ft)	2.3	1.18	7,099	3,498	0.39	0.19	1.57	2.15
MAX.DEPTH	0.94	WIDTH (ft.)	4.0	1.76	9,098	8,247	0.59	0.35	1.76	3.21
FALL	1.18	FACE (in.)	4.00	2.35	11,097	14,174	0.78	0.54	1.96	4.55
NO. DAMS	3	WIDTH (ft.)	0.0	2.94	13,096	21,278	0.98	0.75	2.16	6.17
DROP?	NO	WEIR (ft.)	0.0	3.53	15,095	29,557	1.18	0.99	2.35	8.06

Figure 14: NJURMM Bioswale Spreadsheet

presents an excerpt of the NJURMM BMP Design spreadsheet displaying the design criteria for Bioswale 1-1 check dam pool and its outlet. The green cells represent user input, while the blue cells displays results of the analysis.

Due to the extremely low permeability of the soils on the site, the option of using infiltration according to the bioretention approach discussed in Part 1 was not viable. As a result, there is no exfiltration component in these bioswales, which is unusual. Note how the DURMM stage /area entries are adjusted to correspond to the design elevations, with a multiplier applied to adjust the total volume for each pond's percentage. The corresponding entries on check

Pond 4P: SWALE 1-1

```
Inflow Area =    264,126 sf, Inflow Depth =  0.13"   for 1.25 inch event
Inflow      =    1.14 cfs @   1.11 hrs, Volume=       2,954 cf
Outflow     =    0.58 cfs @   1.78 hrs, Volume=       2,954 cf, Atten= 49%, Lag= 40.0 min
Primary     =    0.58 cfs @   1.78 hrs, Volume=       2,954 cf
Secondary =      0.00 cfs @   0.00 hrs, Volume=           0 cf

Routing by Dyn-Stor-Ind method, Time Span= 0.00-30.00 hrs, dt= 0.01 hrs
Peak Elev= 304.22' @ 1.78 hrs   Surf.Area= 1,520 sf   Storage= 677 cf
Plug-Flow detention time= 18.7 min calculated for 2,954 cf (100% of inflow)
Center-of-Mass det. time= 18.7 min ( 107.9 - 89.2 )

 #     Invert    Avail.Storage  Storage Description
 1     303.40'        9,902 cf  Custom Stage Data (Prismatic)Listed below x 0.33

Elevation      Surf.Area       Inc.Store       Cum.Store
   (feet)        (sq-ft)   (cubic-feet)    (cubic-feet)
  303.40             10              0               0
  303.99          3,050            903             903
  304.58          7,099          2,994           3,897
  305.16          9,098          4,697           8,594
  305.75         11,097          5,958          14,551
  306.34         13,096          7,137          21,688
  306.93         15,095          8,316          30,005

 #  Routing     Invert   Outlet Devices
 1  Primary     303.40'  Special (user-defined)
                         Head (feet) 0.00 0.20 0.39 0.59 0.78 0.98 1.18 1.37 1.57 1.76 1.96 2.16 2.35
                         Disch. (cfs) 0.00 0.07 0.19 0.35 0.54 0.75 0.99 1.38 2.15 3.21 4.55 6.17 8.06
 2  Secondary   306.10'  150.0 deg x 12.0' long Sharp-Crested Vee/Trap WeirC= 2.47

Primary OutFlow Max=0.58 cfs @ 1.78 hrs  HW=304.22'  TW=302.93'  (Dynamic Tailwater)
└─1=Special (user-defined)(Custom Controls 0.58 cfs)

Secondary OutFlow Max=0.00 cfs @ 0.00 hrs  HW=303.40'  TW=302.20'  (Dynamic Tailwater)
└─2=Sharp-Crested Vee/Trap Weir( Controls 0.00 cfs)
```

Figure 15: HydroCAD® Bioswale Summary

dam and bioswale design parameters are presented in a HydroCAD® summary in Figure 15.

Routing Analysis

Figures 16 displays the hydrographs for the 1.25" water quality event and the 2-year event. Note that the flow and elevation axes change scales as the flows are reduced. While providing better visualization of the dynamics of each pond, this display understates the decline in flows through the entire bioswale in the quality event. If the particular setting were better drained, the reductions in the quality event would be substantial, due to exfiltration.

Furthermore, note the effect of the long swale upon hydrograph timing, with peak flows being delayed by well over an hour, resulting in a very broad hydrograph. In this manner, the post-development flows more closely approach pre-development flow dynamics, where much of the runoff response is mediated by slower flow pathways.

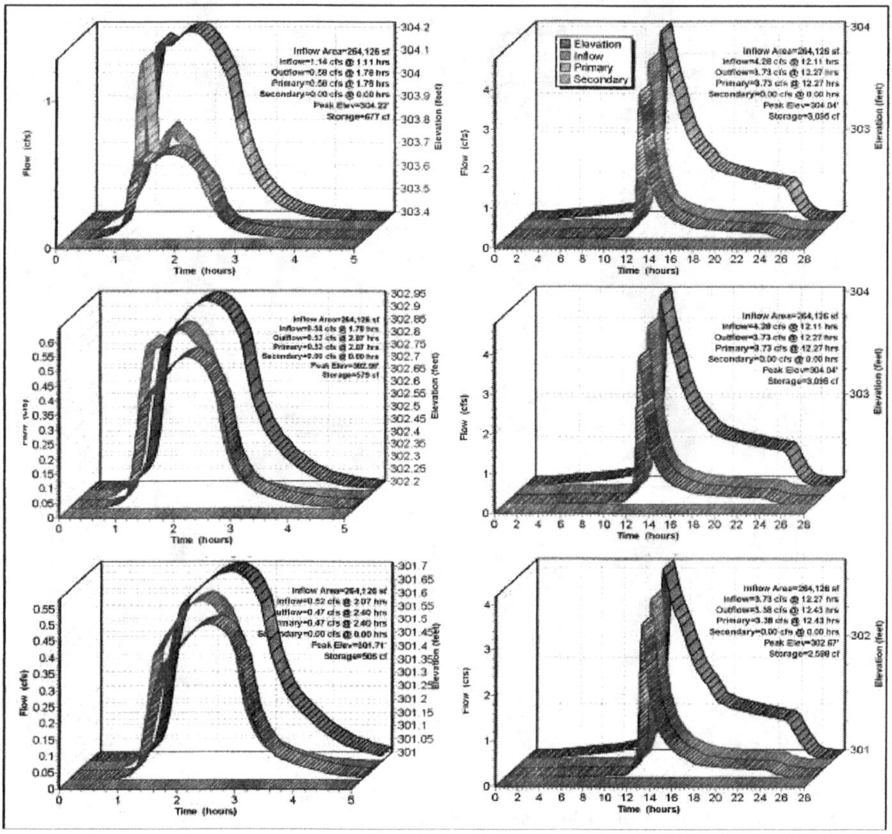

Figure 16: 1.25 inch and 2-year Hydrographs

While the peak reduction in the 2-year event is smaller, there is still a substantial delay in the 2-year event, with peak flows being delayed by 0.32 hours. These dynamics result in meeting the requirements for a 50% reduction in predevelopment flows, which are quite high due to the poorly drained site conditions.

Figures 17 displays the hydrographs for the 10-year and the 100-year events. Even though the check dams do not substantially reduce incoming peak flows in these larger events, they delay the peaks enough that they are no longer synchronized with undetained offsite peak flows. In this manner, the overall routing dynamics meet the required peak release rate criteria.

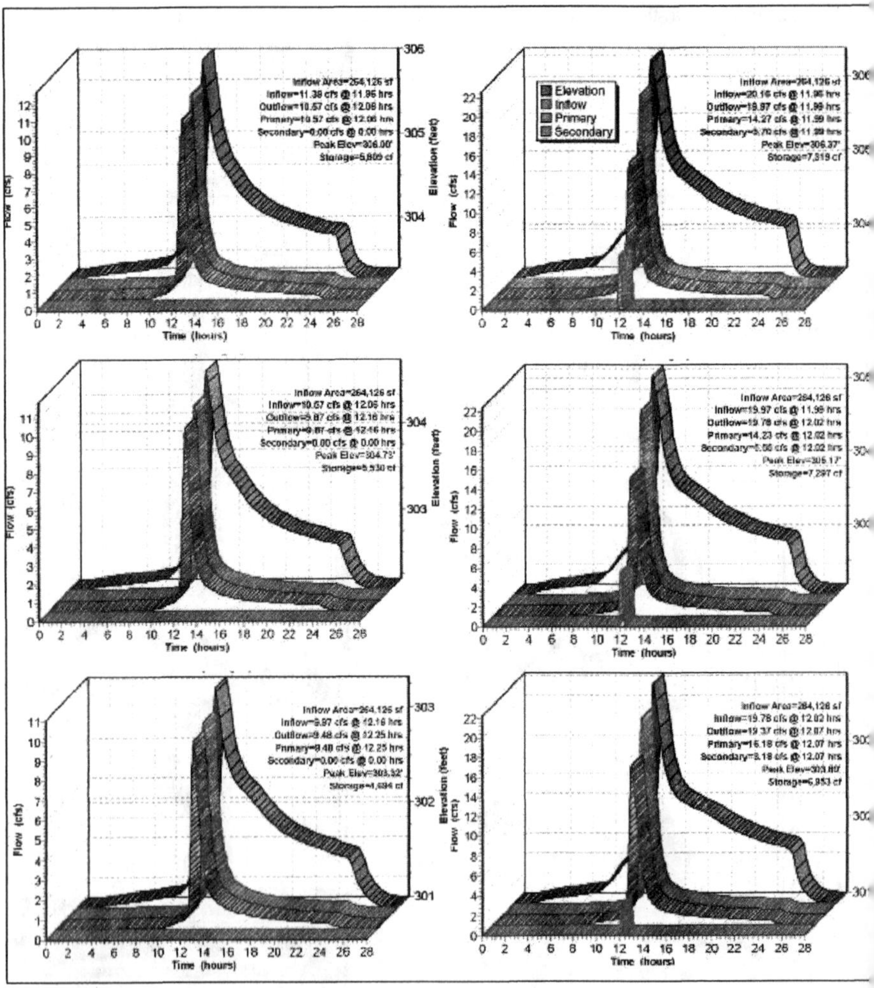

Figure 17: 10-year and 100-year Hydrographs

Conveyance/Overtopping Analysis

The flows over and through the check dams actually percolate down through the dams, as opposed to cascading over the top. The fall within the dams is generally similar to the drop between pools, so there is little potential for scour in smaller events. Even during the 100-year event, the maximum depth of flow over the top of the dams is only several inches. Flowing over 4-inch stone for a length of three feet, this flow will actually emerge from the stone some distance below the top. The difference in peak ponding elevations is no more than 1.3 feet between pools, which occurs when there is at least a foot of tailwater to disperse potential scour.
As a result, flows through and over the check dams should have minimal potential for scour. However, since there is a fall of one foot between the swales in Bioswale 2, the check dam designs include a gabion grade drop structure, which is placed on a turf reinforcement matting (TRM). In Bioswale 1, there is no fall between dams, so there is no drop structure, but the TRM is provided. Notwithstanding the low erosive aspect of flow through the dams, TRM is also placed along the entire length of the bioswales. This TRM will also aid in establishing a good stand of vegetation.

Since the gabion dams have such a large capacity, a foot of freeboard is provided in the extreme event. Note that runs with the openings clogged (an extremely unlikely event for a full width of large stone) indicate that the maximum increase in depth would be less than 6 inches, leaving at least 6 inches of freeboard.

Conclusions

The preceding analysis shows how an effective design for an in-line bioretention facility can provide excellent treatment of the water quality event, while addressing a wide range of flows without excessive ponding in terms of depth and duration. Likewise, effective stormwater management for peak flow rate reduction can also be provided in bioswales utilizing check dams to provide detention storage and routed outflows. However, to accomplish these design tasks, it takes a thorough hydrologic, hydraulic and engineering approach. This level of analysis is necessary to ensure that not only do the facilities address the required water quality and quantity management criteria, but that they can also be constructed with available off-the-shelf technology.

References

Davis, A.P., C. Minami, H. Sharman, and M Shokouhian. 2000. "Laboratory Study Bioretention for Urban Stormwater Management." *Water Environ. Res.*, 72(1), 1-10.

Davis, A.P., Shokouhian, M., Sharma, H. and Minami, C., 2001. "Laboratory Study of Biological Retention (Bioretention) for Urban Storm Water Management," *Water Environ. Res.*,73(1), 5-14.

A.P. Davis Home Page: **http://www.ence.umd.edu/~apdavis/Biodata.htm**.

Horner, R. R, 1988. "Biofiltration Systems for Storm Runoff Water Quality Control". Metro, Seattle, WA.

Hsieh, C.H. and Davis, A.P. 2002. "Engineering Bioretention for Treatment of Urban Storm Water Runoff," *Watersheds 2002, Proceedings on CDROM Research Symposium*, Session 15, Ft. Lauderdale, FL.

Hsieh, C.H. and Davis, A.P. 2003. "Evaluation of Bioretention for Treatment of Urban Storm Water Runoff," presented at *ASCE EWRI Engineering Conference*. Philadelphia, PA.

HydroCAD® Stormwater Modeling System. **http://www.hydrocad.net/support.htm**.

Lucas, W.C., 2004a. "DURMM: The Delaware Urban Runoff Management Model. A Users Manual for Designing Nonstructural BMPs to Minimize Stormwater Impacts from Land Development" 2004 Draft. Dover, DE.

Lucas, W.C., 2004**b** "Green Technology, Delaware's Urban Runoff Management Approach, A Technical Manual for Designing Nonstructural BMPs to Minimize Stormwater Impacts from Land Development". January 2004 Draft. Dover, DE.

McIntyre, C., G. Aron, J.H. Willenbrock, and M. Diemler. "Analysis of Flow Through Porous Media as Applied to Gabion Dams Regarding the Storage and Release of Storm Water Runoff," NAHB/NRC H.R.C. Research Series, Report No. 10, University Park, PA, 1992.

Natural Resource Conservation Service (NRCS). 1992. "Engineering Field Manual" Washington, DC.

Pitt, R.E. 2000. The Integration of Water Quality and Drainage Design Objectives. SLAMM Model Documentation. Chapter 2.

Chapter 10

BMP Technology in Pacific Rim Region

Introduction

By Shaw L. Yu, University of Virginia

The past two decades have witnessed a great deal of progress in the application of BMP technology for nonpoint source pollution control in the United States and other parts of the world. In the Pacific Rim Region, the importance of nonpoint pollution control has been gaining widespread recognition in recent years as a result of more point source pollution being gradually put under control. In Taiwan, for example, field-testing of BMPs began in early 1990s and a few BMP guidebooks have been written. However, the implementation of BMPs is still voluntary because there is no regulatory framework dealing with nonpoint or stormwater runoff pollution at the present time. In China, on the other hand, there has been a strong movement towards mandating nonpoint pollution control since a few years ago. The phenomenal economic growth and urban-industrial development in China have made worse the environmental situations, but at the same time, made more resources available for mitigating the environmental impacts of progress.

A special session was organized for the ASCE Salt Lake City Conference to highlight the nonpoint pollution control efforts in the Pacific Rim Region, especially in China and Taiwan. The three papers selected for the present chapter represent well the work done recently in the Region. The first paper, by Kuo et al., deals with the optimal placement of BMPs in a watershed, which is a subject receiving a broad attention in the States. Jan-Tai Kuo is a professor at National Taiwan University and has been active in water resources and environmental engineering research for many years. The second paper, by Yu et al., gives an overview of the recently initiated nationwide effort in China dealing with nonpoint pollution control. Gang Yu is a senior professor at Tsinghua University, which has the leading environmental engineering program in China. The third paper, by Zhang et al., presents a case study on nonpoint pollution control in Shenzhen, China. Xihui Zhang is a professor at the Shenzhen Tsinghua Graduate School.

Optimal Best Management Practices (BMPs) Placement Models

Application to Fei-Tsui Reservoir Watershed in Taiwan

J. T. Kuo[1], C. D. Hsieh[2], S. K. Chiu[3] and P. H. Hsieh[4]

[1] Professor, Department of Civil Engineering and Hydrotech Research Institute, National Taiwan University, Taipei 106, Taiwan; PH +886-2-2362-1841; FAX +886-2-2366-1725; kuoj@ ntu.edu.tw

[2] Division Director, Water Resources Agency, No. 41-3, Sec. 3, Hsin-Yi Road, Taipei 106, Taiwan; cdhsieh@wra.gov.tw

[3] Graduate Student, Department of Civil Engineering, National Taiwan University, Taipei 106, Taiwan; r91521316@ntu.edu.tw

[4] Ph. D. Candidate, Department of Civil Engineering, National Taiwan University, Taipei 106, Taiwan; d90521002@ntu.edu.tw

Introduction

The purpose of this research work is to propose a model to search for the optimal BMPs placement strategy which considers the minimal cost and reservoir water quality standards. Similar to several related research, such as Yeh and Labadie (1997), Behera et al (1999), Zhen (2002) and Harrell and Ranjithan (2003), the first step of pollution control is to figure out how much pollutant loads from a watershed. Therefore, field data sampling and water quantity and quality modeling are needed to calculate each pollutant load under different landuses. Second, a preliminary procedure is required for finding suitable sites for various BMPs construction in the watershed or subbasin. This work may be carried out by a experienced engineer going out to the field for investigation. With the tool of modern geographic information system (GIS), we can consider many conditions (slope, area, land ownership, etc.) for each kind of BMPs construction. Using those tools, we can find all suitable locations quickly. The optimization model considers total minimum cost for BMPs construction as an objective function. The constraints are to satisfy the water quality standards of the reservoir water quality. Thus, a reservoir water quality model is required to couple with the optimization model. The complete model of this study consists of two interacting components, i.e. an optimization model based on dynamic programming (DP) and a zero-dimensional reservoir water quality model (Vollenweider type).

The Fei-Tsui Reservoir watershed is chosen as an example for this study. Fei-Tsui Reservoir is the major water supply source of over four million people in metropolitan Taipei, Taiwan. Two major pollutants, total phosphorus (TP) and total suspended solid (TSS), are taken into consideration in the model. Fei-Tsui Reservoir has a drainage area of 303 km^2 and a water surface of 10.24 km^2. In order to estimate pollutant load and stormwater discharge, Fei-Tsui Reservoir watershed is divided into 14 subbasins. Furthermore, each subbasin is further divided into 2 to 5 small regions according to their topography and hydrologic characteristics. The Fei-Tsui reservoir watershed and its landuse types distribution are shown in Figure 1. Due to excessive landuses and significant loads of pesticide and fertilizer, various well-designed BMPs should be built to treat stormwater runoff before it flows into the reservoir.

Dynamic Programming

The traditional optimization method, dynamic programming (DP), is used in this study. The objective function is to minimize the BMPs construction and operation, maintenance, and repair (OMR) cost. It can be express as:

$$\underset{\lambda_{i,j},X_{i,j}}{Min.} \sum_{j=1}^{14} \sum_{i=1}^{N} \lambda_{i,j} C_{i,j}(X_{i,j}) \tag{1}$$

where, j = the jth subbasin ;

i = the ith site that allow BMPs to be considered;

N = number of possible BMPs in subbasin j;

$\lambda_{i,j}$ = binary variable , $\lambda_{i,j} \in \{0,1\} \quad \forall\, i$ and j;

$C_{i,j}$ = the cost function of jth subbasin and i^{th} site;

$X_{i,j}$ = decision variable of the jth subbasin and i^{th} site; and 14 denotes the total number of subbasins.

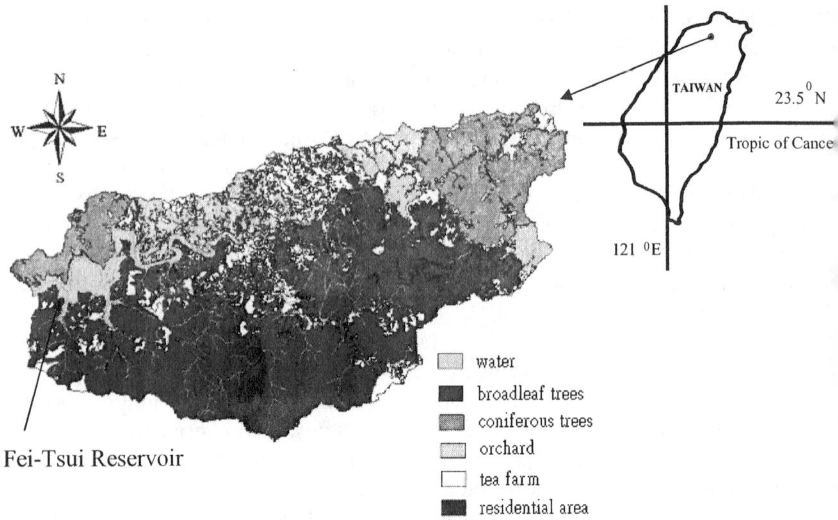

Figure 1. Fei-Tsui reservoir watershed and its landuse types distribution

Reservoir inflow containing TSS and TP should satisfy the reservoir water quality standards, so constraints can be written as:

$$C_d^{TSS} \leq C_c^{TSS} \tag{2}$$

$$C_d^{TP} \leq C_c^{TP} \tag{3}$$

where C_d^{TSS} and C_d^{TP} represent the desired TSS and TP concentrations after BMPs construction in the watershed. C_c^{TSS} and C_c^{TP} are the water quality standards of TSS and TP concentrations which are shown in Table 1 and 2.

Table 1. Water quality standards of TSS in water bodies(Taiwan Environmental Protection Agency, 1998)

Category	Max. allowable concentration (*mg/l*)
One (for drinking water supply)	25
Two (for drinking water supply)	25
Three (for industrial use)	40
Four (for irrigation)	100

Table 2. Water quality standards of TP in reservoirs (Administration National Academy of Science and National Academy of Engineering, 1972)

Classification	Max. allowable concentration (*mg/l*)
Oligotrophic	0.01
Mesotrophic	0.015
Eutrophic	0.02

In this research, the background data includes: pollutant load of different landuses, discharge of each region and subbasin topographic information. The forward moving discrete DP is used to solve the optimization problem. There are eight stages and four kinds of BMPs to be considered in the optimization model. The decision variable, $D_{n,i,j,k}$, is defined as for the kth kind BMPs, the pollutant removal efficiency in the jth subbasin, at i^{th} site and at n^{th} stage. The state transformation function $S_{n+1,i,j,k} = G(S_{n,i,j,k}, D_{n,i,j,k})$ can be expressed as:

1. Combination of concentrations from left and right subbasins:

$$S_{junction,n} = \frac{S_{L,n} \cdot Q_{L,n} + S_{R,n} \cdot Q_{R,n}}{Q_{L,n} + Q_{R,n}} \tag{4}$$

where, $S_{junction,n}$ = TSS or TP concentration at a junction for stage n;

$S_{L,n}$ = TSS or TP concentration of left subbasin for stage n;

$S_{R,n}$ = TSS or TP concentration of right subbasin for stage n;

$Q_{L,n}$ = discharge of left subbasin for stage n; and

$Q_{R,n}$ = discharge of left subbasin for stage n.

2. Pollutant concentration variation from upstream to downstream:

$$S_{junction,n+1} = S_{junction,n}(1 - e^{-K\frac{l}{u}}) \tag{5}$$

where, $S_{junction,n+1}$ = TSS or TP concentration at the next junction;

K = TSS settling rate or TP decay rate;

l = channel length; and

u = averaged flow velocity in the channel.

3. total cost function: $C_n(S_{n,i,j,k}, D_{n,i,j,k}) = \sum_{n=1}^{M} \sum_{i=1}^{I} \sum_{j=1}^{J} C_{n,i,j,k}(S_{n,i,j,k}, D_{n,i,j,k})$ (6)

where, M = number of stages;

I = number of subbasins;

J = number of regions or sites in each subbasin; and

$C_{n,i,j,k}$ =the kth BMPs cost at stage n, subbasin j, site i.

In this study, N =8, I=14 and J=2-5.

4. recursive equation :

$$f_n(S_{n,i,j,k}) = Min.[C_n(S_{n,i,j,k}, D_{n,i,j,k}) + \sum_{i=1}^{I}\sum_{j=1}^{J} f_{i-1}(S_{n-1,i,j,k})] \quad (7)$$

Seasonal Inflow and Pollutant Load

For a planning purpose, average inflow of the storm season and pollutant loads are used for the model to keep the problem small and easier for solution. After knowing the inflow and pollutant loads, a seasonal average reservoir model is used to estimate the reservoir's pollutant concentrations. The streamflow and reservoir inflow data have been monitored since 1987 by Taipei Fei-Tsui Reservoir Administration Bureau. The average inflow volume from June to October (rainy season) during 1987-1997 is 50×10^6 m^3.

According to formal studies (Kuo et al, 1998; Yu et al, 2001), the annual pollutant loads can be estimated as shown in Table 3. From previous measurements, we can know that 70% of the annual pollutant loads occur during June to October.

Table 3. Annual pollutant loads of different landuse in Fei-Tsui Reservoir watershed

Landuse / Pollutant	Average pollutant load (*kg/ha/yr*)			
	Residential	Tea farm	Orchard	Forest
TSS	5,200	5,000	2,000	1,000
TP	1.54	2.94	0.42	0.3

BMPs and its Installing Conditions

There are three kinds of BMPs considered in this study. They are detention pond, grass swale and buffer strip. Each of them has different construction cost function, OMR cost, pollutant removal rate and installing condition.

In Fei-Tsui Reservoir watershed, the pre-selecting procedure is needed before using the optimization model to find suitable sites for BMPs in each region. The installing conditions of BMPs are:

1. Slope: smaller than 30%
2. Land requirement: detention pond needs more than 1 ha. of land; grass swale and buffer strip need at least 61 m for each.
3. Land ownership: government or wild land

By setting the installing conditions as mentioned above, there are only 23 possible sites allowed to construct detention pond. The construction cost function and OMR cost are shown in Table 4 (Sample et al, 2003). The pollutant removal efficiency is shown in Table 5 (Yu et al, 2001; Strecker et al, 2001).

Table 4. Construction cost function and OMR cost of different BMPs

	Construction cost function	Parameters	OMR cost/year
Detention pond	$Cost_d = 35 \times 2.195 \times 10^4 V_d^{0.69}$	V_d = 20,000 m^3	20% of the construction cost
Grass swale	$Cost_g = P_g L_g$	P_g = NT$ 1,700 L_g = 61 m	10% of the construction cost
Buffer strip	$Cost_b = P_b L_b$	P_b = NT$ 3,400 L_b = 61 m	5% of the construction cost

Note: V_d is volume of detention pond; P_g and P_b are prices of grass swale and buffer strip; and L_g and L_b are lengths of grass swale and buffer strip.

Table 5. Pollutant removal efficiency of different BMPs (%)

	Detention pond	Grass swale	Buffer strip
TSS	90	60	40
TP	85	65	55

Seasonal Average Reservoir Water Quality Model

After the stormwater is treated by BMPs and flows into the reservoir, the pollutants may settle to the bottom and decay due to biochemical reactions in the reservoir. A zero-dimensional reservoir water quality model, Vollenweider type model (Vollenweider, 1976), is used to simulate the pollutant concentration in Fei-Tsui Reservoir.

The completely mixing and steady state assumption are made in the seasonal average model. The equation can be written as:

$$C = \frac{W}{Q + V_s A_s} \tag{8}$$

where C and W are the average pollutant concentration and load of TP or TSS, Q is outflow discharge, V_s is pollutant settling rate and A_s is the average reservoir surface area. According to the modeling results by Kuo et al (1998), V_s of TP is 0.274 m/day, and average surface area of Fei-Tsui Reservoir is 8617.249×10^3 m^2. The annual average outflow is 1032.56×10^6 m^3 which is 32.74 cms. In this study, V_s of TSS is assumed to be the same as TP, which is also 0.274 m/day.

Optimization Model Results

From Tables 1 and 2, there are four cases of combination of water quality standards considered in this research. Each case has different TP and TSS concentration standards in the reservoir (Table 6). The optimal BMPs placement strategy should be the one with a minimum cost in the final stage (stage 8). The benefits of BMPs construction are water quality improvement and decreasing water treatment cost. The water quality improvement for each case and its optimal placement of BMPs are listed in Table 6.

Table 6. Optimal BMPs placement strategies

Case	TP Standard (*mg/l*)	TSS Standard (*mg/l*)	Minimum BMPs Cost (NT$)	Water Treatment Cost (NT$)	Water Quality Improvement (Benefit)	Number of BMPs Placement	
1	0.01	25	134×10^6	10×10^6	Satisfying drinking water and oligotropic standards	Detention pond Grass swale Buffer strip Without BMPs	18 18 14 0
2	0.015	30	41×10^6	5×10^6	Satisfying industrial use water source and oligotropic standards	Detention pond Grass swale Buffer strip Without BMPs	5 32 12 1
3	0.015	40	1.9×10^6	150×10^6	Satisfying irrigation water source and mesotropic standards	Detention pond Grass swale Buffer strip Without BMPs	0 0 41 9
4	0.028	90	0	250×10^6	No	Without BMPs	

The optimal total minimum cost considering water quality improvement and BMPs construction cost (trade-off curves) is shown in Figure 2. Case 1 is costly in nonpoint source pollution control, but also saves a lot of cost in drinking water treatment. But it is not the best selection considering total minimum cost (BMPs cost plus water treatment cost). Assuming water treatment costs for case 1 to 4 are 0.02, 0.1, 0.3 and 0.5 dollars per ton, for a capacity of water treatment plant of 500×10^6/day, the water treatment costs are shown in fifth column of Table 6. From Figure 2, case 2 is the best selection when we consider both the BMPs construction cost and drinking water treatment cost.

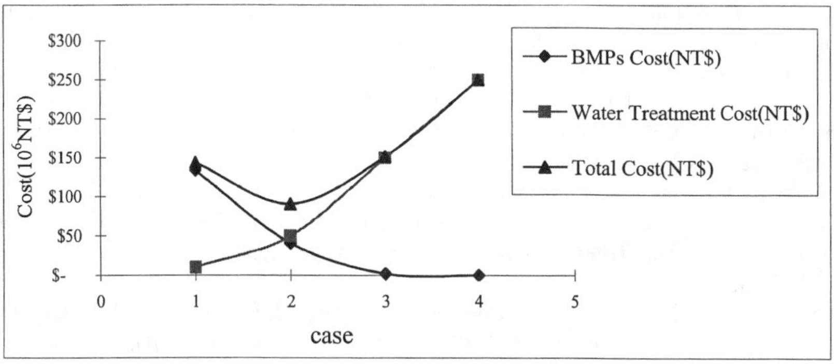

Figure 2. Optimal BMPs construction cost vs. benefit of water quality improvement curves for the four cases in this study

Conclusions

This research develops a forward moving discrete dynamic programming model to find the optimal BMPs placement strategy in Fei-Tsui Reservoir watershed by coupling the optimization model to a seasonal average reservoir water quality model. The optimal BMPs placement policies consider four cases of different water quality standards for total phosphorus and suspended solid. Decision makers can choose a suitable optimal placement strategy according to their budgets and expectation using different water quality constraints. The results obtained by this approach can not only provide practical solutions to nonpoint source pollution control in a watershed, but also give the related agency an optimal facility planning from the whole watershed viewpoint.

Finally, some further research tasks can be followed: (1) solving the optimization problem by evolutionary search methods and (2) life-span consideration of BMPs. Genetic algorithms may be used to solve the same problem by taking the life-span of BMPs into account. A more comprehensive research for policy implementation can be carried out in the future.

References

Administration National Academy of Science and National Academy of Engineering (1972), *Water Quality Criteria,* a Report of the Committee on Water Quality, U. S. A.

Behera, P. K., Papa, F. and Adams, B. J. (1999), "Optimization of regional stormwater management systems", *Journal of Water Resource Planning and Management*, ASCE, Vol. 125, No. 2, 107-114.

Harrell, L. J. and Ranjithan, S. R. (2003), "Detention pond design and land use planning for watershed management", *Journal of Water Resource Planning and Management*, ASCE, Vol. 129, No. 2, 98-106.

Kuo, J. T. et al. (1998), *Water quality simulation and applications for Fei-Tsui Reservoir*, Research report sponsored by Taipei Fei-Tsui Reservoir Administration Bureau, National Taiwan University. (in Chinese)

Kuo, J. T., Yu, S. L. , Kuo, M. C. ,and Pan, C. H. (1998), "Enhancement of the Virginia Storm Model for nonpoint source pollution simulation", Water Resources Engineering 98, Vol. 2, *Proceedings of International Water Resources Engineering Conference*, ASCE, 1177-1181.

Sample, D. J., Heaney, J. P., Wright, L. T., Fan C. Y., Lai, F. H. and Field, R. (2003), "Costs of Best Management Practices and Associated Land for Urban Stormwater Control", *Journal of Water Resource Planning and Management*, ASCE, Vol. 129, No. 1, 59-68.

Strecker, E. W., Quigley, M. M., Urbonas, B. R., Jones, J. E. and Clary, J. K. (2001), "Determining urban storm water BMP effectiveness", *Journal of Water Resource Planning and Management*, ASCE, Vol. 127, No. 3, 144-149.

Taiwan Environmental Protection Agency (1998), *Surface water classification standards*.

Vollenweider, R. A. (1976), "Advances in defining critical loading levels for phosphorus in lake eutrophication", *Mem. Ist. Ital. Idrobiol.* 33, 53-83.

Yeh, C. H. and Labadie, J. W. (1997), "Multiobjective watershed-level planning of storm water detention systems", *Journal of Water Resource Planning and Management*, ASCE, Vol. 123, No. 3, 336-343.

Yu, S. L., Kuo, J. T., Fassman, E. A. and Pan, Henry (2001), "Field test of grassed-swale performance in removing runoff pollution", *Journal of Water Resource Planning and Management*, ASCE, Vol. 127, No. 3, 168-171.

Zhen, X. Y. (2002), *Development of Best Management Practice (BMP) Placement Strategies at the Watershed Scale*, Doctoral Dissertation, Civil Engineering Department, University of Virginia, U. S. A.

To Improve the Quality of Urban Water Environment in China: Technology and Demonstration Projects

Gang Yu[1], Jiuhui Qu[2], Bing Ke[3], and Shaw L. Yu[4]

[1]Department of Environmental Science and Engineering, Tsinghua University, Beijing, 100084, China; PH (86-10) 6278-5685; FAX (86-10) 6279-4006; email: yg-den@tsinghua.edu.cn

[2]Research Center for Eco-environmental Science, Chinese Academy of Science, Box 2871, Beijing, 100085, China
[3]The Administrative Center for China's Agenda 21, Ministry of Science, Beijing, 100089, China

[4]Department of Civil Engineering, University of Virginia, Charlottesville, VA, USA

Serious urban water environmental problems in China

Over the past two decades, China has experienced economic growth at an unprecedented scale. Although the governments at various levels have made relentless efforts and realized significant achievements in urban environmental protection, China's extremely rapid economic development has come at an environmental cost, and pollution of the environment is increasingly threatening the health of its citizens.

In 2002, the wastewater treatment rate by the secondary treatment processes in 47 environmental protection key cities only reached 36.48%. Huge amounts of untreated wastewater discharges have caused serious urban water environmental problems. Major lakes were severely polluted by nitrogen and phosphorus, resulting in a predominant problem of eutrophication. For example, Lake Dianchi is in a state of serious eutrophication, while Lake Tai and Lake Chao are in a state of light eutrophication. The water quality of the lakes in some cities is also relatively poor. Of the major scenic lakes in China, only the water quality of the Kunming Lake in Beijing meets the Grade III standards, while the water quality conditions of the West Lake in Hangzhou, the East Lake in Wuhan and the Lake Daming in Jinan all fail to meet the Grade V standards (SEPA, 2003).

Many Chinese cities face the situation of a worsening water environment. Water pollution prevention, improvement of water environment and protection of aquatic system are the key to the sustainable development of economy and society. In recent years, by focusing on prominent urban water environmental issues, many cities have implemented a number of projects and measures associated closely with issues which people are most concerned about, such as the clean lake projects, clean river projects, clean water projects, new regulations on management of collection and wastewater discharge fees, etc. However, there are many problems in many cities and some local governments have requested for technical supports for their projects.

The ongoing research projects to improve the urban water environment quality in 11 cities

In order to support the improvement of the urban water environmental quality, the Ministry of Science and Technology of China initiated the project "Technologies and demonstrations for improving the urban water environmental quality" in August 2002. Eleven cities , namely, Beijing, Dalian, Daqing, Guilin, Shanghai, Shenzhen, Suzhou, Tianjin, Wuhan, Zhenjiang and Zibo became the demonstration cities of this project (MOST, 2003). Table 1 summarizes the water pollution problems faced by each city and the main research contents.

Table 1 Key problems and main research contents in 11 cities

City	Watershed	Key Problem	Main Research Contents
Beijing	Beihuan Water Channel	✧ Water shortage ✧ Lake and channel pollution	✧ Wastewater reuse technologi ✧ Liangma river restoration technology and demonstratio
Dalian	Malan River	✧ Water shortage ✧ River ecosystem damaged	✧ Rainwater collection practice and demonstration ✧ Wastewater treatment and re technology ✧ Restoration technology and demonstration ✧ Watershe manage systems
Daqing	Liming River	✧ Ecosystem was damaged ✧ Oilfield wastewater pollution	✧ Ecosystem construction technology and demonstratio ✧ Oilfield wastewater treatmen reuse technology
Guilin	Taohua River	✧ Water pollution by agriculture and industry	✧ Water quality improvement technology ✧ Non-point source pollution control technology in Karst a ✧ Watershed management syst
Shanghai	Mengqingyuan District	✧ Water pollution by industry	✧ Small scale wastewater treat technology ✧ River quality improvement technology

City	Watershed	Key Problem	Main Research Contents
Shenzhen	Futian River and Buji River	✧ Ecosystem was damaged by construction ✧ Wastewater pollution	✧ Futian River restoration technology and demonstration ✧ Wastewater treatment and reuse technology
Suzhou	Central Channel System	✧ Sewage did not treated very well ✧ Water pollution	✧ Watershed monitoring systems ✧ Water quality improvement technology and demonstration ✧ Watershed management systems
Tianjin	Taida Area	✧ Water shortage ✧ Wastewater treatment	✧ Sewage treatment and reuse technologies ✧ Lake water quality improvement technology
Wuhan	Hanyang District	✧ Wastewater did not treated ✧ Urban NPS pollution ✧ Lake eutrophication	✧ Urban non-point source pollution control technology and demonstration ✧ Lake water quality improvement ✧ Sewage treatment by ecological technology
Zhenjiang	Neijiang Watershed	✧ Industrial and agricultural pollution ✧ Ecosystem was damaged	✧ Improve the water quality by a series of practices and technologies ✧ Jinshan Lake restoration ✧ Jing-Hang Channel non-point source pollution control technology and demonstration
Zibo	Xiaofu River	✧ Water shortage ✧ Industrial and agricultural pollution	✧ Treat and reuse 3000 m^3/d of wastewater ✧ Improve the water quality to meet the agricultural irrigation demand standard by ecological technologies, such as artificial wetlands ✧ To restore 5 kilometers of a river-riparian ecosystem

BMP in some demonstration cities

Urban NPS pollution characteristics in Wuhan City

A continuous monitoring was carried out in 2003. The maximum storm water was detected on June 23, 2003. Figure 1 to 3 shows the variation of current velocity, COD_{Cr} and total nitrogen with time that day. The maximum current velocity, CODcr and total nitrogen reached 1.28 m/s, 520 mg/L and 44 mg/L, respectively (Peng Jun, 2003)

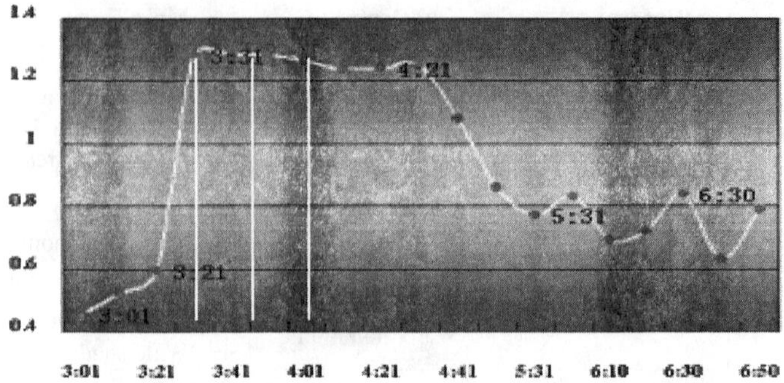

Figure 1. The variation of current velocity (m/S) of storm water with time on June 23, 2003

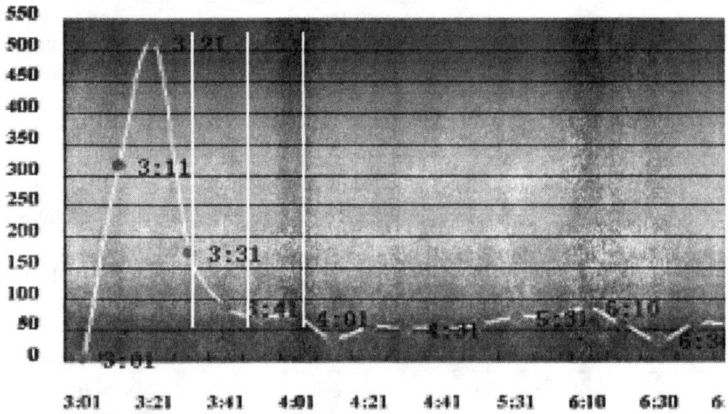

Figure 2. The variation of COD_{Cr} (mg/L) of storm water with time on June 23, 2003

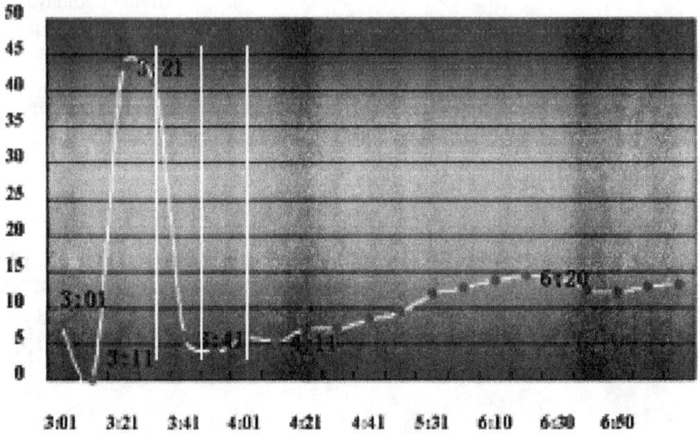

Figure 3. The variation of total nitrogen (mg/L) of storm water with time on June 23, 2003

Urban NPS control in Wuhan City

Laboratory results indicated that ecological engineering technologies are good ways for controlling NPS pollution and restoring the watershed. There are two types of treatment strategies to control the polluted runoff and to reduce the pollutants from the sources. Several control technologies will be used, such as multi-pond systems, grassed filter strips, wetland systems, ecological treatment of wastewater and solid waste. These technologies will be combined systematically in order to form a watershed ecological engineering. Figure 4 shows the ongoing demonstration project to control NPS pollution in the Wanjiaxiang area (Peng Jun, 2003)

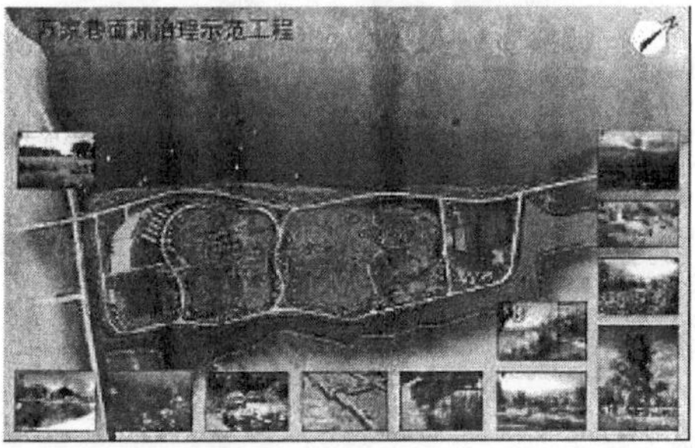

Figure 4. NPS control project in Wanjiaxiang Area

Urban NPS control in Zhenjiang City

Zhenjiang sub-project is designing and constructing 5 kilometers of river-riparian ecosystem to control urban and agricultural NPS pollution in its section of Jing-Hang Channel. Figure 5 is the sketch map of the constructing ecosystem (Zhu Wei, 2003)

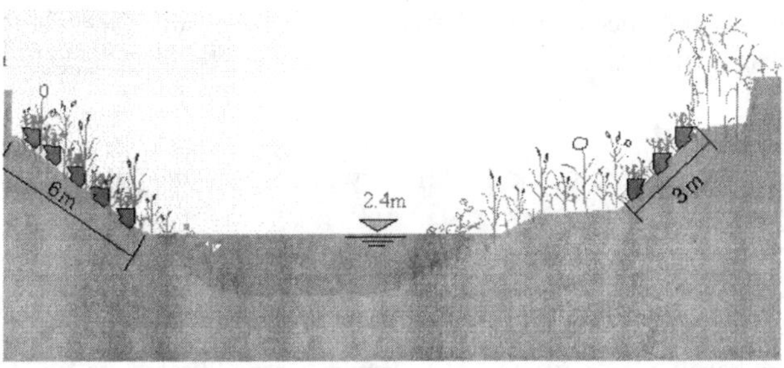

Figure 5. NPS control eco-system in Zhenjiang section of Jing-Hang Channel

Rainwater utilization system in Dalian City

Several main forms of urban rainwater utilization will be demonstrated in Dalian City. Rainwater collection and storage systems, pollutant separators and bio-filtration systems as well as rainwater utilization systems in ecological residential area are main methods to collect rainwater (Yang Fenglin, 2003).

Artificial wetland reduce pollutants from Xiaofu River in Zibo City

As a practical engineering technique for controlling non-point source pollution, an artificial wetland showed in Figure 7 is being constructed to reduce nitrogen, phosphorous and organisms efficiently in Zibo City. The area is 15000 m^2 with depth of 0.5 m. The retention time is 24 h (Hu Hongying, 2003).

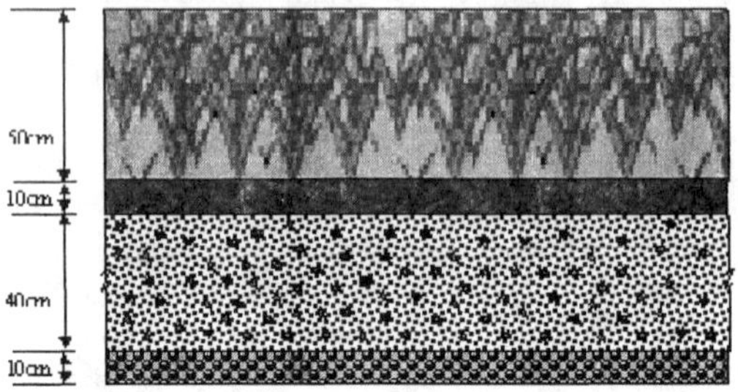

Figure 7. The structure of an artificial wetland in Zibo City

Conclusion

Water shortage, industrial wastewater and sewage pollution as well as non-point source pollution are most serious water problems in many Chinese cities. In order to develop the technologies for improving the urban water quality, the Ministry of Science and Technology supported research and demonstration projects in 11 cities, namely Beijing, Dalian, Daqing, Guilin, Shanghai, Shenzhen, Suzhou, Tianjin, Wuhan, Zhenjiang and Zibo. Some practices and technologies to control NPS pollution are being researched and designed and will be monitored in the near future. The results will provide the needed technical guidance for other Chinese cities to improve urban water quality.

Acknowledgement

The authors would like to thank the Ministry of Science and Technology, China for supporting and funding this work (2003AA601120).

References

SEPA (2003) Annual Report on Urban Environmental Management and Comprehensive Improvement of 47 Environmental Protection Key Cities 2002

MOST (2003) Annual report of project "Water pollution control technology and demonstration"

Peng Jun (2003), Annual report of sub-project "Urban water quality improvement in Wuhan City"

Zhu Wei (2003), Annual report of sub-project "Urban water quality improvement in Zhenjiang City"

Hu Hongying (2003), Annual report of sub-project "Urban water quality improvement in Zibo City"

Yang Fenglin (2003), Annual report of sub-project "Urban water quality improvement in Dalian City"

Nonpoint Pollution Control for the Xili Reservoir in Shenzhen, China

X. H. Zhang[1], T. H. Wang[2], W. Q. Yuan[1], H. L. Duan[2], W. H. Zhang[1], C. X. Li[2]

[1]Research Center for Environ. Eng. & Management, Tsinghua Graduate School at

Shenzhen, Tsinghua University, Shenzhen 518055, China; PH(86-755)-26036707; email: xihuizh@tsinghua.edu.cn[2]Shenzhen Water Resource Bureau, Shenzhen 518055, China; PH(86-755)83072669; email: dhlwater@sina.com

Introduction

Shenzhen is one of the most active cities in China, located in southern China next to Hong Kong. During the past two decades, its population grew from several thousands to now several millions people. Xili reservoir is one of the major resources for water supply in Shenzhen (Latitude 113°57.072E, Longitude 22°35.671N), which is surrounded with small hills. It serves for almost 4 millions residents in the western part of Shenzhen.

The Xili reservoir was constructed in 1960's. Since then, no dredging work has been done for the reservoir deposit. From the last decade of 20th century, the reservoir experienced more and more pollution. With the development of urbanization, nonpoint pollution sources increased sharply due to land reclamation and industrial development, the Xili reservoir was highly polluted at the end of the last century and was under eutrophication condition. Excessive growths of floating plants (e.g. water hyacinth) and cyanobacteria bloom (e.g. *Microcystis sp.*) undertook particularly in the shallow arms of the reservoir, fish-kills took place occasionally. The fish-kills probably resulted from a combination of low dissolved oxygen, high ammonia concentrations and high concentrations of toxic chemicals (e.g. pesticides).

For the security of the water supply, the government carried out a series of restoration projects for the reservoir, and recently a long-distance diversion project as Dongbu water supply project finished at the beginning of the new century to renew its water.

The water quality of the Xili reservoir has been improved gradually. The objective of this paper is to reveal the variation of water quality, to quantify the related nonpoint and point pollution sources, and to identify the controlling strategies for the further improvement of water quality in Xihi reservoir.

Experimental Methods

The map of Xili reservoir was shown in Figure 1. The monitoring sites and sediment sampling sites are also shown. The reservoir has a maximum depth of 15m and a maximum area of 4.6 km$_2$. The capacity of the reservoir is 42×10^6 m^3. Experimental water and sediments were taken from the lake during the period of April 2002 to March 2003. Both online monitoring and lab analysis were conducted with the samples. Online monitoring was conducted a multi-parameter detector (Horiba, Japan). For lab analysis, a number of instruments were used according to requirement (APHA, 1995). For water samples, dissolved oxygen was measured with an Oxi 330 meter and a Cellox 325 electrode (WTW, German). The pH was measured in situ with a pHS-25 meter (Shanghai, China). DO and pH were recorded every day during the experimental period. Ammonia was measured with the Nessler's spectrophotometric method (APHA, 1995), nitrate was measured with an ion chromatography (DX-100, USA, Dionex), SRP was determined with the ascorbic acid method (APHA, 1995), iron and manganese were measured directly with the ICP/MS (Agilent Hp4500 plus). All the measurements were made with triplicate samples.

An aeration experiment was conducted in Polyethylene tubes with 15.4 cm internal diameter and 30 cm length. Well-mixed sediment samples were added into the tubes to 10 cm, then pre-aerated water samples were transferred into the tubes to 15 cm depth. One dissolved oxygen (DO) electrode and one pH electrode were inserted through a rubber stopper to monitor the change of DO and pH in water column. Each tube has one sampling port drilled at heights of 5 cm above the sediment surface, which was sealed with a silicon rubber caulking.

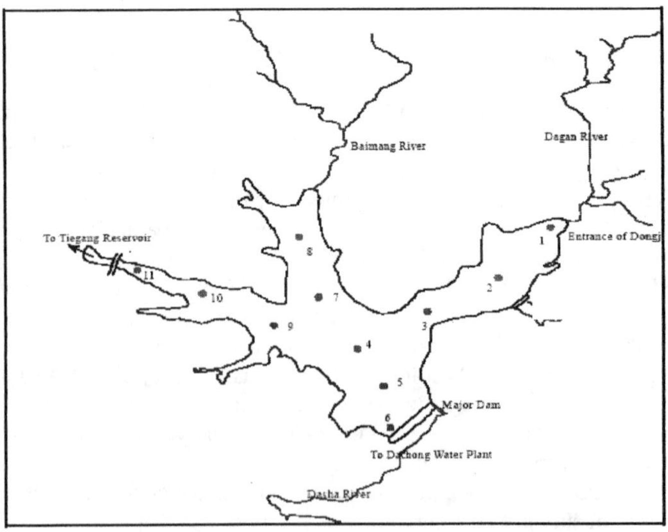

Figure 1. Diagram of Xili reservoir with monitoring site locations.

Results and Discussion

In this study, the weekly changes occurring in the reservoir were investigated, including temperature, dissolved oxygen (DO), pH and turbidity over one year period from April 2002 to March 2003. The number of collected data was over 27000, and the typical results are presented in Figure 2. It was found that stratification occurred in some seasons, which resulted in low concentrations of DO below 6 m in the reservoir.

For turbidity, the concentration near the bottom was obviously higher than close to the water surface, which may reveal that most of the turbidity particles entered into the reservoir through either point sources or non-point sources will settled down as deposits.

Surface water samples were also taken at the same time using plastic bottles, and analyzed for Total-P, NH_4-N, NO_2-N, NO_3-N, Total-Mn and Total-Fe. The concentrations of ammonium, total phosphorus and organic matters are shown in Figure 3. It can be seen than the concentrations of ammonium and phosphorus near the bottom were much higher along the water depth, which may mean that transformations of ammonium and phosphorus are closely associated with the deposit in someway (Reshaw and Riley, 1962; Medina et al., 2003; Burley et al., 2001).

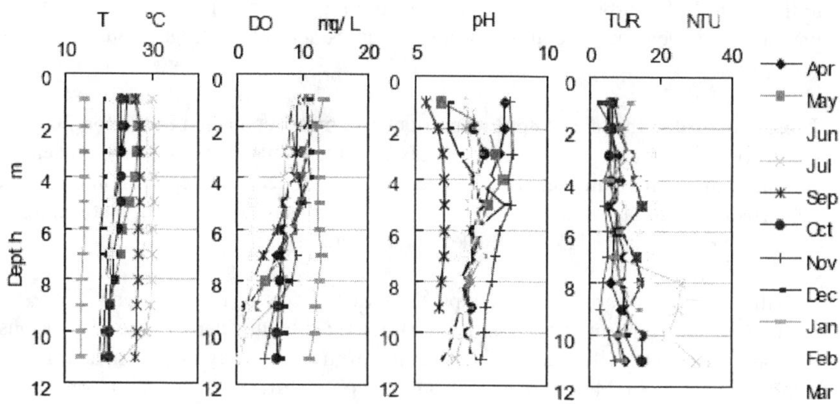

Figure 2. Variation of online monitoring parameters in Xili reservoir.

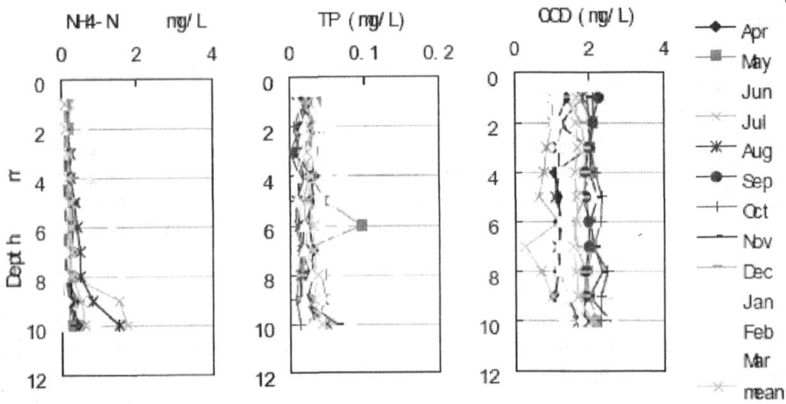

Figure 3. Nutrient concentrations in water samples collected in Xili reservoir.

Deposit samples were collected from the Xili reservoir at eleven sites in August of 2002 using a self-made grabber. The samples were analyzed for pH, Organic Material, Total-N, Total-P and metal species including Fe, Mn, Zn, Pb, Cu, Cr, As, Cd and Hg. Some results are presented in Table 1. The results showed that the contents of Fe and Mn are pretty high in the bottom deposit; the phosphorus content is decreasing while the contents of both organic matter and total nitrogen are still increasing. These phenomena may revealed that some management practices related to the control of phosphorus sources are functioning, however necessary management practices are need for the control of both organics and total nitrogen sources in Xili reservoir.

Table 1 indicated that about 80% of nutrient and metal contained in the top 20cm of the deposit. These nutrients could be released to the overlying water under anaerobic condition or during agitation. The dredge-up of 20cm depth of deposit could remove $103\times10_6$ kg of organic matters, $3.9\times10_6$ kg of nitrogen, $3\times10_6$ kg of phosphorous, $35\times10_6$ kg of iron as well as $1.6\times10_6$ kg of manganese.

In Xili reservoir, the runoff is of typical characteristics of tropical zone. Since the climate in the Xili basin is tropical, it is characterized by distinct wet and dry seasons. The rainy season (April to September), results from the easterly monsoon, and the dry season (October to March) results from the prevailing westerly monsoon. Mean annual rainfall is 1837mm, with almost 80% falling between November and April.

Major non-point sources come from basin runoff. The basin area of Xili reservoir is around 29 km_2. These areas are dominated by traditional agriculture, forestry land uses and small machining factories. The population in this area is around half million, with 300 thousands living in three major villages, most of the others are floating population. It was found that 11.55 km_2 land was occupied by fruit garden, flower garden and vegetable land, the woodland area was less than 1.45 km_2. And more seriously were hundreds of fish pounds distributed around the reservoir with total area of 0.4 km2.

Table 1. Nutrient concentrations in sediment samples.

Depth (cm)	pH	OM (%)	TP (mg/g)	TN (mg/g)	Fe mg/g)	Mn (mg/g)
1	7.1	12.44	2.495	7.652	34.94	1.391
3	7.06	12.51	2.585	7.327	35.49	1.433
5	7.04	11.87	3.38	6.106	37.48	1.419
7	7.21	11.42	3.31	5.491	47.39	1.529
9	7.29	10.74	3.15	4.819	42.16	1.718
11	7.34	10.08	3.31	4.806	42.58	1.721
13	7.52	9.96	2.816	3.657	37.61	1.693
15	7.49	9.08	2.432	3.343	32.18	1.574
17	7.51	8.76	2.326	3.218	29.8	1.473
19	7.58	8.73	2.018	3.163	25.5	1.431
21	7.64	8.09	1.843	3.114	22.1	1.378
23	7.61	7.83	1.768	3.11	21.3	1.361
25	7.6	7.39	1.627	2.989	20	1.363
27	7.59	7.62	1.57	2.948	19.1	1.332
29	7.64	7.04	1.69	2.953	20.9	1.319
31	7.69	6.83	1.35	2.917	18.3	1.259
33	7.72	6.06	1.52	2.931	15.3	1.243
35	7.78	5.08	1.29	2.926	17.9	1.248

The exploited land occupied around 80% of the basin land area. The side hill around the reservoir was planted with mass of litchi. In order to make them growth well, many chemicals including fertilizer, pesticide and herbicide were used. These chemicals could flow into the reservoir with runoff during rainy season. The analyzed results are shown in Figure 4.

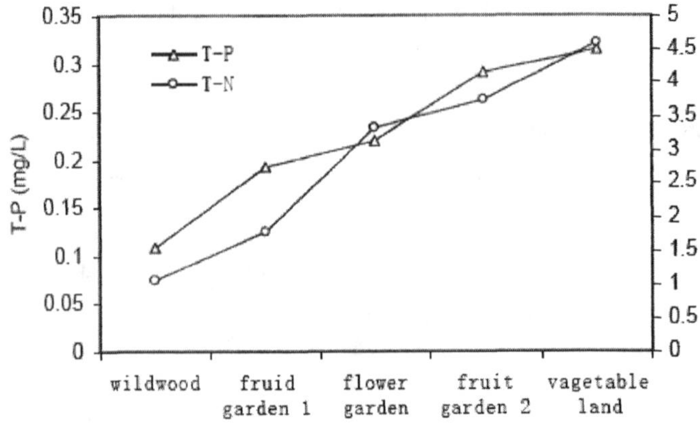

Figure 4. Concentrations of total N and total in runoff flow.

It is estimated that the runoff around the reservoir could bring $1.8\times10_6$ kg of SS, $1.6\times10_6$ kg of nitrogen and $0.2\times10_6$ kg of phosphorus to the Xili reservoir. To control the runoff, it is suggested that an ecological plant zone surround the reservoir should be constructed to prevent the pollution from runoff. In this way, over 80% of SS, N and P can be removed from runoff flows.

In 2002, a long distance diversion channel was built to provide sufficient raw water to Shenzhen city. Now, the diversion project contributes the major inflow to Xili reservoir, also it becomes the major factor that influences the water quality. The inflow was umped from Dongjiang River 136 km away. Before the construction of the diversion channel, the mean residence time of water in Xili reservoir was around 78 days, most of the flow into Xili reservoir came from the surrounding runoff. Now, the inflow was increased enormously to $1\times10_6$~ $1.2\times10_6$ m_3 per day since the diversion project was put into operation in 2002, and the water residence time reduced to about 15 days in Xili reservoir. It is measured that 1000 ~ 1200 Mm_3 of water from Dongjiang river entered the Xili reservoir every day. The water quality of inflow is good for water supply in most of the time. However it changes with seasons and may become worsen at the initial of rainy season. Table 2 shows the results obtained at the intake of the diversion project from 1994-1999.

Table 2. The concentrations of organics and nitrogen in the diversion channel.

Season	COD(mg/L)		NH3-N(mg/L)		NO2-N(mg/L)		NO3-N(mg/L)	
	Range	Mean	Range	Mean	Range	Mean	Range	Mean
Wet Season	1.6-2.2	2.1	0.34-0.52	0.50	0.02-0.14	0.04	0.2-0.84	0.78
Dry Season	1.4-1.8	1.6	0.16-0.32	0.24	0.01-0.03	0.02	0.22-0.48	0.38

The SS and nutrient would increasing greatly at the initial of rainy season because the erosion and runoff at upstream of Dongjiang river. It is estimated that the diversion channel would bring about 3.8×10^6 kg of SS, 4.1×10^6 kg of OM, 3.6×10^6 kg of nitrogen and 0.4×10^6 kg of phosphorus into the Xili reservoir with inflow every year. This will influence the Xili reservoir in both short and long terms. Since most of the turbidity particles will settle down as deposit at bottom, the organics and phosphorus attached on these turbidity particles may be released under some conditions. Therefore, it is necessary to construct a rubber dam at the entrance section, thus form a sedimentation pool. In this way, significant parts of SS and nutrient could be retained in the sedimentation pool. It estimated that 85% of SS, 40% of nitrogen and 30% of phosphorus in the inflow could be captured, preventing from entering the reservoir.

In addition of non-point and point pollution sources, deposit can be consider as an important internal pollution source, which largely formed during gradually accumulation at the reservoir bottom. An experiment on nutrient releasing and control of deposit was carried out in laboratory scale(Beutel, 2001; Tieckelmann, 1991). The deposits taken from Xili reservoir was put into a closed reactor and covered with overlying, then calculating the releasing rate by monitor the concentration of nutrient in water. The experimental results are shown in Figure 5. This experiment indicated that a well designed aeration with addition of either hydrogen peroxide or calcium peroxide could restrain 40% of nitrogen and 60% of phosphorus from releasing, which means $1.2\ 10^6$ kg of nitrogen and $0.06\ 10^6$ kg of phosphorus.

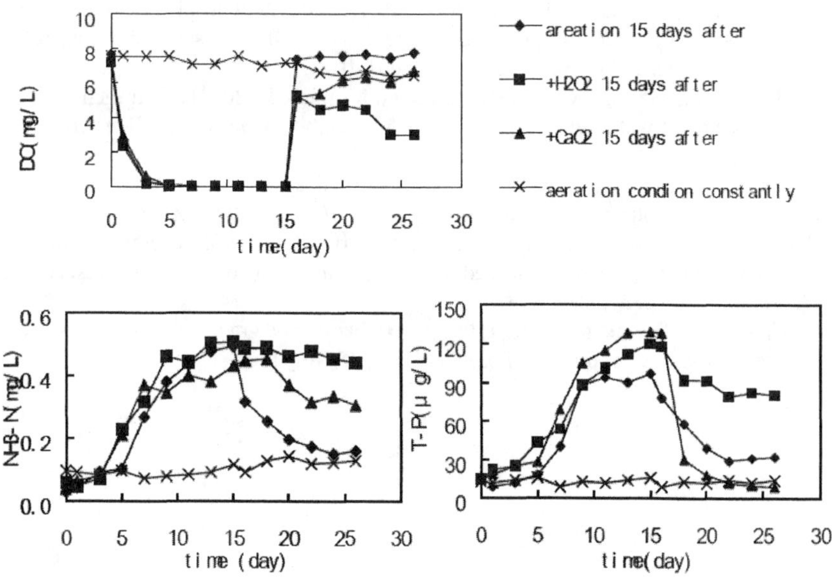

Figure 5. Experimental results for control of sediment releasing.

Conclusion

A year long investigation indicated that the water quality of Xili reservoir is mainly impacted by nonpoint pollution sources, diversion channel and deposit as internal pollution sources. To further improve the water quality of Xili reservoir, following five management practices are necessary: (1) Deposit dredge of 20 cm depth can remove 103 millions kg of organic matters, 3.9 millions kg of nitrogen, 3 millions kg of phosphorous, 35 millions kg iron as well as 1.6 millions kg manganese; (2) Construction of ecological plant zone surrounding the Xili reservoir to remove more than 80% of SS during runoff; (3) Control of non-point pollution sources specifically correlated with organics and nitrogen; (4) Construction of a rubber dam to remove 85% of SS from the diversion channel; (5) Control of the water layer near the reservoir bottom to inhibit the releasing of pollutants from deposits into overlying waters.

Acknowledgment

Support was obtained from grants awarded by the National 863 Project (2002AA601120) and by Shenzhen Water Resources Bureau, P. R. China.

References

American Public Health Association (APHA), American Water Works Association (AWWA) and Water Environmental Federation (WEF). (1995). *Standard methods for examination of water and wastewater*. 18th Edition. Washington, DC.

Beutel, M.W. (2001). "Oxygen consumption and ammonia accumulation in the hypolimnion of Walker Lake, Nevada." *Hydrobiol.*, 466, 107-117.

Burley, K. L.; Prepas, E. E.; Chambers, P. A. (2001). "Phosphorus release from sediments in hardwater eutrophic lakes: the effects of redox-sensitive and -insensitive chemical treatments." *Fresh. Biology*, 46, 1061-1074.

Medina, H. L.; Marín, J. C.; Gutiérrez, E. and Morales, J. (2003). "Nitrogen mobility at the sediment-water interface of Lake Maracaibo, Venezuela." *Wat. Air. Soil. Pollution*, 145, 341-357.

Murphy, J. and Riley, J. P. (1962). "A modified single solution method for the determination of phosphate in natural waters." *Anal. Chim. Acta.*, 27, 31–36.

Redshaw, C.; Mason, C.; Hayes, C. and Roberts, R. (1990). "Factors influencing phosphate exchange across the sediment-water interface of eutrophic reservoirs." *Hydrobiol.*, 192, 233–245.

Tieckelmann, R. E.; Steele, R. (1991). "Higher assay grade of calcium peroxide improves properties of dough." *Food Technol.*, 45, 106-112.

Chapter 11

Low Impact Development (LID) BMP Technology: Planning Tools

Introduction

by Xiao-Yue "Jenny" Zhen

Best management practices for controlling urban runoff pollution are normally designed using the design-storm approach at the site level. In general, the design criteria specify that the post-development peak discharge should be attenuated to the pre development rate, and that a certain water quality volume of runoff should be treated for a selected design storm. The design storm approach is widely accepted and employed because of its simplicity and ease of implementation However, BMPs designed using the design storm approach are not guaranteed for their long-term effectiveness. Also, their water quality and flood control benefits at the watershed scale are often in question. In recent years, owing to the extraordinary advances in computer technology, process-based BMP simulation models have been developed and are becoming an important tool in evaluating BMP performance, as well as facilitating the planning and design of BMPs. When coupled with a watershed model, BMP simulation models can be used not only at the site level, but also for evaluating BMP effectiveness at the watershed scale. In addition, optimization techniques can be applied together with BMP and watershed models to facilitate BMP placement analysis and design for achieving cost-effectiveness.

The first paper "BMP Model for Low-Impact Development" described the Best Management Practices (BMPs) evaluation module developed by Prince George's County to assist in assessing the effectiveness of LID technologies. This module uses process-based algorithms to simulate BMP function and removal efficiency and accepts flow and water quality time series either from observation or generated by runoff models such as the Hydrologic Simulation Program, FORTRAN (HSPF) as input data. The module offers the user the flexibility to design retention style or open-channel BMPs, define flow routing through a BMP network, and compares BMP controls against pre-development conditions. BMP effectiveness can be evaluated and estimated over a wide range of storm conditions, site designs, and flow routing configuration approaches. The BMP Module has been applied to several real land development projects using actual long-term meteorological data recorded at Washington National Airport. An analysis of the simulation results also raised an interesting question: Is using the Natural Resources Conservation Services (NRCS) Rainfall Type II Distribution appropriate for the design of stormwater BMPs?

This paper is authored by Dr. Mow-Soung Cheng, Larry Coffman, Yanping Zhang, John Riverson, and Jenny Zhen. Dr. Mow-Soung Cheng is the Assistant Associate Director of the Environmental Services Division and the Section Head of the Regulatory Programs Section within the Division, Larry Coffman the former Associate Director, and Yanping Zhang a former staff engineer of Programs and Planning Division, Department of Environmental Resources, Prince George's County, Maryland. Prince George's County has been a pioneer in prompting Low Impact Development (LID) concepts and applications for stormwater management in the United States. John Riverson and Dr. Jenny Zhen are environmental engineers at Tetra Tech, Inc, Fairfax, Virginia.

The second paper "A Planning Tool for BMP Implementation at Watershed Scale" describes a planning tool for selecting the optimal spatial BMP placement strategy. The framework of the planning tool includes three components: 1) a watershed model, which simulates hydrology and nonpoint source pollutant loadings; 2) a BMP; and 3) an "optimization" tool, which assists in identifying the cost effective BMP placement strategies in a watershed. A case study application on a tributary watershed of the Swift Creek Reservoir near Richmond was presented to demonstrate how the tool can be used to assess and identify the cost effective BMP placement strategy. Three scenarios that placing potential BMP sites at various spatial scales (i.e., on-site, subwatershed, and watershed scales) were compared, and the results suggest that cost saving benefit can be achieved when stormwater treatment plans are implemented at the watershed scale.

The authors of the second paper are Xiaoyue "Jenny" Zhen, Yanyun Zhai, and Shaw L. Yu. Dr. Xiaoyue "Jenny" Zhen is an Environmental Engineer at Tetra Tech, Inc., Fairfax, Virginia. Yanyun Zhai is a Graduate Assistant at Tsinghua University

Shenzhen Graduate School, Shenzhen, China. Dr. Shaw L. Yu is an Emeritus Professor of Environmental Engineering, Department of Civil Engineering, University of Virginia, Charlottesville, Virginia.

BMP Model for Low-Impact Development

Mow-Soung Cheng[1], Larry S. Coffman[1], Yanping Zhang[1], John Riverson[2] and Jenny Zhen[2]

[1] Prince George's County Government, Department of Environmental Resources, 9400 Peppercorn Place, Largo, Maryland, 20774 (Phone: 301-883-5836; Fax: 301-883-9218)

[2] Tetra Tech, Inc., 10306 Eaton Place, Fairfax, Virginia 22030 (Phone: 703-385-6000; Fax: 703-385-6007)

Introduction

Conventional stormwater structural Best Management Practices (BMPs) are designed to manage runoff from a site with a significant drainage area, therefore the conventional stormwater models are usually oriented to facilitate design of such "centralized" BMPs. A newer concept to minimize stormwater impacts is the application of "source control" techniques such as Low Impact Development (LID). LID is particularly effective in urban settings and is becoming well accepted for controlling urban stormwater impacts from new development and redevelopment. These techniques involve the integration of small, low-cost management practices at the site level that buffer the effects of development and attempt to emulate or improve upon the predevelopment condition. LID uses on-site infiltration and runoff storage and treatment before polluted runoff drains to the sewer system.

The purpose of this paper is threefold. First, it highlights the existing efforts and analytical applications that have been developed for assessing LID-BMP impacts in the County. Second, it discusses the modeling results of a real development site using the BMP Module. Finally, it presents plans for program additions and future enhancements.

Overview of the Existing Modules

In order to assess the stormwater impacts and BMP effectiveness at small-scale drainage areas, Prince George's County, has developed a LID-BMP simulation and assessment model (BMP Module). The BMP module is a predictive tool for evaluating the effectiveness of BMP configurations for stormwater and runoff management from multiple sources under various storm conditions, including critical conditions associated with TMDL development. It is an ideal platform for evaluating the design and placement of non-point source controls necessary for meeting water quality objectives.

The model provides the continuous simulation of hydrographs and pollutant loads and concentrations so that the effectiveness of LID approaches can be simulated within large-scale watersheds models such as HSPF and SWMM. Process-based algorithms include weir and orifice control, storm swale characteristics, hydraulic transport, infiltration and saturation, underdrain outflow, evapotranspiration, general pollutant removal, and stormwater filtration through a soil media. It offers the user the flexibility to design stormwater structural practices such as bioretention cells, rain barrels, roof gardens, vegetated swales, infiltration chambers, wetlands and off-line regional stormwater retention and detention ponds. It also includes the simulation of site design characteristics such as storm drains, building density, road and sidewalk dimensions, disconnection of impervious surfaces, and compares BMP controls against some defined benchmark such as a simulated pre-development condition. Figures 1 and 2 illustrate the basic processes supported by the current BMP module configuration. Figures 3 and 4 show conditions modeled for BMP assessment and the existing BMP Module interface with a sample site represented.

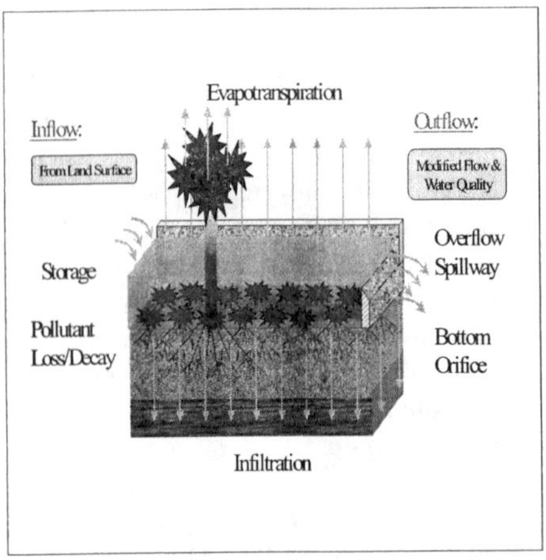

Figure 1. Processes considered in the first BMP –type design.

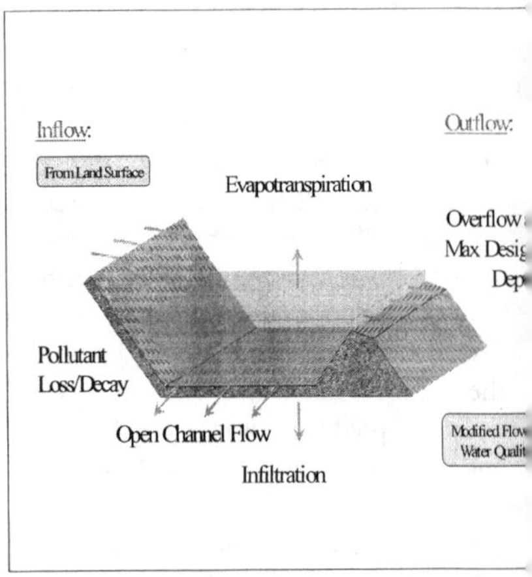

Figure 2. Processes considered in the second BMP-type design.

Since completion of the module, the County validated and reported process-based removal rates and methods for various pollutants and BMP configurations using both

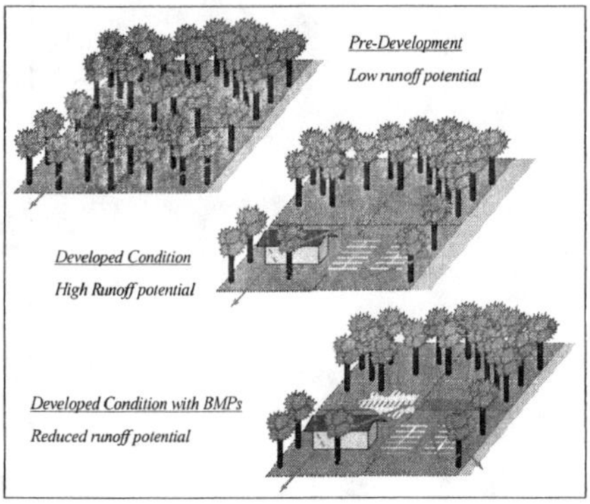

Figure 3. Sample modeling scenarios for BMP assessment.

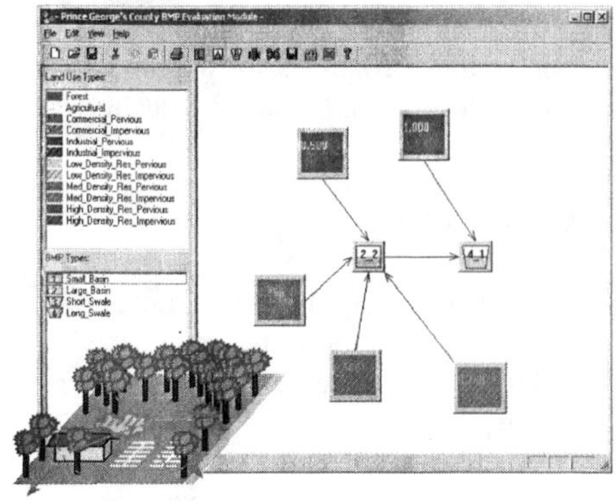

Figure 4. BMP Module interface with example site representation.

published literature values and local experimental site configurations with monitoring data.

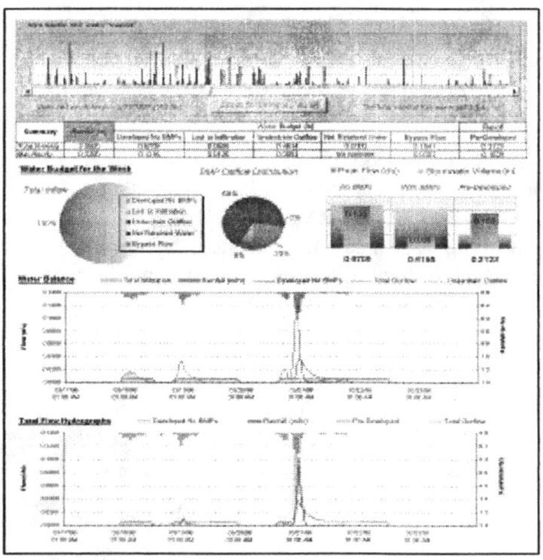

Figure 5. BMP Analysis postprocessor illustrating BMP evaluation technique

The module has been applied to project both existing and proposed development sites with BMPs, and is an active component of the County's planning and assessment work. It includes innovative numerical assessment methodologies, in the form of a Visual Basic Application (VBA) spreadsheet analysis tool, which has been developed to standardize BMP performance evaluation to better support planning and decision analysis. Figure 5 shows how the BMP module evaluates BMP performance against

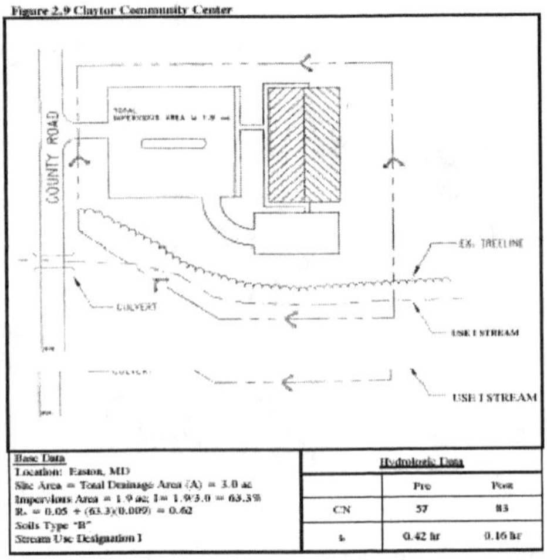

Figure 6. MDE Design Example

both pre-developed condition, and proposed condition with and without BMPs. This tool can assist in testing LID plans for a suite of hydrologic and water quality objectives.

Applications

To demonstrate how the module works comparing to the traditional design, one design example of the Maryland State Stormwater Management Manual was used here. The LID technologies were used to treat stormwater runoff. It can be bioretention ponds, grass swales, roof-top trenches or rain gardens. To simplify the demonstration, only one bioretention pond was used in the example. Figure 6 shows the site data and layout for the site. Claytor Community Center is located in Easton, Maryland, with a total drainage area of 3.0 acres, in which 1.9 acres are impervious area.

The algorithms were evaluated over a range of different storm conditions. Consider the layout shown in Figure 7. One BMP has been installed at the site, called "Bioretention_basin 2_3." It is a Class A BMP bioretention garden that covers an area of 91 by 180 feet, with a relatively large 12-inch upper storage capacity. The BMP-to-drainage area ratio is about 12%. A 12-inch high rectangular overflow weir along the 180-foot length of the basin represents outflow control. The vegetation inside the bioretention basin consists of sod, small trees, and shrubs, which

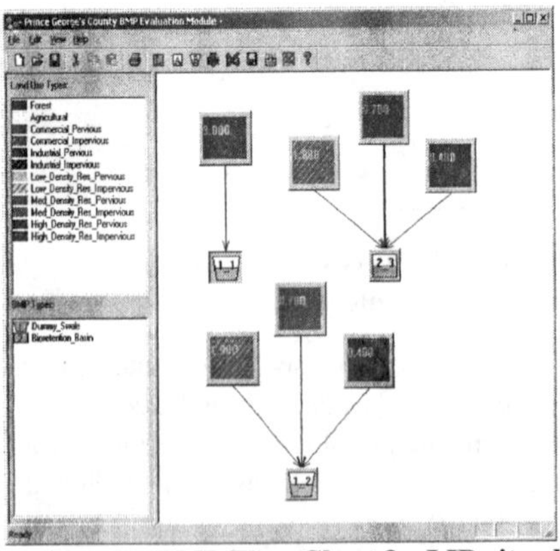

Figure 7. BMP Flow Chart for LID site plan

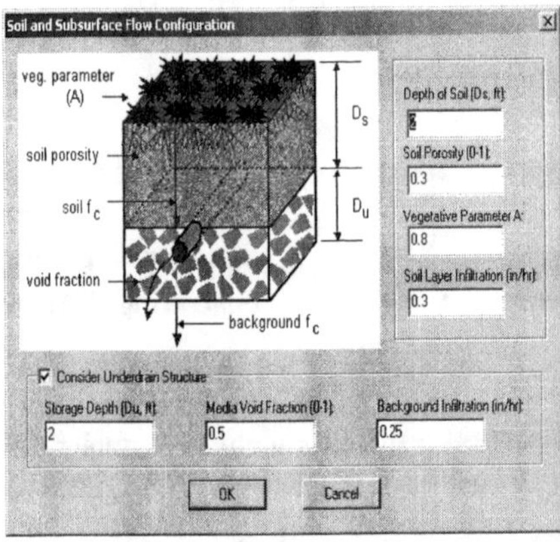

Figure 8. Soil Configuration

corresponds to a Vegetative Parameter-A value of 0.8. Also, as shown in Figure 8, the depth of soil is 2 feet with 30% porosity, and the soil layer infiltration is 0.3 inches per hour. BMP 2_3 also has an additional 2 feet gravel layer above native soil type B soils with a final infiltration rate of 0.25 inch per hour. The other two "BMPs" in Figure 7, 1_1 and 1_2, are dummy cells used to catch and track the total inflows for comparative purposes only.

The module was run with 10 years of rainfall data from January 1, 1989 to December 31st, 1998 collected at nearby Reagan National Airport. Figure 9 shows the 10-year long-term effect of the LID approach comparing to forest condition and developed condition without BMPs. As shown in the bottom chart, the runoff amount was reduced by 94% comparing to developed condition without BMPs. As for pollutant removal rate, it's 98% for sediment, 96% for BOD-5, 96 % for total nitrogen, 99% for phosphorus and zinc. If achieving the modeled pre-developed flow and pollutant condition is the objective, long-term results suggest that the proposed bioretention Basin for this site may be over-designed.

To assess the hydrologic performance of individual storm events, one week of precipitation data was evaluated. Five real storm events have been examined and the model runs have been made. The results are shown in the following Table 1 below:

Table 1. Multiple scenario peak flow comparisons for selected storms.

Date	Rainfall (inches)	Type	Peak Discharge (cfs)		
			Existing	Proposed w/o BMP	Proposed with BMPs
11/24/1993	4.03	Fall	1.80	1.96	0.03
10/13/1995	3.36	Fall	1.46	1.69	0.02
11/6/1996	2.94	Fall	1.14	1.48	0.02
2/2/1998	2.39	Winter	0.11	0.30	0.02
8/1/1993	1.95	Summer	0.00	0.78	0.02

Indicator	Units	Predeveloped	Developed no BMPs	Developed with BMPs
Flow	ft³/yr	31,316	259,272	16,345
Sediment	Tons/yr	0.477	21.203	0.452
BOD-5 day	lb/yr	4.18	81.66	2.90
Total Nitrogen	lb/yr	7.53	22.89	0.97
Total Phosphorus	lb/yr	0.836	2.117	0.030
Total Zinc	lb/yr	0.251	1.218	0.009

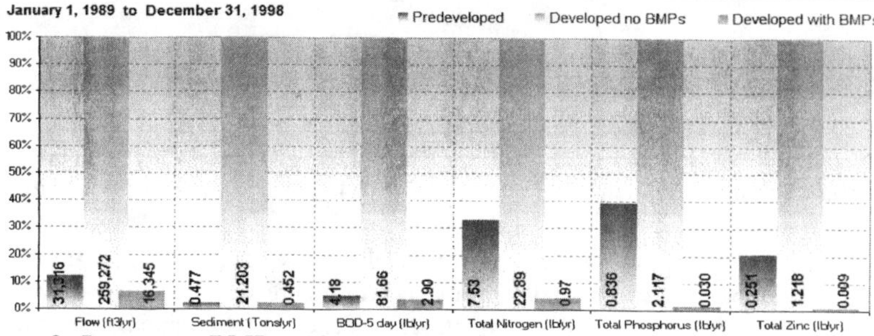

Figure 9. Long-term BMP performance assessment

Table 1 raises some very interesting questions:

- For the extreme storm events, why is there very little difference in runoff peak flow between natural forest and mixed urban areas with high impervious surfaces?
- Does the module over-predict runoff peak flow for forest land?
- Does the module under-predict runoff from urban impervious land?

To answer these questions, a detailed analysis of the rainfall distributions and antecedent moisture conditions (AMC) of these storm events was conducted. It is found that both the rainfall distribution and AMC play very significant roles on the peak flow comparison. The fall storm shown in Figure 10 is a 14-hour duration double-peak storm with 2.94 inches volume. With very short duration and very high intensity, the water is falling faster than it can infiltrate into the ground. This double peak storm-within-a-storm causes a high second-peak flow runoff even for the forested pre-developed condition. Another fall storm shown in Figure 11 is a 16-hour duration double-peak storm. The 16 hours storm duration results in total rainfall of 4.03 inches, with 2 consecutive hours of intense rainfall (0.86 in/hour, 0.8 in/hour) in the second peak. Also since the ground was generally wetter-than-usual during the

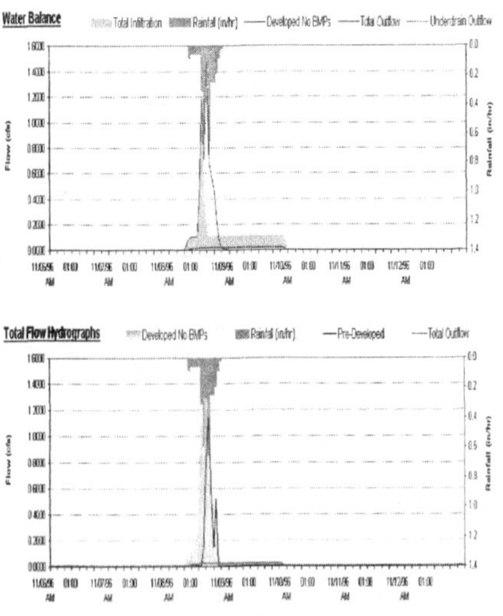

Figure 10. 11/6/1996 Storm
(2.94 inches, 14 hours duration)

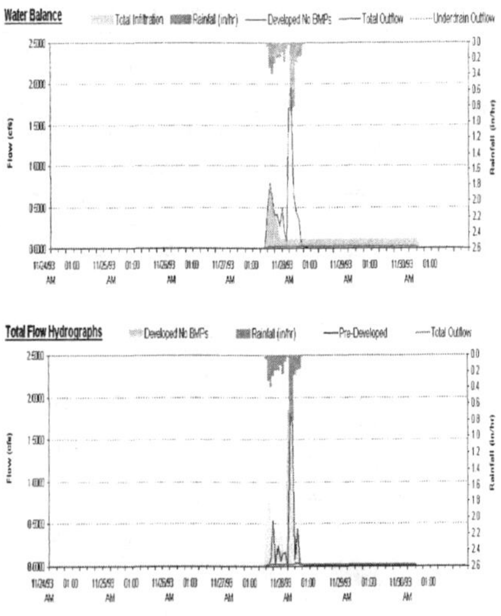

Figure 11. 11/24/1993 Storm
(4.03 inches, 16 hours duration)

fall of 1993, the peak flow of the natural forest is almost the same the peak flow for the urban composite. The last fall storm shown in Figure 12 shows two consecutive events. The first storm lasts 8 hours with 2.21 inches of total rainfall. Four hours later, the second storm comes in with 9-hour duration and 1.15 inches rainfall total.

Since the rainfall is coming faster than the ground can infiltrate within that time period, once again the predeveloped forested condition responds almost exactly like the urban impervious land.

From this evaluation, it is clear that there are multiple competing factors that affect runoff including, but not limited to: storm intensity, storm duration, season that the storm occurs (infiltration potential), antecedent soil moisture conditions and the level of the water table. Although peak flow is highly dependent on the factors listed above, long-term forest runoff volume is generally lower than urban runoff volume due to the attenuation of lower intensity events. Depending on the timing of a storm and antecedent conditions, different factors may have a more significant influence on peak flow than others.

During the week of 2/5/1998, 2.39 inches of winter rain fell within a 24-hour period. The hydrograph is shown in Figure 13. 94% of the peak discharge was reduced. In comparing the LID site design with pre-development conditions, it was predicted that the LID site design, with it's added storage and outflow controls, this bioretention basin was more effective is reducing peak discharge and volume and reducing the nutrient load.

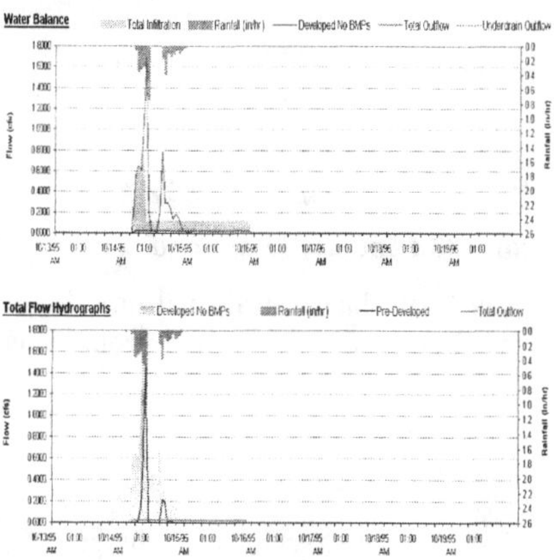

Figure 12. 10/13/1995 Storm (3.36 inches, 17 hours duration)

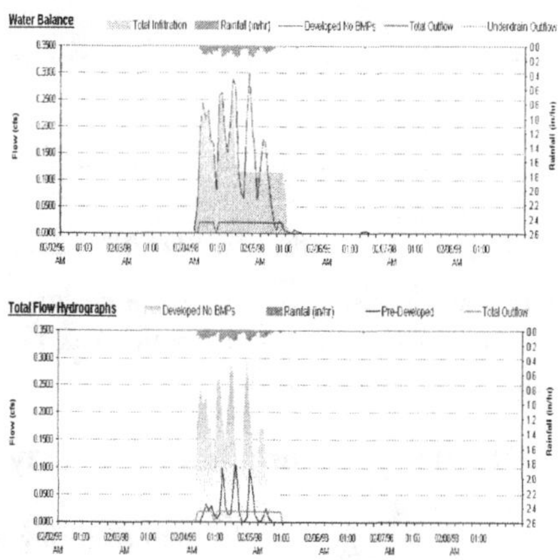

Figure 13. Short-term and Assessment Peak Reduction

After evaluating the natural storm distributions, a synthetic USDA/SCS 2-year design storm was placed in the fall of 1998, with about 12 dry days prior to the start of synthetic storm. The storm has a 24-hour duration with a Type II rainfall distribution as defined by SCS. Under this idealized condition, the model results indicate that the forest peak flow is 94 % lower than the peak associated with the urban composite runoff, as shown in Figure 14. This result show why the natural ebbs and flows of the hydrologic cycle must be considered in BMP design.

Design storms, occurring under idealized conditions, may not always be reflective of the actual natural runoff response. Under certain conditions, when there is added retention and outflow control capability, it is possible for BMPs to attenuate peak flow better than pre-developed conditions. With the BMP module, BMP effectiveness is simulated under variable storm and AMC conditions using actual

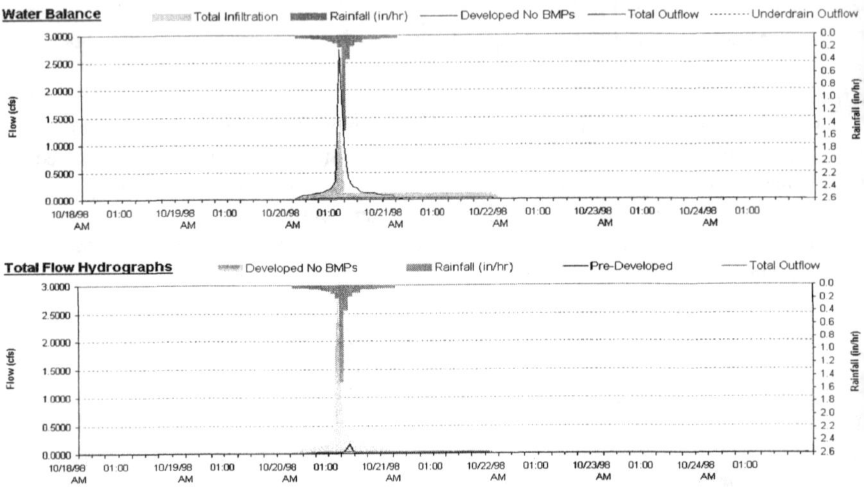

Figure 14. Synthetic USDA/SCS 2-year storm (3.5 inches, 24 hours duration)

rainfall time series, and overall performance can be evaluated evaluated on a long-term basis. Therefore, both peak-flow- and long-term flow-volume-control is considered during BMP evaluation.

Model Enhancements Underway

LID technologies can be applied to highly urbanized areas, such as Washington D. C., to minimize Combined Sewer Overflow (CSO) effects as well. Since LIDs are very site specific, it is also important to know that the same sets of LID technologies at different locations can yield very different removal efficiencies, from a watershed point of view. There is a need to enhance the functionality of the existing BMP Module to address these emerging needs.

The purpose of this section is to outline the required methodology for enhancing and linking the County's LID-based BMP model to the CSO Model, SWMM, and incorporate various optimization techniques to determine the most cost-effective urban retrofit plans. In addition, this effort also evaluates the feasibility of integrating the model with various GIS data layers (e.g. land use, storm drains, sewer lines, etc.) to continue to build a comprehensive assessment tool that can be applied nationwide.

Another key design objective is the application of this strategy for TMDL implementation. Since source contributions from relevant drainage areas are required during model setup, the tool inherently would provide a link between source contributions, BMP influence, and downstream response, which are all required for evaluating TMDL implementation. In light of the watershed approach promoted by EPA, it is envisioned that the goal of stormwater pollution control can be achieved in a more cost effective way by applying a systematic planning/design approach to BMP design and implementation.

This work is of national significance and the EPA Office of Research and Development has also made management practice modeling and optimization a high priority research need to support stormwater and urban TMDL achievement.

Approaches that are typically used to evaluate stormwater BMPs for TMDL implementation are lumped parameter techniques or rely on overall qualitative criteria (e.g., percent reduction). Much of the guidance for selecting removal rates is based on monitoring studies from a wide range of sites, scales, and configurations, resulting in estimated ranges that are sometimes less than useful for management decisions (e.g., total phosphorus reductions between 20 – 80% removal). There is little consideration for the actual processes whereby reduction efficiency may be enhanced or reduced. Furthermore, these approaches typically fall short of predicting BMP response during wet-weather critical conditions when effectiveness may be reduced as a result of antecedent conditions. There was a need to establish a procedure for effectively simulating, testing, and evaluating the effectiveness of BMPs and LID technology.

In order to correctly simulate BMPs, one must examine the underlying processes involved, which include various combinations of storage, infiltration, filtration, biological processes, and hydrologic separation to provide control of hydrology and water quality measures. Besides the physical processes that occur within BMPs, the analysis methodology needs to address watershed-integration issues including: long-term simulation of the landscape, placement of management practices, and general hydrologic and water quality impact. Key considerations for the approach include linkages between the land activities, the management practices, and the stream or hydrologic network. Several global requirements are placed on this system. For example, the system should operate at a short or variable time-step sufficient to represent hydrologic and pollutant loading pollutographs, typically 1 hour or less.

First, the system should be able to address issues of scale such as the simulation of larger watersheds or slightly more detailed subwatersheds and catchment areas. Issues of scale require a flexible computational scheme that is able to correctly handle variable spatial configurations and temporal influences. Second, in light of these analytical options, there arises a need to efficiently (and effectively) visualize, organize, and integrate associated data for modeling purposes. A GIS linkage or platform provides an effective medium for achieving this goal. Third, BMP optimization involves outlining a sample space of potential BMP types, sizes, placement locations and scale, and methodically iterating through the possibilities to arrive at a solution that meets a set of defined objectives. Since BMP simulation is process-based, and the selected results can be numerically represented, the objectives for optimization is defined by measurable physical quantities (e.g., peak-flow, flow volume, sediment load, pollutant concentrations).

1. Link to the CSO Models

The county is building upon existing modeling systems such as SWMM and the County's BMP Module to predict flow and pollutant transport in urban settings. The County adapts and enhances the BMP module algorithms to provide greater flexibility in simulating BMPs and represent the added complexities of CSO processes and impacts, incorporating design requirements related to NPDES and MS4 regulations. The following specific algorithms are added to the BMP module.

- Update BMP module subroutine to handle variable time steps (less-than hourly)
- Update BMP module infiltration algorithm to better handle soil retention
- Update BMP module to better represent evapotranspiration
- Update BMP module water quality to handle adsorption/desorption
- Add BMP module ecological model to simulate plant nutrient uptake & life-cycle
- Add a new stream routing algorithm to BMP module to address larger scale routing issues and incorporate combined sewer overflows (CSOs).

2. GIS Data Layer Linkage

A supplementary interface is added to facilitate the linkage of the BMP module with various GIS data layers (e.g. land use, storm drains, sewer lines, etc.) to form a comprehensive assessment tool that can be applied nationwide. This linkage also provides a visual aid to facilitate the conversion of spatial information into the network system required for simulation. The following activities are performed in the execution of this task:

- Define required formats for associated spatial, physical, and temporal data sets
- Prepare associated computer codes for seamless linkage of the various components
- System testing and debugging
- Perform initial linkage testing and model confirmation against known solutions

3. Optimization of LID/BMP for CSO Management

The system incorporates efficient tools to design and execute optimization techniques to determine cost-effective urban LID/BMP implementation plans with a particular focus on CSO management needs. In order to provide flexible optimization of LID/BMP design for systems including CSOs an interface is needed to facilitate the selection and design of the objective functions and constraints of the optimization problem. The interface guides the user through the formulation and calls the optimization engine to perform the analyses. The optimization occurs at two layers of scale: 1) the site-level, and 2) the watershed level, including both controlled site-level outputs and uncontrolled contributions. The design of the interface defines optimization parameters, benchmarks, and system requirements and integrates performance indicators, acceptable ranges, and other constraints.

The key to the use of the system for management decisions is the interpretation of spatial information (including placement of management practices) into a network and model input deck. This process must be iterative to address the need for multiple simulations of the various management options and search for the optimal solution. Figure 15 below shows the dynamic relationship between the various system components. Beginning with the spatial or GIS view, a simulation network is developed, the network is interpreted into a model input file, the selected modules are executed, external inputs or outputs are generated, data is evaluated, and the model decision matrix is populated. The model decision matrix is maintained as a database to describe the range of options for consideration and the resulting comparison with the identified environmental measures of condition (i.e., hydrology, water quality criteria). The model decision matrix provides the interface with the optimization module. This iterative process is illustrated in three dimensions in Figure 16. Assuming an optimization goal is set based on meeting a target load, the location and type of management practices are varied through a range of conditions through numerous iterations of the ISMDSF. The network manager and decision matrix are essential to providing a streamlined process for achieving evaluating these conditions.

The optimization engine acts as a moderator for the modeling system. It iterates through a defined set of conditions and constraints to arrive at an optimal solution. The optimization system will be set-up and tested in the application case study.

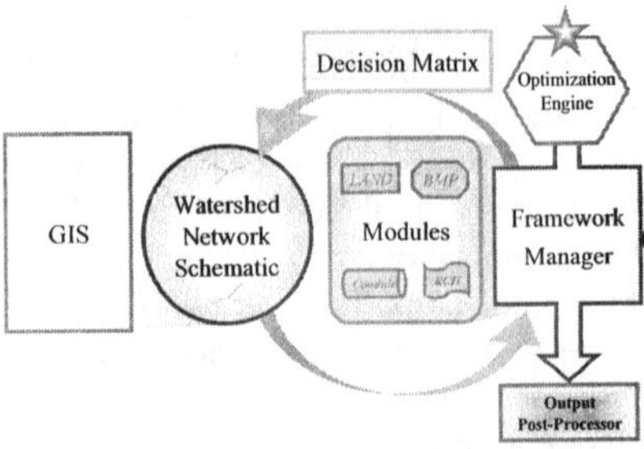

Figure 15. Conceptual design of system components and interactions

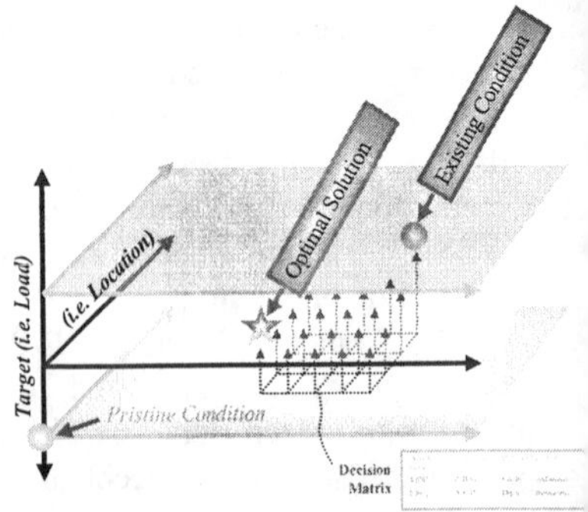

Figure 16. Conceptual diagram of the ite optimization process with three objecti

Conclusion

This BMP module has been proved to be a very powerful tool in assessing the efficiencies of the Low Impact development technologies. Furthermore, this module can also be used for those conventional stormwater management facilities such as ponds and infiltration devices. With the ongoing enhancements, this module will be much more useful for the evaluation of efficiencies for various stormwater management devices. It has also indicated that the commonly used design storm distribution may not always be representative of actual conditions, and may even be overly conservative.

A Planning Tool for BMP Implementation at Watershed Scale

Jenny X. Zhen[1], Yanyun Zhai[2], and Shaw L. Yu[3], LM, ASCE

1 Environmental Engineer, Tetra Tech, Inc. Fairfax, VA;
2 Former Graduate Student, Dept. of Civil Engineering, Univ. of Virginia, Charlottesville, VA.
3 Professor of Civil Engineering, University of Virginia, Charlottesville, VA;

Introduction

Best management practices (BMPs) have long been employed as a standard technology for controlling nonpoint sources of pollution. Since the enactment of the various federal, state and local stormwater management and nonpoint pollution control regulations in late 1980's and early 1990's, there have been tens of thousands of detention ponds constructed in the US. For example, in Charlottesville, Virginia, a relatively small city, there are over 100 detention ponds scattered throughout the landscape at the present time (year 2004). The number of BMPs will continue to grow in light of the announcement of the March 2003 federal NPDES Phase II stormwater regulations. Many small municipalities and construction sites, which were not regulated under the NPDES Phase I program, will now be included.

To date, most BMPs have been designed at the local stormwater management level, instead of the regional, or watershed level. For example, the Virginia Department of Transportation (VDOT) has over the years constructed or installed hundreds of BMPs, most of them detention ponds, in all parts of the state in order to comply with various regulatory requirements. The synergetic impact on water quality of water bodies in the watershed, which can be attributed to these BMPs has not been studied. A key question, therefore, is: How should BMPs be strategically placed in a watershed in order to meet specific water quality goals? Currently, on-site BMPs are designed based on a "performance" expectation such as "80% total suspended solids removal" The new USEPA total maximum daily load (TMDL) program requires a “water quality based” implementation of BMPs. How the hundreds of BMPs will do "collectively" to improve water quality at the watershed level will need to be determined.

The overall goal of this study was to develop an approach to determine the optimal placement of the BMPs so that their combined effects will provide the most cost-effective way of treating stormwater runoff from highways and other sources. GIS-based watershed models were used for simulating the hydrology and water quality changes resulting from various BMP placement scenarios. The effectiveness of a

BMP placement scenario was judged according to flow volume and pollutant reduction, as well as peak flow attenuation by BMPs at the watershed scale. The cost of BMPs is also included in the evaluation.

Methods

Placing BMPs at different spatial levels, ***i.e.*** on-site, sub-regional, and regional, can affect the overall cost-effectiveness of the stormwater control system (Zhen, 2002). Therefore the location of a BMP is an important factor in the optimization framework. Potential BMP locations can be pre-selected based on an inventory of NPS pollution critical source areas, as well as considering the availability of spaces for BMP installation. For each pre-selected location - ***i***, one set of decision variables is assigned. The decision variable vector (BMPi) for location ***i*** represents the existence and the configuration of the BMP. Because the existence of a detention pond can only be represented or described by a discrete variable, the optimization of stormwater control systems naturally falls into the category of an integer or a discrete problem.

Mathematically, the optimization model can be expressed as:

Objective:

$$\text{Minimize} \sum_{i=1}^{n} \cos t(BMP_i) \tag{1}$$

Subject to:

$$L_k \le Lmax_k \tag{2}$$

$$BMP_i \in \text{feasible range} \tag{3}$$

where,
cost = construction cost,
BMP_i = a set of BMP decision variables for pre-selected location i,
L_k = pollutant load at a reference point k, with BMPs,
$Lmax_k$ = maximum pollutant load allowed for the reference point k.

In this study, selected decision variables are the surface area for detention ponds, and length for grass swales. The decision variables are discretized between zero and a maximum value at a constant increment, with a zero value representing that a BMP is not built at the site. The cost is calculated as the construction cost.

The water quality constraint, $Lmax_k$ can be determined based on existing regulations such as federal, state and local watershed pollution control schemes such as a TMDL point and nonpoint source pollution control framework for the watershed being examined. The annual pollutant loads (L_k) at a designated check point (k) determines the pollutant removal effectiveness of the potential BMP alternatives. It is computed by using a watershed model that has BMP simulation module embedded. To solve the above stated problem, the planning tool needs to include three components: 1) a watershed model, which simulates hydrology and nonpoint source pollutant loadings; 2) a BMP module; and 3) an "optimization" tool, which assists in identifying the cost effective BMP placement strategies in a watershed.

For this tool, the Windows version of VirginiA STorm (VAST) model, called WinVAST, was chosen as the watershed model. Also WinVAST has an embedded BMP simulation module, which is capable of simulating detention ponds and grass swales. Recently, a GIS interface was developed to prepare required input data for VAST. The GIS interface was developed as an Arc Info Toolbox in the ESRI ArcGIS environment. The Scatter Search technique (Zhen and Yu, 2004) was selected as the optimization technique for finding optimal solutions. The structure of the optimization framework is illustrated in Figure 1, which consists of three functional components, namely, an initialization run of a watershed model (e.g., the WinVAST model was used in this study), the system evaluator, and the optimizer, i.e., the OptQuest™ Callable Library (OCL), which performs the optimization search process.

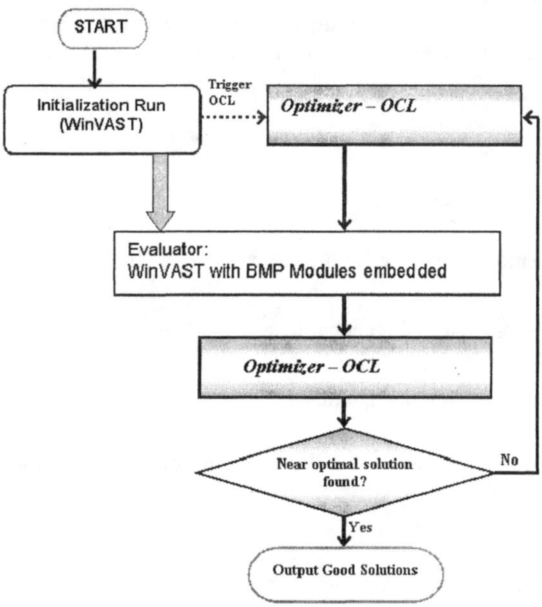

Figure 1: Optimization Framework.

Application of the Planning Tool

The planning tool was applied to the Little Tomahawk Creek, a subwatershed draining to Swift Creek Reservoir, Richmond VA, to demonstrate how it can be used in facilitating BMPs placement for highway stormwater control. For the past few years VDOT has been building the section of Rt. 288 that provides the link between Rt. 95 and Rt. 64 west of Richmond. A stretch of Rt. 288, including the interchange with Rt. 76, passes through the Little Tomahawk Creek, a tributary to Swift Creek Reservoir, which is a major drinking water reservoir for Chesterfield County. To comply with the various stormwater regulations, VDOT is required to implement BMPs to mitigate the stormwater pollution from the highway for the purpose of protection the drinking water reservoir.

Problem Formulation

The Little Tomahawk Creek watershed was divided into fifty-nine (59) sub-watersheds for the case study. The section of Rt. 288 passing through the watershed was segmented into eleven one mile-long segments. It was assumed that for each of the eleven highway segments, one or two potential BMPs (either a detention pond, a swale, or both) could be applied. Also BMPs are assumed to be placed at the outlet of each sub-watershed. For illustration purpose, the water quality requirement was set at 20% removal of sediment at the Little Tomahawk Creek watershed outlet. The goal of BMP implementation was therefore to reduce the sediment loads from the watershed by 20%. In summary, the problem formulation for the case study can be expressed as:

Minimize: BMP Construction Cost;

Subject to: Pollutant removal efficiency at the target point≧ 20%;

BMP sites ∈ pre-selected potential locations

BMP types ∈detention pond, swale, or both.

The BMP construction costs were calculated using the following functions (Schuler, 1987).

- Detention Pond:

$$\text{Cost}(\$) = 18.5\ V^{0.7}$$

where, V is the pond excavation volume (ft3).

- Swale:

$$\text{Cost}(\$) = 3090\ A$$

Where, A is the swale construction area (acre).

Optimization Analysis and Results

In order to compare the cost-effectiveness of various BMP placement spatial scales, three placement scenarios were examined:

- Scenario 1: BMPs can only be placed along the highway within the right-of-way. . (This represents the current VDOT strategy.)
- Scenario 2: BMPs can be placed along the highway sections, and also in sub-watersheds where the highway segments locates . These sub-watersheds therefore are those receiving runoff from the highway section directly.
- Scenario 3: BMPs can be placed either along the highway or in any sub-watershed within the Little Tomahawk Watershed.

Result of Scenario 1: Under this scenario, BMPs can only be constructed at the 11 highway sections. The results show that a detention pond (□) and a grassed swale (▲) are required for all eleven sites, as shown in Figure 2. The total cost was $40,910.

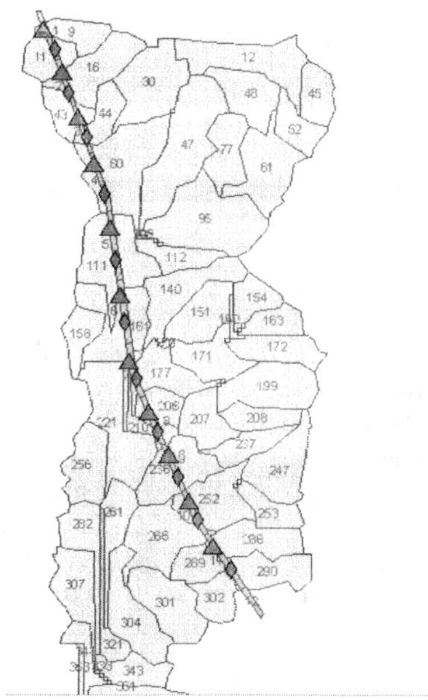

Figure 2: BMP Placement under Scenario 1.

Result of Scenario 2: Under this scenario, it was assumed that BMPs can be placed at the outlet of the eleven sub-watersheds that directly connected to the highway and along the highway. The selected BMP sites are shown in Figure 3. The total cost was #31,350.

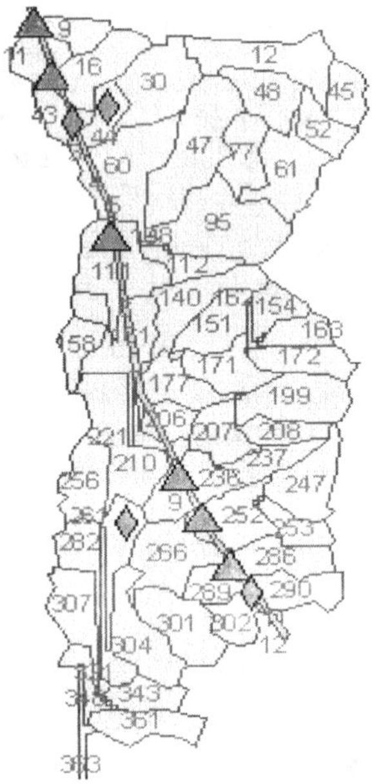

Figure 3: BMP Placement Under Scenario 2.

Results for Scenario 3: Under this scenario, BMPs can be built in any sub-watershed and also along the highway. This scenario has the maximum flexibility in placing the BMPs, i.e., putting them at key locations so that the combined benefit of all the BMPs would be optimized. The results, as shown in Figure 4, gives lower total cost ($25,830) than that of Scenario 1 and Scenario 2. Since some of the BMPs will need to be at locations far outside of the highway right-of-way, VDOT needs to coordinate the implementation of BMPs in the watershed with other stakeholders such as local government and other interested parties. Such a strategy clearly demonstrated the fact that the "watershed approach", promoted by EPA for water quality control in watersheds, may lead to a most cost-effective way for controlling NPS pollution.

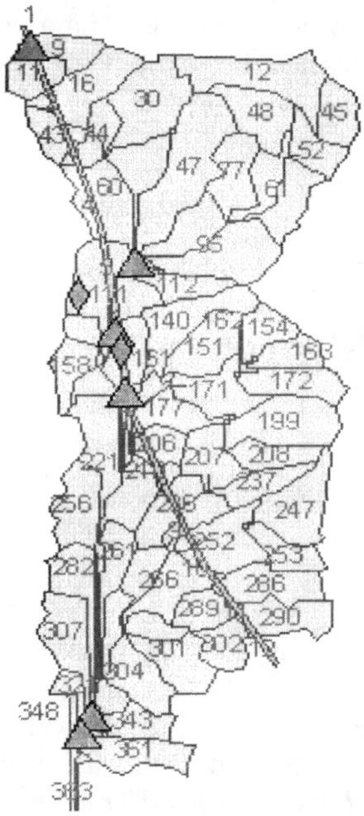

Figure 4. BMP Placement Under Scenario 3.

In summary, by comparing the above three scenarios, it is obvious that by not restricting place BMPs at the on-site scale (in this case the highway sections), or in other words, by allowing placement of BMPs at the watershed scale, more cost-effective strategies can be achieved. Although the current regulations require "on-site" treatment, for the watershed as a holistic target, the best strategy would be for VDOT to work with other stakeholders to develop an optimal placement scheme. By doing so VDOT could save its BMP costs while achieving water quality goals for the watershed as a whole.

Conclusions

A holistic approach for determining the most cost-effective placement and configuration of stormwater best management practices (BMPs) was developed in the present study. The methodology involved the coupling of a comprehensive watershed simulation model with an optimization technique. Specifically, the methodology developed in the study consists of three interacting functional components, *i.e.* a watershed simulation model, a BMP simulation module (the detention pond and grass swale routines), and an optimization model. A highway application case study was done using the Rt. 288 Project in Chesterfield County in Virginia. The results showed that the current VDOT BMP placement approach, i.e., on-site treatment of stormwater runoff from highways, might not be the most cost-effective in terms of protecting the reservoir water quality. The results thus suggest that, by cooperating with other stakeholders in a watershed to plan a BMP implementation strategy at the watershed scale, VDOT can help achieve a more cost-effective reservoir water quality protection goal..

Acknowledgments

The research project, on which the present paper was based, was funded by the Virginia Transportation Research Council (VTRC), Charlottesville, VA. The authors would like to especially thank Mr. G. Michael Fitch of VTRC for his assistance in the GIS applications.

References

Schuler, T. R. (1987). *Controlling Urban Runoff*, Dept. of Environmental. Programs. Metropolitan Washington Council.

Tisdale, T.S., Kaighn, R. J. Jr., and Yu, S. L., (1996). *The VirgiA Storm Model for stormwater management, v6.0 User's Guide*, Dept. of Civil Engineering, Univ. of Virginia

Yu, S. L, (1993). *Stormwater Management for Transportation Facilities*, Transportation Research Board, National Research Council, Washington, D. C.

Zhen, Xiaoyue, (2002). *Development of a BMP Placement Strategy at the Watershed Scale*. Ph.D. Dissertation, Dept. of Civil Eng., Univ. of Virginia, Charlottesville, VA.

Zhen, J. X., and Yu, S. L. (2004) "Optimal Location and Sizing of Stormwater Basins at the Watershed Scale", *J. of Water Resources Planning and Management*, July/August 2004, 130(4), 339-347

Index

About the Editors

Richard Field, who has worked in the environmental engineering field for 43 years, received a bachelor of civil engineering degree from the City College of New York (CCNY) in 1962 and a master's degree in civil engineering (sanitary engineering option) from New York University (NYU) in 1963, graduating first in his class. He is a registered professional engineer (P.E.) in the states of New York and New Jersey, a member of Chi-Epsilon National Civil Engineering Honor Fraternity, a member of the Environmental Water Resources Institute of the American Society of Civil Engineers (ASCE) and chair of its Urban Water Resources Research Council. He has presided as chair of the annual World Water and Environmental Resources Congress' Urban Watershed Management Symposia from 2004 through 2006. He is also a Diplomate by Eminence of the American Academy of Water Resources Engineers.

Since May 1970, Mr. Field has been in charge of the U.S. EPA National Storm and CSO (combined sewer overflow) Technology Research and Development Program located at the National Risk Management Research Laboratory in Edison, New Jersey. He has received numerous outstanding-achievement awards and citations for on-the-job performance and technological contribution including two U.S. EPA bronze medals, the ASCE State-of-the-Art of Civil Engineering Award, two New York Water Pollution Control Association awards for excellence in technological advancement, three U.S. EPA Scientific and Technological Achievement awards including a first-level award, and a first place U.S. EPA National Award in the CSO category. He has authored and coauthored, presented, or published a combination of more than 800 peer-reviewed articles, conference proceedings, papers, U.S.EPA reports, books and book chapters, some of which are internationally recognized publications in his field. He has been invited to lecture and present seminars throughout the world and has presented more than 300 times. Mr. Field is a U.S. EPA expert and an internationally recognized expert in urban wet weather discharge impacts and control technology including the areas of CSO, SSO, and infiltration/inflow (I/I), urban stormwater, diffuse or nonpoint sources, and watershed management. He is listed in Who's Who in Engineering, Who's Who in Technology Today, Who's Who in Science and Engineering, Who's Who in Finance and Industry, Who's Who in the East, International Who's Who of Professionals and Who's Who in the World.

Scott D. Struck received his Ph.D. in Environmental Science in 2003 from Indiana University. He also holds two masters degrees, one in Water Resources and one in Environmental Policy and Natural Resource Management. His background is in aquatic ecosystem ecology, restoration, habitat evaluation, and natural resource management. Through his doctoral work he gained significant understanding in biogeochemical, ecological, and hydrologic interactions of aquatic systems and their interface with terrestrial systems.

Scott has worked with the U.S. EPA's Office of Research and Development in the Urban Watershed Management Branch for the past three years. In doing so, he designed monitoring strategies and coordinated collection efforts that measured chemical, physical, and biological parameters in both dry- and wet-weather flow conditions. The monitoring data was often combined with land use and landscape information to evaluate the impacts of changing environments at the sub-watershed level for assessing improvements with more intensive watershed management. His research has also looked at the unit-processes affecting wetland and retention pond best management practice performance so that practitioners can design these practices to optimize possible benefits. Other projects he has worked on include: evaluating the effectiveness of stream restoration and additional stormwater BMPs implemented within the upper watershed of an impaired urban aquatic ecosystem to improve in-stream water quality and quantity; how sustainability and land use planning may prevent further degradation in urbanizing and suburbanizing areas; and effectiveness of grassed swale and bioretention management techniques for improving receiving water quality and volume.

Scott participates in many committees and professional organizations within and outside of the U.S. EPA. With the Agency he was a member of the Water Quality Leadership Team and the ORD Leadership Development Team. He was also co-chair of the Urban Watershed Management Symposium (formerly the BMP Technology Symposium) of the Environmental Water & Resources Institute's (EWRI) and American Society of Civil Engineers' (ASCE) World Water and Environmental Resources Annual Conference for three years, and served on EWRI's Urban Watershed Resources Research Council, the National Guidelines for Low Impact Development committee (co-chair), and the Urban Stream Restoration committee. He is also a member of the Society of Wetland Scientists and currently serves as the treasurer of the Mid-Atlantic Chapter.

Anthony N. Tafuri received a bachelor of science degree in engineering from Hofstra University in 1964, and master's degrees in civil engineering from Columbia University in 1966 and (sanitary engineering option) from New York University (NYU) in 1971. He is a registered professional engineer (P.E.) in the states of New York and New Jersey and a member of Chi-Epsilon National Civil Engineering Honor Fraternity.

Mr. Tafuri has worked in the environmental field for 38 years. He has been involved in several program areas since his research career started at the U.S. EPA in 1971. From 1971 to 1980 he was an environmental engineer in the U.S. EPA's Storm and Combined Sewer Technology Program and worked on the development and demonstration of technologies to abate pollution from storm and combined sewer discharges and non-sewered urban runoff. From 1980 to 1995, he researched technologies to clean up and abate pollution from uncontrolled hazardous waste sites. For the last 10 years of this period he managed the U.S. EPA's Underground Storage Tank Program and was instrumental in providing technical support for the development and implementation of regulations for the prevention, detection and cleanup of contamination from leaking tanks and pipelines containing petroleum products. Since 1995, Mr. Tafuri has been working in the U.S. EPA's Urban Watershed Management Program and has been chief of the Urban Watershed Management Branch since September 2004, addressing the problems associated with the collection, control, and treatment of wet-weather flow within the urban watershed. He is also involved in research activities associated with aging and obsolete water and wastewater infrastructure and in the U.S. EPA's Environmental Technology Verification (ETV) Program.

Mr. Tafuri has received numerous awards and citations for on-the-job performance and technological achievements, including four U.S. EPA Bronze Medals for outstanding contributions to the U.S. EPA's storm water runoff pollution abatement program, underground storage tank program, and hazardous waste program. He is the recipient of Vice President Gore's Hammer Award for his teamwork in developing the environmental technology verification program. He is also the recipient of a U.S. EPA Scientific and Technological Achievement Award for his research publication on locating leaks with acoustic technology and was nominated for U.S. EPA's Science Achievement Award for exceptional leadership and administration of U.S. EPA's Leaking Underground Storage Tank Program. During his research career, Mr. Tafuri has authored, coauthored, presented and published a combination of more than 250 peer-reviewed technical documents, including books and book chapters on wet-weather flow technology, drinking-water distribution systems, hazardous waste control, and corrective actions for leaking underground storage tank sites.

As Director, Urban Watershed Management, **Michael A. Ports** is responsible for integrating the the various aspects of water resources and environmental management for Black & Veatch's municipal clients. Michael has over thirty-five years of planning, analysis, design, and construction experience in a broad spectrum of water resources engineering applications, including surface water hydrology and hydraulics, storm water management, master planning, soil and water conservation, urban drainage and flood control, river training works, fishery and wildlife habitat mitigation, stream channel rehabilitation, erosion and sediment control, environmental impact assessment, water law, sediment transport modeling, bridge scour analysis, and environmental regulatory compliance.

Michael has authored more than 120 technical papers, reports, articles, chapters, and notes. He is a licensed professional engineer, certified professional hydrologist, certified professional in erosion and sediment control, and Diplomate Water Resources Engineer. Michael is past president of ASCE's Environmental and Water Resources Institute. He also is president-elect of the American Academy of Water Resources Engineers.

Michael Clar is a water resources engineer with 30 years of experience in the evaluation, development and application of innovative stormwater management technology. His education includes a B.S. in Civil Engineering from the University of Maryland and an M.S. in Mining Engineering from the Pennsylvania State University. He is a registered professional engineer in the States of Maryland and Pennsylvania and is also a Diplomate of the American Academy of Water Resources Engineers of the American Society of Civil Engineers.

Currently Mr. Clar serves as president and owner of Ecosite, Inc., a minority owned small business that specializes in many aspects of watershed protection technology. Ecosite is a national leader in innovative stormwater management technology, specially the development and application of ecologically sensitive design techniques such as Low Impact Development (LID), and the protection and restoration of urban streams.

Mr. Clar is the past president of the Maryland Society of Professional Engineers, and past chair of the Urban Water Resources Research Council (UWRRC) of the Environmental and Water Resources Institute (EWRI) of the American Society of Civil Engineers. He is currently the co-chair of the LID technical committee of the UWRRC and chair of the technical committee for the manual of Practice for Erosion and Sediment Control of the Standards Council of EWRI.

Shirley Clark received a Bachelor of Science in Chemical Engineering from Washington University in 1987, a Master of Science in Civil Engineering (environmental engineering major) from the University of Alabama at Birmingham (UAB) in 1996 and a doctoral degree in Environmental Health Engineering from UAB in 2000. Dr. Clark has worked in the environmental field since 1987 and in water resources and stormwater since 1992. Dr. Clark is a registered professional engineer in the state of Alabama. Dr. Clark is a member of the Water Environment Federation (WEF), the American Water Resources Association (AWRA) and the Environment and Water Resources Institute (EWRI) of the American Society of Civil Engineers (ASCE). She is currently committee chair of the Urban Wet Weather Flows Literature Review Committee for WEF, a member of the Management Practice for Control of Erosion and Sediment Committee of EWRI, and a member of the Control Group of the Urban Water Resources Research Council of EWRI. In addition, she serves on the EWRI Task Committees on Low Impact Development and on Urban BMP Technology.

Dr. Clark currently is an Assistant Professor of Environmental Engineering at Penn State – Harrisburg (PSH). Prior to joining PSH in 2003, she was an Assistant Professor at UAB in the Department of Civil and Environmental Engineering. Between her Ph.D. and joining the faculty at UAB, she was a Post-Doctoral Research Engineer for the U.S. EPA's Urban Watershed Management Branch in Edison, New Jersey. Her current research focus is on stormwater treatment and infiltration, with an expertise in stormwater filtration.

Betty Rushton received her M.S. and Ph.D. degrees from the University of Florida in Environmental Engineering Sciences in 1985 and 1988. Her graduate research tested tree species suitable for reclaiming phosphate clay settling areas and she related the results to theories about natural succession. Between the years 1983 and 1987 she supervised the planting of over 3000 tree seedlings in phosphate clay settling areas using various experimental designs to try to determine conditions that would be most conducive for swamp forest reclamation. She also tested tree species planted in clay areas that utilized various reclamation methods such as sand/clay mix and sand cap on top of clays.

Betty went to work for the Southwest Florida water Management District (SWFWMD) in mid-1988 where her main task was to develop a stormwater research program. The goal of the research was to evaluate how well SWFWMD's stormwater rules met state water quality objectives as well as to determine methods to improve water quality from storm runoff. In addition, research from the program has provided stormwater monitoring for projects constructed under the Surface Water Improvement and Management (SWIM) Department. Since the stormwater program was initiated, over sixteen research projects have evaluated how well wet-detention, wetland treatment, alum injection, low-impact designs, effluent filtration, hydrodynamic separators, drop boxes and other methods have reduced pollution to receiving waters. Results from this research have been published in SWFWMD reports, professional journals and many are available on the internet. The stormwater program is recognized by the Environmental Protection Agency (EPA) and the Florida Department of Environmental Protection (FDEP), which has provided funding for many of the research efforts. An outgrowth of the program has resulted in SWFWMD sponsoring a statewide stormwater research conference, held every two years, which attracts scientists from throughout the United States. Betty retired from SWFWMD in 2005 and now has a part-time appoint with the University of Florida's Environmental Engineering Department to continue her research efforts.

She has received certification as a Professional Wetland Scientist by the National Society of Wetland Scientists and as a Qualified Environmental Professional by the Institute of Environmental Practices. She is a member of the American Water Resources Association, the Society of Wetland Scientists, Florida Lake Management Society and the American Society of Civil Engineers (ASCE). She also serves as the Chairman of the Gross Solids technical committee for ASCE. She is an advisory board editor for "Stormwater", a Journal for Surface Water Professionals. She provides peer review for research articles submitted for publication in the American Water Resources Association Bulletin and for various ASCE journals.